Mechanisms and Machine Science

Volume 32

Series editor

Marco Ceccarelli, Cassino, Italy

More information about this series at http://www.springer.com/series/8779

Andreas Pott · Tobias Bruckmann
Editors

Cable-Driven Parallel Robots

Proceedings of the Second International Conference on Cable-Driven Parallel Robots

Springer

Editors
Andreas Pott
Fraunhofer Institute for Manufacturing Engineering and Automation IPA
Stuttgart
Germany

Tobias Bruckmann
University of Duisburg-Essen
Duisburg
Germany

ISSN 2211-0984 ISSN 2211-0992 (electronic)
ISBN 978-3-319-38209-8 ISBN 978-3-319-09489-2 (eBook)
DOI 10.1007/978-3-319-09489-2

Springer Cham Heidelberg New York Dordrecht London

Softcover reprint of the hardcover 1st edition 2015

Printed on acid-free paper

Springer is part of Springer Science+Business Media (www.springer.com)

Committee

General Chair

Andreas Pott, Fraunhofer IPA, Stuttgart, Germany
Tobias Bruckmann, University Duisburg-Essen, Germany

Scientific Committee

Sunil Agrawal, University of Delaware, United States
Clément Gosselin, Laval University, Québec, Canada
Marc Gouttefarde, LIRMM, Montpellier, France
Jean-Pierre Merlet, INRIA, Sophia Antipolis, France
Dieter Schramm, University Duisburg-Essen, Germany
Alexander Verl, University Stuttgart, Germany

Preface

First application ideas and concepts for cable-driven parallel robots were presented in the late 1980s. Due to the unique properties of these robots, like huge size of the workspace, high payload, and outstanding dynamic capacities, the potential advantages became obvious and successful application projects seemed to be within grasp.

During the following years it became clear that the mechanical simplicity is accompanied by practical issues and theoretical challenges. Accordingly, the realization of applications on a reliable and industrial level did not broadly succeed.

Thanks to extensive research—also massively driven by many of the contributors to this book—in the recent years numerous questions were answered and several prototypes were realized. Even more, projects in close cooperation with industry or directly funded by industrial companies are currently testing cable-driven parallel robots in productive environments and first products are expected soon.

In 2012, leading experts from three continents gathered during the "First International Conference on Cable-Driven Parallel Robots" in Stuttgart, Germany. This conference initiated a forum for the cable robot community that is continued by the "Second International Conference on Cable-Driven Parallel Robots" at the University Duisburg-Essen in 2014. This book summarizes the contributions of the participants of this event.

During the lectures it became obvious that practical investigations as well as the stable and reliable control of cable-driven parallel robots are attracting the focus of research teams around the world. We are sure that this pioneers future applications where cable-driven parallel robots enable outstanding solutions in the domains of logistics, handling, production, maintenance, and physical therapy.

We are most grateful to the authors for their significant contributions, to the reviewers for their careful feedback, and for the support of the scientific committee that enabled this. We also thank the people at Springer for their efficient support and help.

The conference was organized by the University of Duisburg-Essen and the Fraunhofer Institute for Manufacturing Engineering and Automation IPA under the patronage of International Federation for the Promotion of Mechanism and Machine Science (IFToMM). It is supported by the Förderverein Ingenieurwissenschaften Universität Duisburg-Essen e.V. and the Duisburger Universitätsgesellschaft e.V. as well as by the Rectorate and the Faculty for Engineering of the University Duisburg-Essen. We would like to express our gratefulness to these institutions for their valuable sponsorship.

June 2014

Andreas Pott
Tobias Bruckmann

Contents

Part I
Modeling

The Forward Kinematics of Cable-Driven Parallel Robots with Sagging Cables

Jean-Pierre Merlet and Julien Alexandre-dit-Sandretto

Abstract Solving the forward kinematics (FK) of parallel robots is known to be a difficult task and the problem is even more complex for cable driven parallel robot (CDPR): the system of equations that has to be solved is larger than with rigid legs as first the static equations have to be taken into account and, second because the deformation of the cables because of their elasticity and their mass may play a role, while being described by a relatively non algebraic complex model. We consider in this paper any arbitrary CDPR whose cables may present a significant deformation due to their elasticity and own mass and we present for the first time an interval analysis based generic algorithm that allows to calculate in a guaranteed manner all the FK solutions and illustrate its use for a CDPR with 8 cables.

1 Introduction

Cable-driven parallel robot (CDPR) have the mechanical structure of the Gough platform with rigid legs except that the legs are cables whose length may be controlled. We will assume that the output of the coiling system for cable i is a single point A_i, while the cable is connected at point B_i on the platform (Fig. 1). Classical kinematics problem are the inverse kinematics (find the lengths of the cables for a given pose of the platform) and the forward kinematics (FK) (find the pose(s) of the platform for given cable lengths). Solving the FK of parallel robots is one of the most challenging problem in modern kinematics. Two categories of FK may be distinguished:

J.-P. Merlet (✉) · J. Alexandre-dit-Sandretto
INRIA, 2004, Route des Lucioles, Sophia-Antipolis, 06902 Cedex, France
e-mail: Jean-Pierre.Merlet@inria.fr

J. Alexandre-dit-Sandretto
e-mail: julien.alexandre-dit-sandretto@ensta-paristech.fr

A. Pott and T. Bruckmann (eds.), *Cable-Driven Parallel Robots*,
Mechanisms and Machine Science 32, DOI 10.1007/978-3-319-09489-2_1

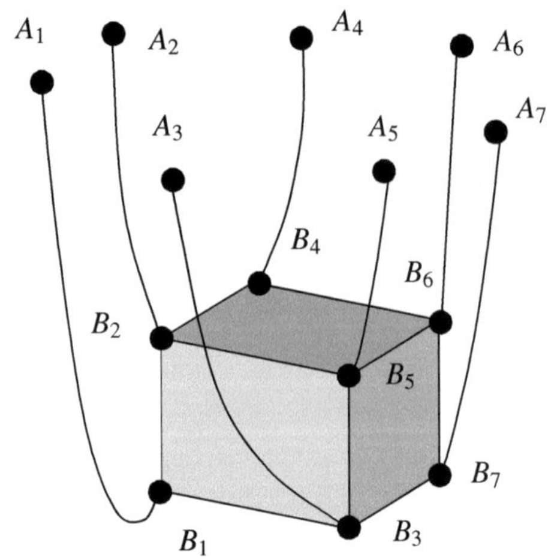

Fig. 1 Cable driven parallel robots with sagging cables

- *real-time FK*: used for control purposes with the objective of determining the current pose of the robot. It need to be fast (running within a sampling time of the robot controller) and uses the knowledge that the solution is "close" to a known pose. It may also be used for simulation purposes
- *full FK*: the purpose is to determine all solutions of the FK. It is used off-line for determining the possible initial states of the robot when running a simulation

We will see later on that the real time FK is not a problem and we will address only the full FK. This problem has been addressed for CDPRDs in a very preliminary stage only very recently and under restrictive assumptions on the behavior of the cables. Indeed we may consider:

- *non-deformable cables*: they are aligned along the direction A_iB_i
- *elastic cables*: they are also aligned along A_iB_i but their lengths depend upon the tension to which they are submitted
- *catenary cables*: they exhibit elasticity and their own mass leads to a deformed shape

Most of the kinematic works have assumed non-deformable cables. For robots having at least 6 cables the FK is equivalent to the one of classical parallel robots (for more than 6 cables at most 6 will be under tension simultaneously [1]). With less than 6 cables the problem is still open as the geometrical constraints relating the length of the cables to the pose leads to less equations than unknowns, which imposes to add the 6 additional statics equations and the cable tensions as additional unknowns. For a CDPR with m cables the minimal system has $6+m$ equations in $6+m$ unknowns, to be compared with the system of 6 equations for the Gough platform. Although there has been progress recently to solve these problems [2–4] there are still a lot of progress to be made in order to determine the maximal number of solutions according to m, solutions that should have only positive tensions and are stable.

If we assume elasticity in the cables there has been some works for the IK [5, 6] but to the best of our knowledge the FK has never been addressed in the general case.

This paper addresses the most general case of FK for CDPR having catenary type cables. Very few works have addressed the IK and FK of such a robot [7–9] and none have considered the general case.

2 Problem Statement

2.1 Cable Model

We will assume cables with linear density μ, cross-section A_0 and E will denote the Young modulus of the cable material. A reference frame $O, \mathbf{x}, \mathbf{y}, \mathbf{z}$ will be used and the coordinates of the A_i points are known in this frame. In the vertical plane of the cable we may assume that the cable is attached at point A with coordinates (0,0) while the other extremity is attached at point B with coordinates (x_b, y_b) (Fig. 2). The vertical and horizontal forces F_z, F_x are exerted on the cable at point B and the cable length at rest is L_0. With this notation the coordinates of B are related to the forces F_x, F_z [10] by:

$$x_b = F_x\left(\frac{L_0}{EA_0} + \frac{\sinh^{-1}(F_z) - \sinh^{-1}(F_z - \frac{\mu g L_0}{F_x})}{\mu g}\right) \quad (1)$$

$$z_b = \frac{\sqrt{F_x^2 + F_z^2} - \sqrt{F_x^2 + (F_z - \mu g L_0)^2}}{\mu g} + \frac{F_z L_0}{EA_0} - \frac{\mu g L_0^2}{2EA_0} \quad (2)$$

2.2 The FK Problem

We consider a spatial robot with m cables whose lengths will be denoted $L_0^1, \ldots, L_0^m$. Without loss of generality we will assume that the Young modulus, linear density and cross-section of the cables are identical. The problem we have to solve is to determine all the possible poses of the platform being given the L_0^j and the location of the A_i, together with the external forces/torques $\mathcal{F}$ that act on the platform. In terms of unknowns we will assume a minimal representation of the pose with 6 parameters

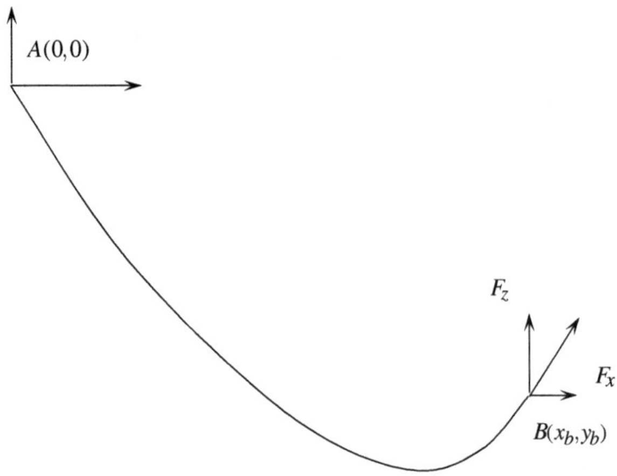

Fig. 2 A deformed cable

and we have also the $2m$ unknown forces F_x^i, F_z^i, for a total of $6+2m$ unknowns. Note that to express these forces in the reference frame we need to establish a rotation matrix **Rot** that rotates the planar frame around the z axis by an angle u_i so that in the reference frame we get the force acting on the platform **F** as $\mathbf{F} = \mathbf{Rot}(-F_x, 0, -F_z)^T$ and hence for cable j $\mathbf{F_j} = (-F_x^j \cos u_j, F_x^j \sin u_j, F_z^j)$. If we assume that the external force acting on the platform is the gravity and that the platform mass is M the mechanical equilibrium imposes that:

$$\sum_{j=1}^{j=m} F_x^j \cos u_j = 0 \quad \sum_{j=1}^{j=m} F_x^j \sin u_j = 0 \quad \sum_{j=1}^{j=m} F_z^j = Mg \tag{3}$$

If C is the center of mass of the platform we get also

$$\sum_{j=1}^{j=m} \mathbf{CB_j} \times \mathbf{F_j} = \mathbf{0} \tag{4}$$

In terms of equations we have the $2m$ Eqs. (1), (2) and the 6 static equations that express the mechanical equilibrium of the platform. Hence we end up with $6+2m$ equations so that solving the FK requires to solve a square system, which will usually have a finite number of solutions. It may also been seen that the FK in that case is much more complex than the FK of the Gough platform (it has $2m$ additional equations) and that the classical methods used to determine an upper bound of the maximum number of solutions (Bezout number, elimination, Gröbner basis) cannot be applied here as Eqs. (1), (2) are not algebraic.

3 Solving the FK

As a theoretical solving appears to be difficult to be used we will have to resort to a numerical solving method, that has to provide all the solutions. We will use an interval analysis (IA) approach, which guarantees to find all the solutions lying within some given ranges. The basis of IA is the *interval evaluation*: being given a function $f(x_1, \ldots, x_n)$ in n variables and assuming that each variable x_i lies in the range $[\underline{x_i}, \overline{x_i}]$ the interval evaluation of f is a range $[A, B]$ such that $\forall x_i \in [\underline{x_i}, \overline{x_i}], i \in [1, n]$ we have $A \leq f(x_1, \ldots, x_n) \leq B$. There are multiple ways to define an interval evaluation but the most simple is the *natural evaluation*: each mathematical operator has an interval equivalent (for example the addition interval operator $\underline{+}$ is defined as $[a, b] \underline{+} [c, d] = [a + c, b + d]$) and transforming any function by using the interval operators allows to calculate the interval evaluation. One of the property of the interval evaluation $[A, B]$ is that if $A > 0$ of $B < 0$, then f cannot cancel whatever is the value of the variables in their ranges. Note that an interval evaluation may be *overestimated*: there may not be value of the variables in their respective range such that $f(x_1, \ldots, x_n) = A$ or $f(x_1, \ldots, x_n) = B$. Indeed overestimation may occur because of multiple occurrences of a given variable that are considered as independent: for example the evaluation of $x - x$ when $x \in [-1, 1]$ is $[-2, 2]$. But the size of the overestimation decreases with the width of the variable ranges.

The second key ingredient of IA is the *branch and bound* algorithm. A *box* $\mathcal{B}$ is defined as a set of ranges for the variables. If for a given box we have $f(\mathcal{B} = [A, B]$ with $A < 0, B > 0$, then we select one of the variable, x_i bisect its range in two and create two new boxes $\mathcal{B}_1, \mathcal{B}_2$ that are identical to $\mathcal{B}$ except for the range for x_i which result from the bisection. These boxes are stored in a list and will be processed later on. We will see in the next section that if a box is tiny enough we may determine if it includes a single solution and compute this solution with an arbitrary accuracy.

However we will not use the minimal set of equations for the FK. Indeed the pose of the platform will not be represented by the coordinates of C and three orientation angles. The motivation is that coordinates x_b, z_b in (1, 2) will be obtained after using the rotation matrix and will include several occurrences of the rotation angles, possibly leading to large overestimation. We prefer to represent the pose of the platform by the 12 coordinates in the reference frame of 4 of the points B_i, which are not coplanar (we assume here that B_1, B_2, B_3, B_4 are chosen). With this choice the coordinates of any point M on the platform may be obtained as $\mathbf{OM} = \alpha_1\mathbf{OB_1} + \alpha_2\mathbf{OB_2} + \alpha_3\mathbf{OB_3} + \alpha_4\mathbf{OB_4}$ where the α are known constants. Such a representation allows one to obtain the coordinates of the $B_j, j > 4$ and of the center of mass C. As we have now 12 unknowns for representing the pose of the platform instead of 6 with the minimal representation we need 6 additional equations that are obtained by stating that the distance between a pair of points in the set $B_1, \ldots B_4$ is a known constant. Note that these equations are not sufficient to fully describe the geometry of the platform (e.g. the equations does not allow to differentiate if a B_j point is over or under the plane that includes the three other points). Another test is needed and we will use the fact that for any point M of the platform there exist constants β_j such that

$$\mathbf{B_jM} = \beta_1 \mathbf{B_jB_k} + \beta_2 \mathbf{B_jB_l} + \beta_3 \mathbf{B_jB_k} \times \mathbf{B_jB_l}, \quad j, k, l \in [1, 4] \tag{5}$$

It remains to manage the angles u_i: formally they can be obtained using the equation

$$x_b^j \sin(u_j) + y_b^j \cos(u_j) = 0 \tag{6}$$

where x_b^j, y_b^j are the coordinates of B_j in the reference frame. But by so doing their interval evaluation will be relatively large even for small width for the x_b^j, y_b^j intervals, especially if $j > 4$. Hence we prefer to add them as additional unknowns and to use (6) as additional equations. Hence we end up with $12 + 3m$ unknowns for $12 + 3m$ equations. Note that we have checked that solving this system in the context of a real-time FK is not difficult as soon as a certified strategy is used [11]: this strategy allows one to determine the current pose of the robot or eventually that the pose is too close to a singularity (in which case the FK has an infinite number of solutions).

An IA approach impose to determine a domain in which are located all the solutions. This can be easily done for the x_b^j, y_b^j variable that are restricted to lie within the convex hull of the A_i points. This can also be done for z_b^j, the z coordinate of B_j that cannot be greater than the highest z coordinate of the A_i points and cannot be lower than the length obtained if we assume that the cable is vertical and bears the platform. As for the u_i as the cable have to lie within the convex hull of the A_i we can also get bounds for these variables. It remains the variables F_x, F_z which have no natural bounds except that F_x cannot be negative. We will first define m new variables λ_j such that $F_z^j = \lambda_j F_x$ (which allows one to have simpler expressions for (1, 2). We take then as upper bound for F_x 10 times the value of mg and for the λ a range of $[-10, 10]$ (at the extremities of this range the cables are almost vertical).

3.1 Determining Exact Solutions

The classical IA branch and bound algorithm assume that if the width of a box is smaller than a small value and the interval evaluations of all equations include 0, then we have found an approximate solution of the system. Here we proceed in another way: for each box of the algorithm we run a few iterations of the Newton-Raphson scheme with as estimate of the solution the center of the box H. Note that even if the NR algorithm converges there is no guarantee (1) that the result is indeed an approximate solution of the system, (2) that the solution lie within the box or even within the search space, (3) that the result satisfies the constraint (5). In order to check if the result is really a solution of the system we use Kantorovitch theorem [11] that allows one to verify that there is indeed a single solution of the system in a ball centered at H with a known radius. If this test succeed we have furthermore the property that the NR scheme, initialized with H as guess point, converges toward the solution. We will see in the implementation section that this property will allow us to compute an approximation of the solution with an arbitrary accuracy.

As soon as a solution H_0 is found it is stored and in a first step we will assume that there is no other solution in a ball centered at H_0 with a given radius, this being applied only on the 12 coordinates of the point $B_1, \ldots, B_4$. Any box that is fully included in this ball will be eliminated and if a box has an intersection with the ball, then the intersection part will be removed from the box. Our purpose in this first step is to determine balls that include a solution and possibly others. In a second step we will run the algorithm on this ball and this check will be faster because the search domain will be drastically reduced. With this approach the IA algorithm is guaranteed to complete.

3.2 Heuristics

A drawback of the usual IA branch and bound algorithm that eliminates boxes only according to the interval evaluation of the equations is that is not efficient as soon as we have complex equations with multiple occurrences of the variables. But several heuristics allows one to drastically improve the efficiency of the algorithm. A first set of heuristics is called the *consistency* approach, which is based on a rewriting of the equations. Consider for example the equation that described that the distance between the points of a pair of B_i point is constant. This equation is written as

$$(x_b^i - x_b^j)^2 + (y_b^i - y_b^j)^2 + (z_b^i - z_b^j)^2 = d_{ij}^2 \tag{7}$$

which may be rewritten as

$$(x_b^i - x_b^j)^2 = d_{ij}^2 - (y_b^i - y_b^j)^2 - (z_b^i - z_b^j)^2$$

Let $[A, B]$ denote the interval evaluation of the right hand-side of this equation. We deduce that

- if $B < 0$ the equation has no solution
- if $A \leq 0, B \geq 0$ then $-\sqrt{B} + x_b^j \leq x_b^i \leq -x_b^j + \sqrt{B}$
- if $A \geq 0$ then x_b^i belongs to $[-\sqrt{B} + x_b^j, -\sqrt{A} + x_b^j] \cup [\sqrt{A} + x_b^j, \sqrt{B} + x_b^j]$

With this approach we may improve the range for any variable in the equation or even eliminate a box without having to use the bisection process. It is important to note that if the set of variable is denoted x and we are able to write an equation under the form $g(x_i) = G(x)$ the consistency requires an inverse operator of g in order to be able to update x_i. This also motivate our choice not to use the minimal representation of the pose but a more algebraic formulation whose inverse is trivial. In our implementation the consistency is applied on all equations of the system and for all variables. It is also used on Eq. (5) and on the equations $\mathbf{B_iB_j.B_iB_k} = d_{ij}d_{ik}\cos(\theta)$ where θ is the known angle between the lines going through (B_i, B_j) and (B_i, B_k).

Another efficient heuristic is the *3B method*. Assume that we have a box and select one of the variable x_i whose range is $[\underline{x_i}, \overline{x_i}]$. We change this range to $[\underline{x_i}, [\underline{x_i} + \varepsilon]$, where ε has a small value. Interval evaluation of the equations and the consistency are used to determine if this new box may include a solution. If the answer is negative we can safely modify the initial box by setting the range for x_i to $[\underline{x_i} + \varepsilon, \overline{x_i}]$. This process is applied for all variables but also on the right side of the interval for x_i.

Another approach is used for the equations that have multiple occurrences of the same variable. We calculate the gradient of these equations and it interval evaluation for the current box. If this evaluation has a constant sign we set the variable to the appropriate lower or upper bound of the range to improve the interval evaluation. This process has to be recursive: as soon as a variable is set to an extremity of its range, then the interval evaluation of the gradient for another variable may become of constant sign.

Another important issue is the choice of the variable that will be selected for the bisection process. Our strategy is to bisect in priority the 12 coordinates of the $B_1, \ldots B_4$ until the width of their interval is lower than a given threshold. Indeed these variables play an important role in the equations: if they are fixed the Eqs. (1), (2) admit a single solution that correspond to the minimal potential energy of the cables.

3.3 Implementation

The algorithm is implemented using the interval arithmetics of BIAS/PROFIL[1] while the higher level uses the functions of our `ALIAS` library[2] that mixes a C++ library and a Maple interface. The Maple interface has allowed to generate most of the C++ code for the algorithm and includes an arbitrary accuracy Newton scheme which allow us to calculate an approximation of the solution with n digits, the n-th digit being guaranteed to be exact, n being a number given by the end-user.

Another property of the Maple interface is that it allows one to implement the algorithm in a *distributed manner*, i.e. running the algorithm on several computers. Indeed it must be noticed that in the solving scheme the treatment of a given box is independent from the treatment of the other boxes to be processed. This allow to have a master program that manages the list of boxes to be processed and the list o solutions and an arbitrary number m of slave computers. The master computer process the initial box until it has a fixed number of boxes in its list. Then it sends the top boxes to the slave computer that a few iterations of the solving algorithm and send back to the master the eventual solution and the boxes that remain to be

[1] http://www.ti3.tuhh.de/keil/profil/index_e.html.

[2] http://www-sop.inria.fr/coprin/developpements/main.html.

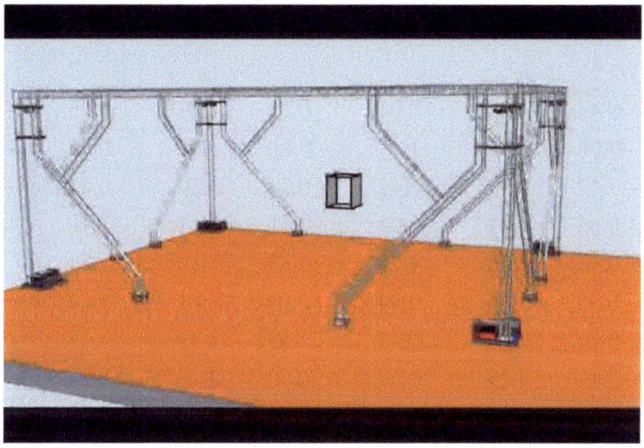

Fig. 3 The robot developed for the ANR Cogiro project. Although the robot is real we present a CAD drawing that allows one to better figure out the CDPR

Table 1 Coordinates of the attachment points on the base (in meters)

x	y	z	x	y	z
−7.175	−5.244	5.462	−7.316	−5.1	5.47
−7.3	5.2	5.476	−7.161	5.3	5.485
7.182	5.3	5.488	7.323	5.2	5.499
7.3	−5.1	5.489	7.161	−5.27	5.497

treated. As the communication overhead is small compared to the computing time of the algorithm the distributed version allows to divide the processing time by m.

4 Example

We consider the large scale robot developed by LIRMM and Tecnalia as part of the ANR project Cogiro [7] (Fig. 3). This robot is a suspended CDPR (i.e. there is no cable pulling the platform downward) with 8 cables, whose A_i coordinates are given in Table 1. The cables characteristics are $E = 100^9$ N/m^2, $\mu = 0.346$ kg/m and their diameter is 10 mm. The mass of the platform of 10 kg. The value of the L_0s (Table 2) are the non-deformed cable lengths for the pose (1, 0, 2) in meters and for an orientation such that the reference frame and the mobile frame are aligned.

Table 2 Lengths of the cable at rest (in meters)

1	2	3	4
10.48215	9.838952	10.16035	8.96827
5	6	7	8
10.310003	8.421629	8.663245	8.655556

Table 3 The coordinates of $B1$, B_2, B_3 for the 19 solutions (meter)

Sol	x_1	y_1	z_1	x_2	y_2	z_2	x_3	y_3	z_3
1	1.55	0.43	4.26	0.04	0.21	4.89	1.05	−0.35	4.71
2	0.57	−0.08	3.53	1.47	−0.80	4.71	0.44	−0.89	4.15
3	1.45	−0.47	2.37	0.42	0.34	3.39	0.43	−0.27	2.38
4	1.27	0.65	4.41	−0.04	−0.31	4.11	1.10	−0.37	4.39
5	0.37	0.47	3.96	1.27	−0.33	5.10	1.34	0.11	4.01
6	0.91	0.98	3.45	0.94	0.04	4.81	1.64	0.64	4.09
7	1.53	−0.74	3.40	0.47	0.53	3.49	1.13	−0.18	4.16
8	1.46	−0.42	4.29	0.52	0.33	3.16	0.94	0.46	4.25
9	1.06	0.64	4.06	1.00	−0.60	2.97	0.93	−0.38	4.12
10	1.66	−0.07	3.45	0.46	0.77	4.20	0.94	−0.30	4.16
11	0.71	0.64	4.10	1.08	−0.64	3.11	0.58	−0.39	4.14
12	1.65	−0.32	4.99	0.73	−0.07	3.64	0.83	−0.78	4.57
13	0.29	0.69	3.64	1.11	−0.67	3.17	0.12	−0.32	3.71
14	1.39	0.45	3.81	0.55	−0.23	2.55	0.88	0.81	2.99
15	1.28	0.53	3.78	0.11	0.01	4.82	1.22	−0.20	4.50
16	1.04	0.90	3.63	0.54	−0.12	4.82	1.52	0.33	4.34
17	1.37	0.52	3.59	0.19	−0.64	3.63	0.44	0.40	3.15
18	0.71	0.92	3.69	0.53	−0.69	4.01	−0.08	0.31	3.94
19	1.86	−0.15	4.12	0.27	−0.60	4.19	1.18	−0.31	4.88

With 8 cables we have to solve a system of 36 equations and this is probably the most challenging FK task that has even been considered. The solving algorithm has been implemented using 10 computers and nineteen solutions were found in the search domain in a computation of about 24 h. They are presented in Table 3, while the cable tensions are given in Table 4. The solutions are depicted in Fig. 4. It is interesting to note that the solution poses are distributed all over the possible workspace: for example the x, y, z coordinates of B_1 are in the ranges [0.29, 1.86], [−0.74, 0.98], [2.37, 4.99]. The F_x forces exhibit also a very large range. For example for cable 1 this force ranges from 20.08 to 417 N. We observe the same variation for the F_z force: for the same cable its ranges from −30.25 to 15.91 N. In 15 cases on 19 the F_Z tension in cable 1 is positive, meaning that the cable exert a downward force on the platform. The number of cables that exert an upward force to support the load is either 2, 3 or 4, meaning that only a small subset of cables contributes to this support: this may be an useful information for dimensioning the cable.

Table 4 Cable tensions for the 19 solutions (Newton)

1 F_x	417.00	22.12	78.64	421.40	14.78	17.25	48.67	98.43
1 F_z	−30.25	15.05	11.19	−43.41	14.86	12.21	4.61	12.48
2 F_x	20.08	138.03	48.65	160.06	292.98	26.46	361.66	11.28
2 F_z	12.79	5.90	10.45	3.60	−16.58	7.61	−38.66	11.63
3 F_x	53.41	51.34	48.02	57.21	55.95	56.22	53.79	61.07
3 F_z	0.49	5.01	1.03	−6.48	5.28	−0.82	−6.66	−1.12
4 F_x	63.89	17.85	429.43	41.10	14.32	73.80	129.88	305.89
4 F_z	11.08	13.29	−29.11	6.18	14.03	5.90	−13.06	−11.58
5 F_x	21.49	242.38	298.60	26.28	19.78	251.91	315.03	21.29
5 F_z	13.77	7.43	−26.40	10.61	16.49	−1.49	−36.82	13.14
6 F_x	76.90	68.82	245.02	15.36	17.30	197.57	129.34	101.20
6 F_z	2.37	11.95	−16.54	11.72	14.32	−16.60	−17.55	7.06
7 F_x	30.99	134.64	90.89	119.67	31.69	63.39	20.24	114.91
7 F_z	10.63	−11.15	5.19	−4.05	7.42	−7.84	8.26	−11.74
8 F_x	29.99	112.30	27.31	37.41	93.70	117.12	19.86	119.86
8 F_z	13.87	−10.68	13.34	9.01	−4.54	−17.91	10.88	−17.24
9 F_x	43.32	78.79	68.24	21.31	66.92	108.11	19.48	125.50
9 F_z	11.47	−4.42	7.80	10.23	−1.06	−18.03	9.69	−18.95
10 F_x	103.16	170.17	57.39	269.24	21.03	13.32	42.69	51.20
10 F_z	−2.67	−5.56	9.40	−36.83	11.86	10.12	1.69	8.73
11 F_x	28.80	99.44	33.76	24.86	90.54	109.52	22.93	114.64
11 F_z	13.35	−8.23	12.24	10.64	−4.98	−18.65	10.11	−17.74
12 F_x	37.36	48.75	270.98	94.64	107.81	16.26	11.05	384.16
12 F_z	15.91	6.93	−7.06	−2.23	−0.10	11.65	12.15	−40.52
13 F_x	23.40	95.81	23.16	54.19	112.26	76.11	74.82	75.04
13 F_z	12.62	−6.67	12.17	6.41	−8.11	−11.21	1.71	−10.19
14 F_x	85.67	52.69	26.80	44.85	89.57	98.61	91.56	42.53
14 F_z	3.87	−0.39	9.28	2.15	−1.29	−12.44	−2.41	−2.02
15 F_x	70.67	21.20	442.79	124.43	14.12	26.91	145.75	307.32
15 F_z	5.97	14.92	−25.45	−2.25	14.31	9.76	−19.27	−1.26
16 F_x	96.61	27.71	316.92	20.56	15.64	175.25	131.51	167.57
16 F_z	0.35	14.57	−18.38	11.45	14.25	−12.87	−17.14	4.51
17 F_x	227.68	18.81	22.23	124.03	52.72	123.33	67.08	30.85
17 F_z	−23.60	11.93	10.69	−9.49	10.80	−12.77	4.35	4.83
18 F_x	38.88	22.14	16.06	59.80	304.33	39.15	209.62	104.55
18 F_z	10.48	12.55	13.66	9.15	−28.04	3.34	−12.75	−11.67
19 F_x	142.07	19.07	208.49	104.84	101.23	68.68	11.11	333.35
19 F_z	−0.72	13.37	4.85	6.58	−3.75	−2.84	11.52	−32.30

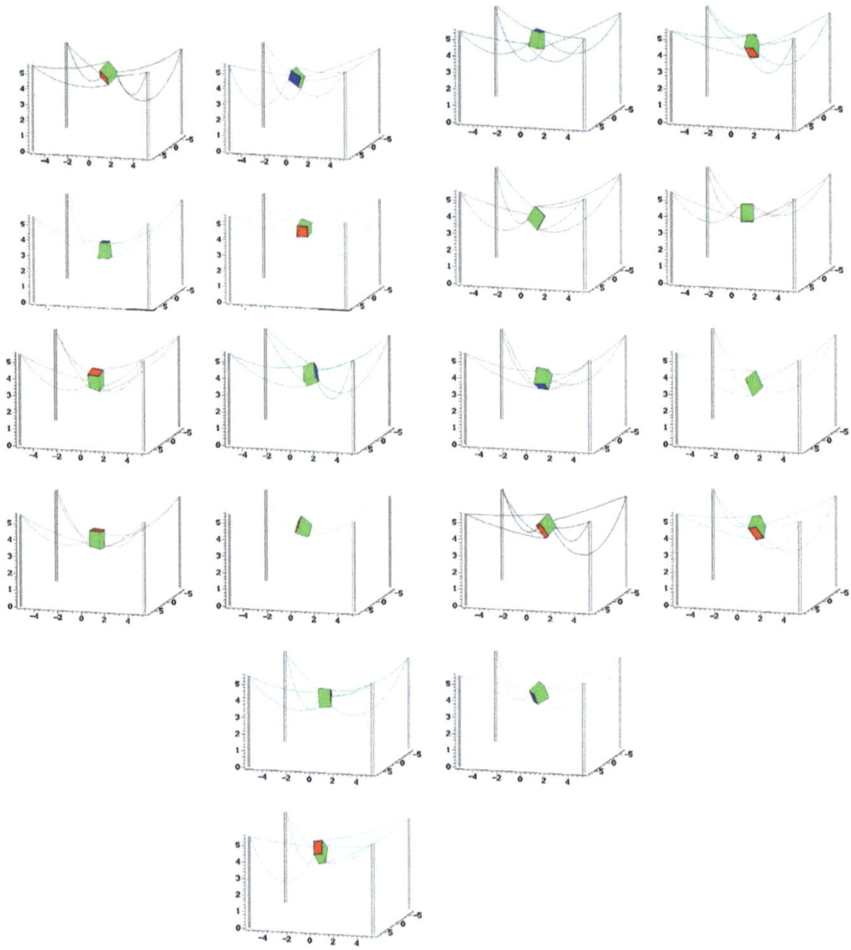

Fig. 4 Solutions 1–19

5 Conclusions

We have presented for the first time a generic algorithm to solve the FK for CDPR with sagging cables. This a computer intensive algorithm (because of the complexity of the problem), that is however guaranteed to provide all solutions. A test case of a robot with 8 cables (probably one of the most complex that has been studied) has shown that we may obtain surprising poses. As prospective our objective is to determine a better balance between the various heuristics that are used in the solving. We will also study the stability of the solutions, possibly introducing stability condition as an additional solving heuristic in order to speed up the computation.

Acknowledgments This research has received partial funding from the European Community's Seventh Framework Program under grant agreement NMP2-SL-2011-285404 (CABLEBOT).

References

1. Merlet J-P (2012) The kinematics of the redundant N-1 wire driven parallel robot. In: IEEE international conference on robotics and automation, Saint Paul, 14–18 May 2012, pp 2313–2318
2. Carricato M, Merlet J-P (2013) Stability analysis of underconstrained cable-driven parallel robots. IEEE Trans Robot 29(1):288–296
3. Carricato M, Abbasnejad G (2012) Direct geometrico-static analysis of under-constrained cable-driven parallel robots with 4 cables. In: 1st international conference on cable-driven parallel robots, Stuttgart, 3–4 Septembre 2012, pp 269–286
4. Jiang Q, Kumar V (2010) The inverse kinematics of 3-d towing. In: ARK, Piran, June 28– July 1, 2010, pp 321–328
5. Merlet J-P (2012) Managing the redundancy of N-1 wire-driven parallel robots. In: ARK, Innsbruck, 25–28 June 2012, pp 405–412
6. Such M et al (2009) An approach based on the catenary equation to deal with static analysis of three dimensional cable structures. Eng Struct 31(9):2162–2170
7. Gouttefarde M et al (2012) Simplified static analysis of large-dimension parallel cable-driven robots. In IEEE international conference on robotics and automation, Saint Paul, 14–18 May 2012, pp 2299–2305
8. Kozak K et al (2006) Static analysis of cable-driven manipulators with non-negligible cable mass. IEEE Trans Robot 22(3):425–433
9. Riehl N et al (2009) Effects of non-negligible cable mass on the static behavior of large workspace cable-driven parallel mechanisms. In: IEEE international conference on robotics and automation, Kobe, 14–16 May 2009, pp 2193–2198
10. Irvine HM (1981) Cable structures. MIT Press, Cambridge
11. Merlet J-P (2004) Solving the forward kinematics of a Gough-type parallel manipulator with interval analysis. Int J Robot Res 23(3):221–236

An Elastic Cable Model for Cable-Driven Parallel Robots Including Hysteresis Effects

Philipp Miermeister, Werner Kraus, Tian Lan and Andreas Pott

Abstract Experimental results indicate that time invariant linear elastic models for cable-driven parallel robots show a significant error in the force prediction during operation. This paper proposes the use of an extended model for polymer cables which allows to regard the hysteresis effects depending on the excitation amplitude, frequency, and initial tension level. The experimental design as well as the parameter identification are regarded.

1 Introduction

A cable-driven parallel robot, in the following simply called cable robot, is a parallel kinematic machine mainly consisting of a platform, cables, and winches as shown in Fig. 1. The cables connect the platform to the winches which in turn control the platform pose by changing the cable length. The control inputs for the winches usually are computed using a simplified kinematic robot model which regards the platform and frame geometry under the assumption that the attachment points at the platform and the contact points at the winch are time invariant. Methods for design and workspace computation of such systems can be found in [3, 4, 10, 12]. Extended kinematic models also include the pulley geometry at the winches and result in higher positioning accuracy [9]. The cable robot IPAnema at Fraunhofer IPA uses Dyneema cables instead of conventional steel cables which brings the advantage of the lower weight but at the same time introduces a more complex elastic behavior in the most relevant force transition element of the cable robot. It showed that the Dyneema polyethylene cables have a changing elastic behavior over time, are subject to settling effects, are sensitive to overload, and show hysteresis effects. Since it is very difficult to build the models and identifying the related parameters using models from different fields such as tribology, viscoelasticity, and multibody systems, here we propose a black box approach to model the drive chain. While white box modeling demands a

P. Miermeister (✉) · W. Kraus · T. Lan · A. Pott
Fraunhofer IPA, Stuttgart, Germany
e-mail: philipp.miermeister@ipa.fraunhofer.de

A. Pott and T. Bruckmann (eds.), *Cable-Driven Parallel Robots*,
Mechanisms and Machine Science 32, DOI 10.1007/978-3-319-09489-2_2

Fig. 1 Overconstrained cable-driven parallel robot

very good knowledge of the inner relations of a system, black box modeling tries to identify the system behavior by observing the input output behavior of a system.

In the fist part of the paper the modeling of the cables is shown. The second part deals with the parameter identification, while the third part shows the evaluation of the model for different input signals.

2 Robot and Cable Elasticity Model

The minimal robot model usually running in a numerical control to operate a robot is based on a solely geometrical model as shown in Fig. 2 including the m platform and winch attachment points $\mathbf{b}_i$ and $\mathbf{a}_i$ with $i = 1 \ldots m$, respectively. The cables can be modeled in different ways depending on the demanded degree of accuracy. The most simple model is just the geometrical model without elastic behavior, meaning that the inverse kinematic equation

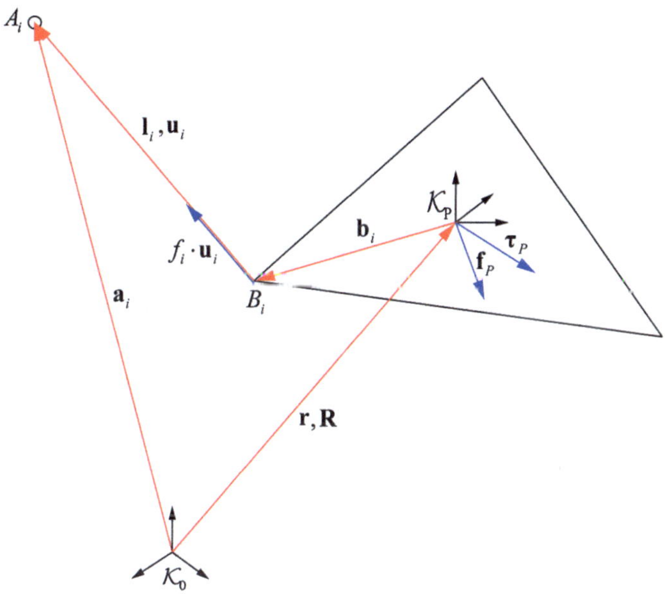

Fig. 2 Kinematic loop for one cable

$$\mathbf{l}_i = \mathbf{a}_i - (\mathbf{r} + \mathbf{R}\mathbf{b}_i) \tag{1}$$

can be used to compute the cable length $l_i = \|\mathbf{l}_i\|_2$ from a given platform position $\mathbf{r}$ and rotation $\mathbf{R}$. More complex models deal with pulleys by introducing time variant anchorage points and handle non negligible cable mass by utilizing catenary equations to model the cable sagging [5, 11].

Since the mass of the cables of the IPAnema robot is small in comparison to the applied tension level it is not necessary to regard the cable sagging in the model. For cable force control, admittance control, and parameter identification it is necessary to predict the cable force very accurately. The force and torque equilibrium of a cable robot is described by the structure equation

$$\underbrace{\begin{bmatrix} \mathbf{u}_1 & \cdots & \mathbf{u}_m \\ \mathbf{b}_1 \times \mathbf{u}_1 & \cdots & \mathbf{b}_m \times \mathbf{u}_m \end{bmatrix}}_{\mathbf{A}^\mathrm{T}} \underbrace{\begin{bmatrix} f_1 \\ \vdots \\ f_m \end{bmatrix}}_{\mathbf{f}} + \underbrace{\begin{bmatrix} \mathbf{f}_\mathrm{p} \\ \tau_\mathrm{p} \end{bmatrix}}_{\mathbf{w}} = \mathbf{0} \tag{2}$$

where $\mathbf{A}^\mathrm{T}$ is the well known structure matrix [12], $\mathbf{u}_i = \mathbf{l}_i \, \|\mathbf{l}_i\|_2^{-1}$ is a unit vector in direction of the cables, $\mathbf{f}$ is the vector of cable forces and $\mathbf{w}$ is the external wrench. Going from the kinematic model to a linear elastic model of type

$$f_i = \begin{cases} f_i = c_i \Delta l_i & \text{for } \Delta l_i > 0 \\ 0 & \text{for } \Delta l_i < 0 \end{cases} \tag{3}$$

with $\Delta l_i = l_{SP,i} - l_i$ where c_i is the cable stiffness and $l_{SP,i}$ is the set point for the cable length while l_i is the actual cable length in the workspace one can achieve a more accurate prediction. Modeling the cable forces in this way gives acceptable results for positive cable tensions, but leads to numerical problems in case of $\Delta l_i < 0$ due to the discontinuity and the zero area.

Comparing the linear elastic model with experimental data showed that cables also come along with a hysteresis behavior which depends on the actuation frequency, amplitude, tension level and amplitude as well as the current angle of the cables with regards to the redirection pulleys of the robot. Different models such as the Bouc–Wen-Model, Bilinear Model, Bingham Model, or Polynomial model can be used to describe the hysteresis behavior [1, 2, 6]. A single tensioned cable of the robot can be seen as single degree of freedom system with an elastic–plastic behavior whose hysteresis by means of the cable elongation x can be described by two polynomial functions, one for the upper part and one for the lower part

$$F_U(x) = \sum_{i=0}^{n_P} a_i x^i, \quad \dot{x} > 0 \tag{4}$$

and

$$F_L(x) = \sum_{i=0}^{n_P} (-1)^{i+1} a_i x^i, \quad \dot{x} < 0, \tag{5}$$

respectively. Under the assumption of displacement anti-symmetry one can compute the hysteresis function by combining Eqs. (4) and (5) to

$$F(x, \dot{x}) = g(x) + h(x) sgn(\dot{x}) \tag{6}$$

with the polynomials

$$\begin{aligned} g(x) &= a_1 + a_3 x^3 + \cdots + a_m x^{n_g} \\ h(x) &= a_0 + a_2 x^2 + \cdots + a_n x^{n_h} \end{aligned} \tag{7}$$

where n_g is an odd number and $n_h = n_g - 1$. The degree of the polynomials can be chosen according to the expected accuracy. The sum of the polynomials $g(x)$ and $h(x)$ can be interpreted as the superposition of an anhysteretic nonlinear elastic part and hysteretic nonlinear damping part as shown in Fig. 3. The expression in Eq. (6) allows to describe the hysteresis for a certain amplitude A independent of the current velocity state. Experiments showed that the hysteretic behavior of the cable is different comparing very small an very large amplitudes. Extending the model in order to deal with variable amplitudes and replacing the solely polynomial description by a velocity depended damping expression one can write

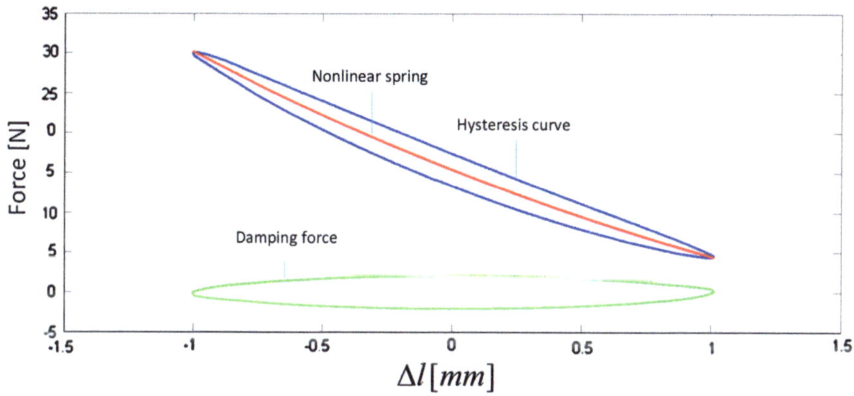

Fig. 3 Decomposition of a hysteresis curve into its anhysteretic nonlinear elastic part and its nonlinear damping part

$$F = K_0 + K_1(A)x + K_2(A)x^2 + K_3(A)x^3 + \frac{S_a|v|^{\alpha(A)}\mathrm{sgn}(v)}{A(A\omega)^{\alpha(A)}l\pi} \tag{8}$$

where the first amplitude dependent polynomial part, now describes the nonlinear non hysteretic spring behavior and the second part of the equation deals with the velocity depended hysteretic damping forces. The hysteretic part is parameterized by the area of the hysteresis S_a and the viscoelastic damping factor α. In case of $\alpha = 0$ the second part of Eq. (8) gets independent of the velocity representing dry friction behavior. The energy dissipation in one cycle for a given friction behavior is

$$l = \frac{\int_0^{T/2} \sin{(\omega t)}^{\alpha(A)+1}\, d\omega t}{\int_0^{T/2} \sin{(\omega t)}^2\, d\omega t} = 2\pi \int\limits_0^{T/2} \sin{(\omega t)}^{\alpha(A)+1}\, d\omega t \tag{9}$$

where T is the time of oscillation and ω is the angular frequency (Fig. 3).

3 Design of Experiments and Parameter Identification

Considering Eq. (8) for a given amplitude level, one has a five dimensional parameter space where the first parameter K_0 determines the pretensioned cable state with $A = 0$ and $v = 0$ reducing the identification problem to four dimensions

$$\mathbf{p_M} = \begin{bmatrix} K_1 & K_2 & K_3 & S_a \end{bmatrix}^{\mathrm{T}}. \tag{10}$$

Table 1 Parameter sensitivity

	K_1	K_2	K_3	S_a
Sensitivity	1.1	0.1	0.03	0.3

The amplitude model is considered separately. While the parameter identification problem for p_M has to deal with the problem of finding the inverse mapping function from four to one dimensions $\mathbf{p}_M = f_M^{-1}(r_M)$, the identification of $p_{M,i} = f_{A,i}(A)$ is just a one dimensional problem. For an efficient measurement and identification process it is important to have a good selection of model parameters and measurement samples. A good selection of model parameters means a high sensitivity with regards to the objective function and a good selection of samples means maximized information gain with a minimal number of samples. To obtain the optimal number of parameters necessary for the model identification, the sensitivity of the parameters is computed by

$$S = \frac{O_2 - O_1}{\bar{O}_{12}} \left(\frac{I_2 - I_1}{\bar{I}_{12}} \right)^{-1}. \tag{11}$$

This sensitivity metric gives the relative normalized change of the output O with regards to the input I using the averages $\bar{O}$ and $\bar{I}$. Parameters with a low sensitivity do not have to be regarded in the model. As can be seen in Table 1, elements with an order higher than three can be neglected in the model. The influence of the parameters K_1 and S_a are visualized in Figs. 4 and 5.

The selection of measurement samples was done according to a D-optimal design of experiments giving a set of tuples $(\mathbf{p}_{M,i}, \mathbf{r}_{M,i})$, $i = 1 \ldots n_D$ where n_D is the number of measurements and $\mathbf{r}_{M,i}$ is the ith error function for the parameter identification problem given by

$$r_i(\mathbf{p}) = F(\mathbf{p}, x_i) - F_{M,i} \tag{12}$$

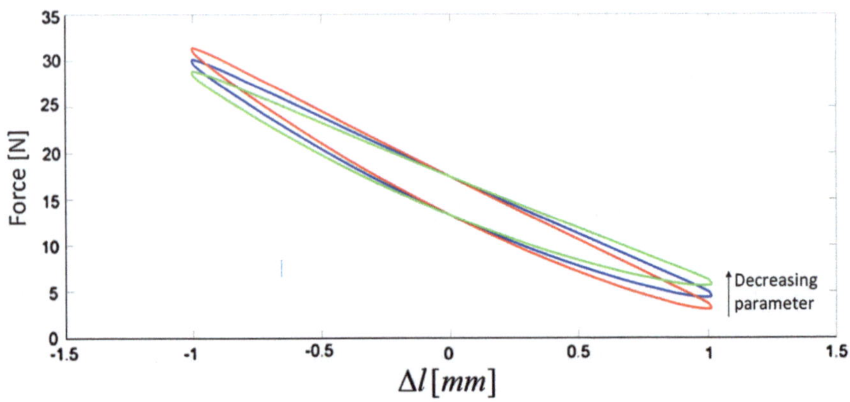

Fig. 4 Parameter sensitivity of K_1

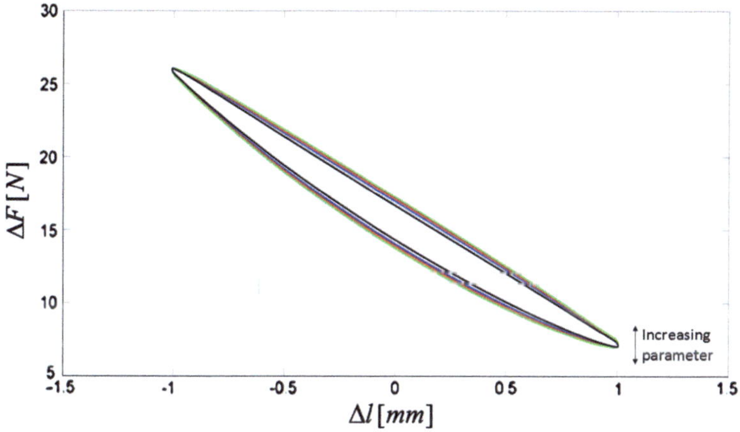

Fig. 5 Parameter sensitivity of S_a

Assuming only small deviations $\Delta\mathbf{p}$ it is possible to use linearization around $\mathbf{p}_0$ which gives the Jacobian $\mathbf{J}_{\mathrm{rp}}$. From that the optimum parameter set can be found by minimizing the least squared error using the objective function

$$\mathbf{p}_{\mathrm{opt}} = \min\left(\frac{1}{2}\sum_{i=0}^{n_{\mathrm{D}}}\left(F(\mathbf{p}, x_i) - F_{\mathrm{M},i}\right)^2\right). \tag{13}$$

Using the linearization one can compute the minimal solution for Eq. (13) by solving the well known normal equation

$$\mathbf{J}_{\mathrm{rp}}^{\mathrm{T}}\mathbf{J}_{\mathrm{rp}}\Delta\mathbf{p} = \mathbf{J}_{\mathrm{rp}}^{\mathrm{T}}\Delta\mathbf{r}. \tag{14}$$

In case that the error of the model is small, a local optimization scheme such as the Levenberg Marquardt algorithm can be used to find the optimal parameter set. Having a good initial guess $\mathbf{p}_0$ is essential for a successful parameter identification. It showed that local optimization is a good choice for long term parameter tracking or repeated adjustment after certain time periods, but did not work well for the very first identification process. Nonlinearities and the huge initial errors demanded for more global optimization techniques such as simulated annealing or genetic optimization procedures which can be used to identify the global optimum.

4 Experimental Results

For measurement, the platform is fixed at the origin such that no interaction between the cables can occur. The excitation function for the cable elongation and it's associated velocity is chosen as

Table 2 Parameter sensitivity

	K_1	K_2	K_3	S_a
Sensitivity	−16.8863	2.411	1.0472	1.1698

$$\Delta l = -A\cos(\omega t)$$
$$\Delta \dot{l} = v = A\omega \sin(\omega t) \tag{15}$$

For this excitation function the damping force can be calculated as

$$F_{\mathrm{h}} = c|A\omega \sin(\omega t)|^{\alpha} \mathrm{sgn}(\sin(\omega t)) \tag{16}$$

where c is damping function

$$c = S_a \left(2A(A\omega)^{\alpha} \int_{0}^{T/2} \sin{(\omega t)}^{\alpha+1}\, d\omega t \right). \tag{17}$$

with S_a describing the area of the hysteresis function. The cables were actuated with an amplitude ranging from 0.1 mm up to 1.5 mm. The pretension level was chosen at 60 N. Running the test was done by generating the cable length set points according to the experimental design using Matlab. The length for each cable in the joint space was commanded by a TwinCat3 controller connected to industrial synchronous servo motors. Force sensors between the cables and the platform were used to measure the cable forces. The force data together with the actual cable lengths were stored in a csv file by TwinCAT and used for model identification in Matlab. Using the actual cable lengths instead of the commanded cable lengths is important to reduce errors introduced by system dead time and controller delay. The results of the parameter identification process can be found in Table 2. The comparison of measurements and simulation results after the identification process gave an average model error of 0.4 N for a static pose. Checking the model prediction after a few days of operation, the model and the real system already started to diverge as can be seen in Fig. 6 where the cable force already shows an offset of 0.4 N to the previously created model. This may be caused by temperature changes, cable settlement and high tension states resulting in lasting changes in the cable's elastic behavior. Using the model for a sinusoidal excitation, a randomly shaped signal can be approximated by a Fourier decomposition. A trapezoidal and a triangular signal where used as test signals to verify the model for different shaped inputs. The Fourier decomposition of the triangular function for example is given by

$$f(t) = \frac{8A}{\pi^2}(\sin(\omega t) - \frac{1}{3^2}\sin(3\omega t) + \frac{1}{5^2}(\sin(5\omega t)) - \cdots). \tag{18}$$

The comparison of the curve progression of the Fourier based force model and the actual measurements are shown in Fig. 7 for an approximation with the first three

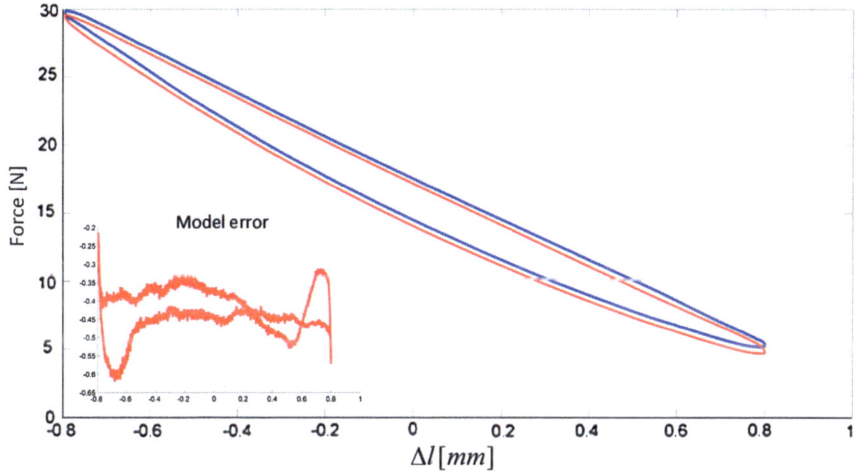

Fig. 6 Comparison of hysteresis model and measurements after approximately for days of operation

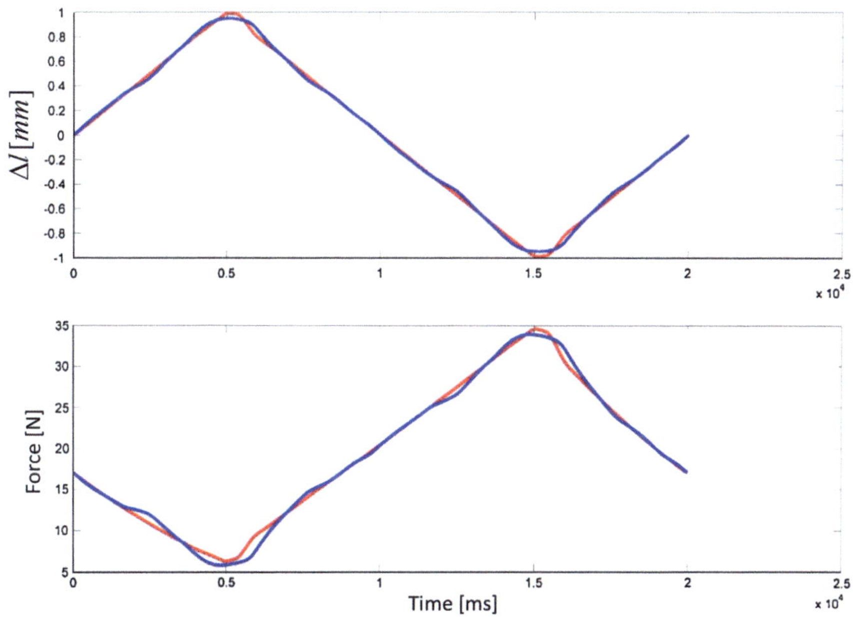

Fig. 7 Triangular test signal with Fourier approximation

Fourier coefficients. The related error shown in Fig. 8 has a maximum magnitude of ±1.5 N at the peak points. The model prediction is based on the whole set of measurement samples which provides knowledge about the past and future of the signal at a certain time stamp. Using the model in a real scenario one has to use

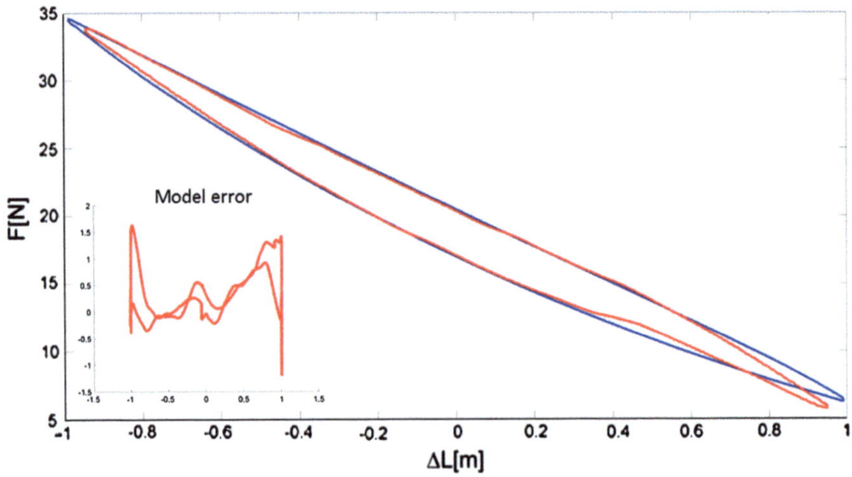

Fig. 8 Hysteresis function and error for a triangular test signal

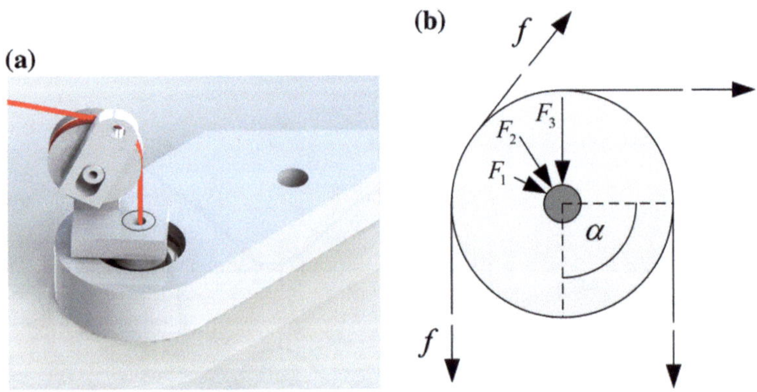

Fig. 9 Pulley assembly of the cable-robot demonstrator. **a** Pulley for cable redirection. **b** Cable and bearing forces

a continuous-time Foruier transformation to deal with the unknown future of the signal, or the numerical control has to feed forward the positioning set points of the cables. While the prediction for a small area around the initial pose was accurate, experimental results showed that the model error increased depending on the platform pose in the workspace. This is is caused by the elasticity of the pulley mounting and the damping behavior of the pulley bearings. To get better model prediction for the whole workspace, these influences have to be regarded in the elasticity model of the power trains. Introducing the angle α to measure the wrapping length of the cable as shown in Fig. 9b, its influence on the stiffness of a single powertrain is experimentally determined as can be seen in Fig. 10. The influence of angle α on the hysteresis behavior is shown in Fig. 11.

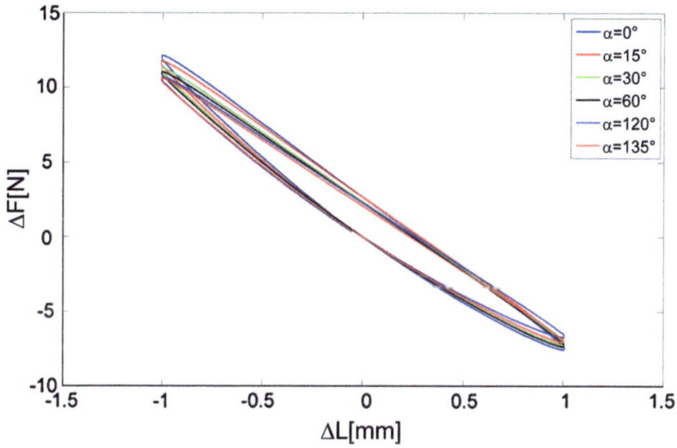

Fig. 10 Hysteresis and stiffness behavior in relation to the pulley angle α

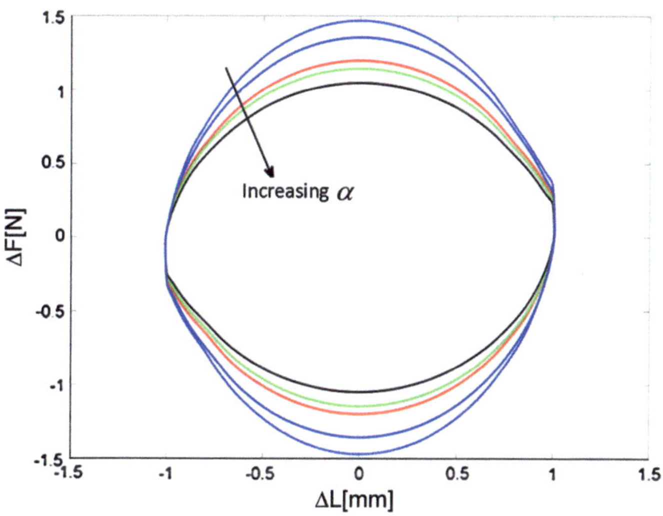

Fig. 11 Damping behavior in relation to the pulley α

The effect may be caused by the increased wrapping angle and the direction of the changing force vector as shown in Fig. 9b, increasing the reaction and friction force at the pulley bearing according to

$$F_i = f\sqrt{2\cos(\alpha_i + 1)}. \tag{19}$$

The angle of attack also influences the torque applied to the bearing on the frame and therefore influences the observed elasticity in the cables.

5 Conclusion and Outlook

In this paper an improved cable model was presented which allows to regard the hysteresis effect during force computation. The approach can be used to improve force algorithms and the identification of geometrical robot parameters in an auto calibration procedure which relies on the force sensors for data acquisition. The simplicity of the model allows to compute the force values in a deterministic time slot, meeting the demand of real-time algorithms. While the model gives significantly better results than the linear elastic model it showed that the cable behavior not only depends on local parameters such as the amplitude, but also on more global and time variant parameters such as the platform pose. Beside that, the actual robot and the model tend to diverge over time depending on the operating load and environmental conditions. Further experiments will be executed to evaluate the long term parameter stability of the robot parameters depending on operational time. It also would be interesting to investigate the influence of overload on the cables alone and in interaction with the surrounding support structure.

References

1. Badrakhan F (1987) Rational study of hysteretic systems under stationary random excitation. Non-Linear Mech 22(4):315–325
2. Bouc R (1971) Modle mathmatique dhystrsis. Acustica 24:16–25
3. Gouttefarde M, Merlet JP, Daney D (2007) Wrench-feasible workspace of parallel cable-driven mechanisms. In: IEEE international conference on robotics and automation, Roma, Italy, pp 1492–1497
4. Hiller M, Fang S, Mielczarek S, Verhoeven R, Franitza D (2005) Design, analysis and realization of tendon-based parallel manipulators. Mech Mach Theory 40(4):429–445
5. Irvine M (1981) Cable structures. MIT Press, Cambridge
6. Ismail M, Ikhouane F, Rodellar J (2009) The hysteresis Bouc–Wen model, a survey. Arch Comput Methods Eng 16(2):161–188
7. Kossowski C, Notash L (2002) CAT4 (Cable actuated truss—4 Degrees of Freedom): a novel 4 DOF cable actuated parallel manipulator. J Rob Syst 19:605–615
8. Merlet J-P, Daney D (2007) A new design for wire-driven parallel robot. In: 2nd international congress, design and modelling of mechanical systems, 2007
9. Pott A (2012) Influence of pulley kinematics on cable-driven parallel robots. In: Advances in robot kinematics, Austria, pp 197–204
10. Pott A, Bruckmann T, Mikelsons L (2009) Closed-form force distribution for parallel wire robots. In: Computational kinematics. Springer-Verlag, Duisburg, Germany, pp. 25–34
11. Thai H, Kim S (2011) Nonlinear static and dynamic analysis of cable structures. Finite Elem Anal Des 47(3):237–246
12. Verhoeven R (2004) Analysis of the workspace of tendon-based Stewart platforms. PhD thesis, University of Duisburg-Essen, Duisburg

On the Improvement of Cable Collision Detection Algorithms

Dinh Quan Nguyen and Marc Gouttefarde

Abstract This paper presents several algorithms to detect the cable interferences for a general spatial Cable-Driven Parallel Robot (CDPR). Two types of cable interferences are considered. The first type is the collisions between cables and cables. The second type is the interferences between cables and the CDPR mobile platform. In each case, an algorithm is proposed to efficiently verify the cable interferences. The use of the proposed algorithms is then illustrated by a verification procedure of the collision free condition over a given Cartesian workspace and orientation workspace of a CDPR. These tools can be used in the design or planning phase of a general CDPR.

1 Introduction

For cable-driven parallel robots, collision detections happen in several cases:

- Interferences between cables and cables
- Interferences between cables and mobile platform
- Interferences between mobile platform and surrounding environment
- Interferences between cables and surrounding environment.

Efficient methods to detect such collisions become necessary, especially for spatial CDPR having a large number of cables (e.g. $m \geq 6$) such as the NIST robot crane [1], the Marionet CDPR [2, 3] and the CoGiRo prototype [4]. These methods could be used in two main situations:

- Design/planning: required to check the capability of CDPR (e.g. compute the bounds on the orientation and Cartesian spaces within which there is no cable interference).
- Control: required to guarantee safety issues in operating CDPR in real-time.

D.Q. Nguyen (✉) · M. Gouttefarde
Laboratoire d'Informatique, de Robotique et de Micro-électronique de Montpellier (LIRMM-CNRS-UM2),161 rue Ada, 34392 Montpellier Cedex 5, France
e-mail: dinhquan.nguyen@lirmm.fr

M. Gouttefarde
e-mail: marc.gouttefarde@lirmm.fr

A. Pott and T. Bruckmann (eds.), *Cable-Driven Parallel Robots*,
Mechanisms and Machine Science 32, DOI 10.1007/978-3-319-09489-2_3

In most situations, the latter case can be avoided if all the safety constraints are dealt with from checking the capability of CDPR over a desired workspace. In this paper, we mainly discuss the verification of collision free conditions for a CDPR in the design or offline planning phases.

In fact, CDPR cable interference problem has not extensively been addressed. Studies on this aspect can be listed as [5–10].

In term of the collisions between cables or CDPR mobile platform with surrounding environment, one can use AABB or OOBB tree methods [11]. These methods are fast and effective for large and complex shape objects (triangulations of the mobile platform and obstacles may consist of a lot of vertices).

For CDPR with light-weight cables, the cables can be considered as straight lines. For CDPR using hefty cables, the mobile platform weight is expectedly large, the cable sagging effects may not really affect the algorithms, thus it may also be sufficient enough to consider the cables as straight lines. In either cases, the interferences between cables and cables can be treated as interferences between straight line segments. The interferences between cables and CDPR mobile platform can be considered as collisions between straight line segments and triangles (the latter triangulating the surface of the CDPR mobile platform). Usual methods [11] can be applied to detect cable interferences. However, these methods are only suited for real-time situations and are not satisfactory enough to our application of interest which is verifying the cable interference free conditions for a given desired Cartesian workspace and set of orientation ranges. In [5], Merlet discussed algorithms to detect interferences between cables and cables as well as between cables and mobile platform. However, the proposed methods were only applied to fixed orientation cases and did not applied for a range of orientations. In [9], Perreault presented an analysis of the cable interference-free workspace of CDPR. The analysis was also mainly applied to the cases of CDPR with constant orientation.

In this paper, we aim to develop algorithms that could improve the efficiency of the verification of collision free operation with respect to given CDPR Cartesian workspace and orientation workspace. These algorithms concern only the interferences between cables and cables, and cables and CDPR mobile platform.

The paper is organized as follows. Section 2 presents the algorithm to detect interferences between cables and cables. The collision detection algorithm for the interferences between cables and CDPR mobile platform is discussed in Sect. 3. An illustrating example of using the proposed algorithms to verify the collision free operation of a CDPR over a given Cartesian workspace and orientation workspace is given in Sect. 4. Finally, Sect. 5 gives an illustration of the performance of the presented methods.

2 Interferences Between Cables and Cables

Let us consider a m-cable CDPR as shown in Fig. 1. We will consider the interference between two cables A_iB_i and A_jB_j $(i \neq j)$.

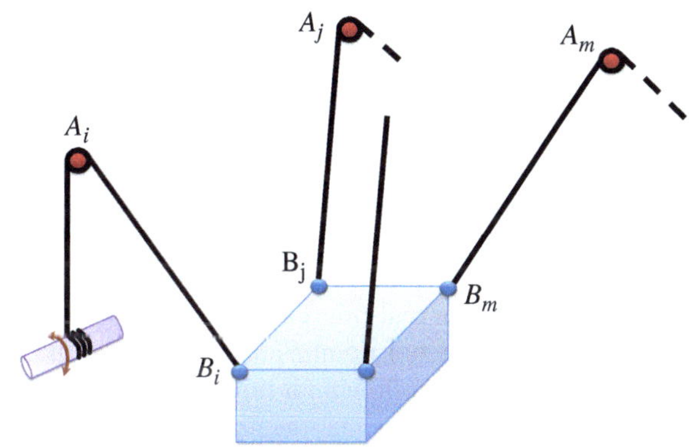

Fig. 1 A general m-cable CDPR

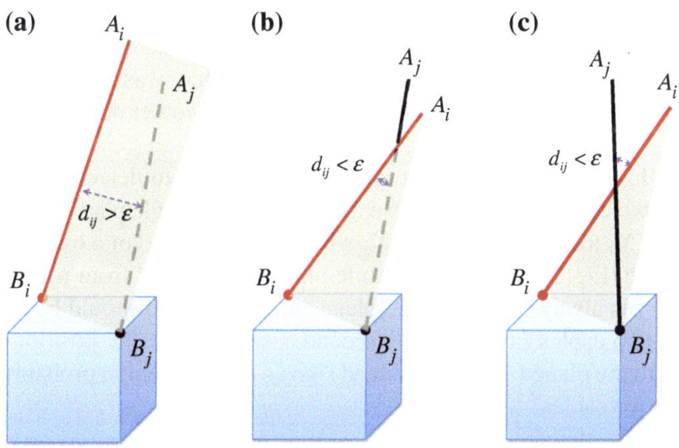

Fig. 2 Usual method of checking interference between two cables. **a** Not colliding, **b** (Going to) collide, **c** Colliding

Figure 2 illustrates a general method to detect the collision between two cables. In the first case (Fig. 2a), the two cables i and j are not colliding since the distance between the two cables d_{ij} is greater than a given small value ε (this value can be chosen as the cable diameter). In the second case (Fig. 2b) when the cable i moves toward the cable j, according to the collision condition $d_{ij} < \varepsilon$, the two cables collide. Note that, in this state, the cable j is "behind" the plane $(A_iB_iB_j)$. In this case, one can say that the two cables are going to collide but a real collision has not yet happened. In the third case (Fig. 2c), the same collision condition is valid. However, the cable j is "in front of" the plane $(A_iB_iB_j)$. In this case, a real collision between the two cables i and j has occurred. This algorithm can be formalized as follows [6, 7, 9]:

A real collision between the two cables i and j will occur if the two following conditions are met

(i) the distance between the two cables is very small: $d_{ij} < \varepsilon$
(ii) the position of the cable j (or the cable exit point A_j) with respect to the plane $(A_iB_iB_j)$ changes sign (e.g. switch from "behind" position to "in front of" position)

It is enough to use this algorithm for real-time collision detection where the positions of cable exit point A_i and cable anchor point B_i are updated online in each sample time (while the mobile platform is following a trajectory). The computation of d_{ij} can be found in [11].

However, in the design phase, where the collision free conditions need to be verified with respect to a range of orientations and a volume of Cartesian space, this usual method may not be really effective.

In fact, to check the cable interferences, it could be enough to consider the second condition (ii) (in the usual algorithm) while neglecting the first condition (i) ($d_{ij} < \varepsilon$). The two cables i and j can be far away (the distance d_{ij} can be large) but their relative positions will tell us whether or not there was a collision when the mobile platform "moved" from an arbitrary pose X_p to another pose X_q in the CDPR workspace.

Figure 3 illustrates the method proposed in this paper to detect the interference between the two cables i and j. Suppose that the CDPR mobile platform moves from an initial pose X_p to an arbitrary pose X_q where a rotation and/or a translation occur. In the first case (Fig. 3a), when the mobile platform "moves" from pose X_p to pose X_q, the cable j is always "behind" the plane $(A_iB_iB_j)$. There should be no collision between the two cables i and j. In the second case (Fig. 3b), the position of cable j with respect to the plane $(A_iB_iB_j)$ changed sign so that a collision probably occurred between the two cables.

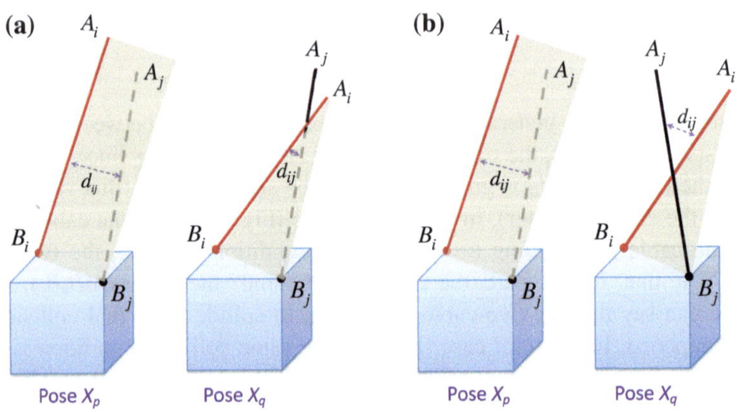

Fig. 3 Checking interference between two cables. **a** Not colliding, **b** Colliding

The collision detection algorithm between the two cables used in this paper consists of the following steps:

- **Step 1.** At pose X_p, compute the position of A_jB_j with respect to plane $(A_iB_iB_j)$ and store it in the variable $s_{ij}(X_p)$:

$$\begin{cases} s_{ij}(X_p) = 1, & \text{if } A_jB_j \text{ is "in front of" the plane } (A_iB_iB_j) \\ s_{ij}(X_p) = 0, & \text{if } A_jB_j \text{ lies on the plane } (A_iB_iB_j) \\ s_{ij}(X_p) = -1, & \text{if } A_jB_j \text{ is "behind" the plane } (A_iB_iB_j) \end{cases} \quad (1)$$

with

$$s_{ij} = \text{sign}\left(\left(\overrightarrow{A_iB_i} \times \overrightarrow{A_iB_j}\right)^T \cdot \overrightarrow{A_iA_j}\right) \quad (2)$$

Here, the two cable exit points A_i and A_j must not be coincident: $A_i \neq A_j$.

- **Step 2.** At pose X_q, compute the projection image $A_j^{'}B_j$ of cable A_jB_j onto the plane $(A_iB_iB_j)$:

$$A_j^{'} = A_j - t * \mathbf{n} \quad (3)$$

with

$$\mathbf{n} = \overrightarrow{A_iB_i} \times \overrightarrow{A_iB_j}$$

$$t = \frac{\mathbf{n}^T \cdot \overrightarrow{A_iA_j}}{\mathbf{n}^T\mathbf{n}}$$

- **Step 3.** At pose X_q, compute the position of A_jB_j with respect to plane $(A_iB_iB_j)$ and store it in the variable $s_{ij}(X_q)$ by using (2).

  If $s_{ij}(X_q) == s_{ij}(X_p)$, *then* no collision should have occurred.
  If $s_{ij}(X_q) \neq s_{ij}(X_p)$ and $A_j^{'}B_j$ is not intersecting A_iB_i, *then* no collision should have occurred.
  If $s_{ij}(X_q) \neq s_{ij}(X_p)$ and $A_j^{'}B_j$ is intersecting A_iB_i, *then* we consider that a collision between the two cables i and j occurred.

In this algorithm, for a m-cable CDPR, step 1 requires to compute s_{ij} in (2) N_{cc} times:

$$N_{cc} = \frac{m \cdot (m-1)}{2} \quad (4)$$

The algorithm stops if there exists any i, j for which a collision occurs (thus, the times of performing steps 2 and 3 is $N_{2,3} \leq N_{cc}$).

The presented algorithm considers that given the two arbitrary poses X_p and X_q, if a collision is detected, then there exists no collision free trajectory that allows the CDPR mobile platform to move from pose X_p to pose X_q (regardless of any trajectory planning method).

3 Interferences Between Cables and CDPR Mobile Platform

3.1 Method 1

Suppose that the CDPR mobile platform is triangulated into $N_{\triangle}$ triangles. The first approach to detect the interferences between cables and CDPR mobile platform is quite straightforward using the method to detect collision between line segments and triangles [11].

At pose X_p, for each cable i, we check for the interferences between cable i and all the triangles that do not belong to the planes which contain the cable end point B_i. If there is a collision then we stop the checking process and give out a warning.

Figure 4 shows an example of the collision between cables i and the mobile platform.

The computational time in this case depends on the number of vertices of the mobile platform as well as the number of cables. This method is quite "heavy" and not really effective if the mobile platform has a complex shape (triangulated with a large number of triangles $N_{\triangle} \gg 1$). To avoid excessive computational time, we can approximate the mobile platform shape by a more simple convex shape whose number of triangles is reduced considerably, e.g. in Fig. 5 (the simplified shape should enclose the CDPR mobile platform).

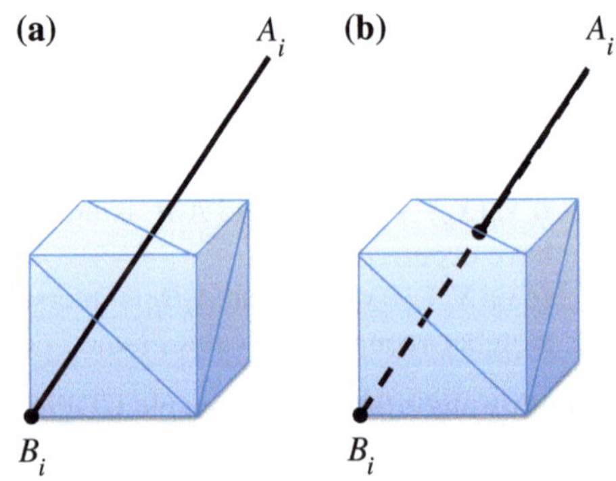

Fig. 4 Interferences between cables and CDPR mobile platform. **a** Not colliding, **b** Colliding

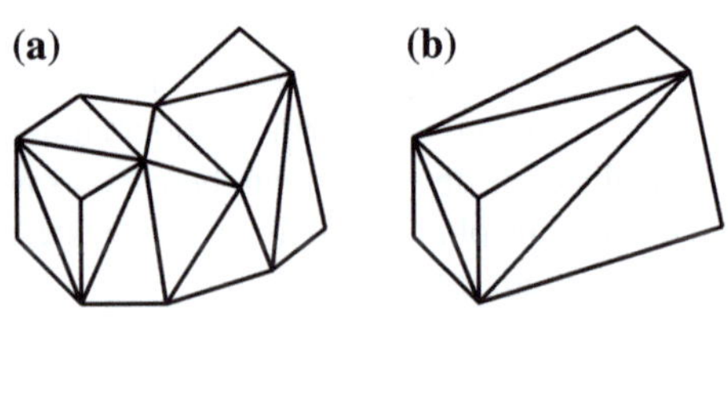

Fig. 5 Simplification of the mobile platform shape. **a** Real shape, **b** Simplified shape

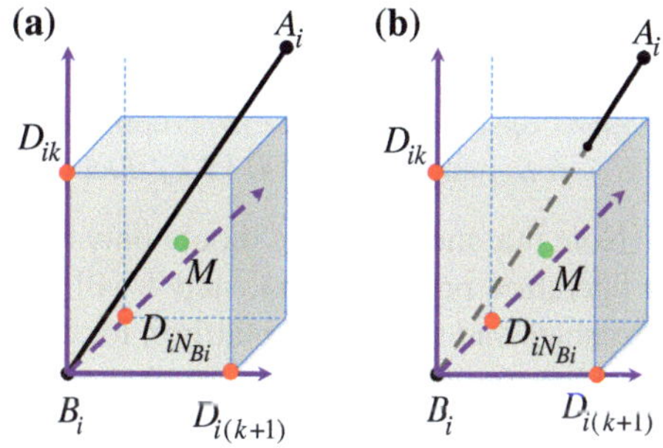

Fig. 6 Detecting collision between cable i and the mobile platform. **a** Not colliding, **b** Colliding

3.2 Method 2

Although the first method (Sect. 3.1) to detect the interferences between cables and mobile platform is simple, the issue of heavy computational time may remain if the mobile platform has a complex shape and the simplification procedure cannot significantly reduce the number of its vertices.

We propose a second heuristic method which consists in checking *whether or not the cable A_iB_i belongs to the subspace (convex cone) spanned by its nearest edges.* Fig. 6 shows an illustrating example of this approach.

The algorithm is given in the following steps:

- **Step 1.** Perform a simplification of the mobile platform to transform it into a simpler convex shape while keeping important vertices. This simplified convex shape should enclose the CDPR mobile platform.
- **Step 2.** Determine the nearest neighbor vertices D_{ik} $(k = \overline{1, N_{Bi}})$ of anchor points B_i in such a way that the convex cone spanned by the vectors $(\overrightarrow{B_iD}_{i1}, \overrightarrow{B_iD}_{i2}, \ldots, \overrightarrow{B_iD}_{iN_{Bi}})$ includes the CDPR mobile platform. N_{Bi} should be the minimum number of such nearest neighbor vertices of B_i.
- **Step 3.** Compute the positions (or the signs) of an arbitrary point M lying within the mobile platform shape with respect to the planes $(B_iD_{ik}D_{i(k+1)})$ and store them into vector S_{Bi} of size $[N_{Bi} \times 1]$:

$$S_{Bi}(k) = \text{sign}\left(\left(\overrightarrow{B_iD_{ik}} \times \overrightarrow{B_iD_{i(k+1)}}\right)^T \cdot \overrightarrow{D_{ik}M}\right) \tag{5}$$

  For instance, the point M can be chosen as the origin of the local frame attached to the mobile platform or as its center of mass. Note that all the cases where $S_{Bi}(k) = 0$ in (5) are considered invalid (the point M must lie strictly inside the mobile platform shape).
- **Step 4.** At an arbitrary pose X_p, compute the signs S_{Ai} of the cable exit point A_i with respect to the N_{Bi} planes $(B_iD_{ik}D_{i(k+1)})$ (vector S_{Ai} is of size $[N_{Bi} \times 1]$):

$$S_{Ai}(k) = \text{sign}\left(\left(\overrightarrow{B_iD_{ik}} \times \overrightarrow{B_iD_{i(k+1)}}\right)^T \cdot \overrightarrow{D_{ik}A_i}\right) \tag{6}$$

If $\exists\, i$ and $\exists\, k$ ($i = \overline{1, m},\ k = \overline{1, N_{Bi}}$) such that $S_{Ai}(k) = 0$, *then* the cable A_iB_i is considered to be colliding with the mobile platform.
If $S_{Ai} \equiv S_{Bi}$ *then* the cable A_iB_i is considered to be colliding with the mobile platform.
If $S_{Ai} \neq S_{Bi}$ and $S_{Ai}(k) \neq 0$ ($\forall\, k$) *then* there is no collision.

Note that the steps 1–3 should only be done at the initial step of an optimization or operation process. Then, step 4 will be used to check the interferences between the cables and the mobile platform for each considered robot configuration.

This approach utilizes the fact that the positions of a point lying within the mobile platform with respect to the planes $(B_iD_{ik}D_{i(k+1)})$ never change. One only need to evaluate (6) N_{cp} times to check the collision, where:

$$N_{cp} = \sum_{i=1}^{m} N_{Bi} \tag{7}$$

This method is fast and reliable. However, there are still a few limitations to this approach. The algorithm only works under the condition that the mobile platform has a convex shape. In case the mobile platform shape is concave, a pre-process (Step 1) is needed to convert it into a convex object (with a number of vertices as small as possible) in order to apply the algorithm. Currently, we are not aware of an efficient (fast) method of selecting the right number of the nearest neighbor vertices D_{ik} of anchor point B_i. One still has to manually select the vertices D_{ik}. The process of simplifying the mobile platform shape to reduce its complexity can be done with available CAD softwares e.g. [12].

4 Verification of Collision Free Condition for a Given Workspace

Let us consider an application where one want to verify the cable interferences of a CDPR with respect to a given Cartesian workspace and orientation range. The CDPR workspace is given as follows:

$$\begin{aligned}
x_{\min} &\le x \le x_{\max}\\
y_{\min} &\le y \le y_{\max}\\
z_{\min} &\le z \le z_{\max}\\
\theta_{x\min} &\le \theta_x \le \theta_{x\max}\\
\theta_{y\min} &\le \theta_y \le \theta_{y\max}\\
\theta_{z\min} &\le \theta_z \le \theta_{z\max}
\end{aligned}$$

where $X = (x\ y\ z,\ \theta_x\ \theta_y\ \theta_z)$ denotes the mobile platform pose. Assume that the Cartesian workspace is discretized into a finite set of N_p points and the orientation workspace is discretized into a finite set of N_q points (these points can be chosen as extreme points which lie on the workspace boundaries). Let us take an arbitrary pose

X_c of the given workspace where we assume that there is no cable interference:

$$X_c = (x_c\ y_c\ z_c,\ \theta_{xc}\ \theta_{yc}\ \theta_{zc}) \tag{8}$$

The verification of collision free condition of the CDPR with respect to the assigned workspace is illustrated in the following pseudocode:

Compute $s_{ij}(X_c)$ *in* (2);
Simplify the CDPR mobile platform shape (if it is necessary);
Determine the N_{Bi} nearest neighbor vertices D_{ik} of B_i;
Compute S_{Bi} (X_c)in(5);
$OK = 1$; (there is no collision)
for $k = 1 : N_p$
  for $l = 1 : N_q$
    $X = (x_k\ y_k\ z_k,\ \theta_{xl}\ \theta_{yl}\ \theta_{zl})$;
    OK = Check the interferences between cables and cables;
    if (OK == 0) *break*;
    OK = Check the interferences between cables and mobile platform;
    if (OK == 0) *break*;
  end;
end;

In short, we perform the verification process at each discrete points in Cartesian space and orientation space. The reference (initial) state of the mobile platform is computed at the pose X_c. In the step checking the interferences between cables and cables, the initial pose is always X_c, and the destination pose is X. This means that when the mobile platform "moves" from pose X_c to pose X, the checking process stops if there is any interference detected. In the step checking the interferences between the cables and the mobile platform, the second approach is used. There is no collision if the returned value of the checking variable is $OK = 1$.

In this way, the collision free condition of the CDPR with respect to a given workspace is ensured in the sense that, when $OK = 1$ is returned, there should always exist one collision free path starting from the home pose X_c to any pose (among the considered discrete set of poses) in the workspace. When $OK = 0$ is returned, there very probably exists no collision free trajectory that allows the mobile platform to move freely within the given workspace (regardless of any path planning method).

Currently, this approach has only been validated on examples. One can select just one home pose X_c to check the collision free conditions with respect to the given workspace (X_c can be chosen as the center pose of the given workspace). To increase the reliability, we can apply the algorithm to a set of N_c home poses X_c and with large numbers of discrete poses (N_p and N_q are large). The computational time is proportional to $N_c \times N_p \times N_q$.

5 Simulation

Let us consider the 8-cable CDPR shown in Fig. 7. The mobile platform is a cube with 8 vertices. In this test, we show the computation time for each function call to check the collisions between cables and cables, and between cables and the CDPR mobile platform (using method 1 and method 2) while assuming that the mobile platform moves from the home pose X_c to the destination pose X_d. The home pose is $X_c = (0\,0\,0,\ 0\,0\,0)$ $(m,\ rad)$. The destination pose is varied. The results are given in Table 1. We use MATLAB to run the simulation on a PC with CPU core i7-2620M 2.7GHz.

In the case of checking the collision between cables and cables, we have the number of cables is $m = 8$. The computation time is quite expensive. The maximum computation time for checking the collision is around $t_{cc} \approx 4.65$ ms.

In the case of checking the collision between the cables and CDPR mobile platform using the first method, the mobile platform surface is triangulated into 12 triangles. For each cable, one need to verify potential collision with a maximum of 6 triangles. The maximum number of calls of the primitive test used to detect the collision

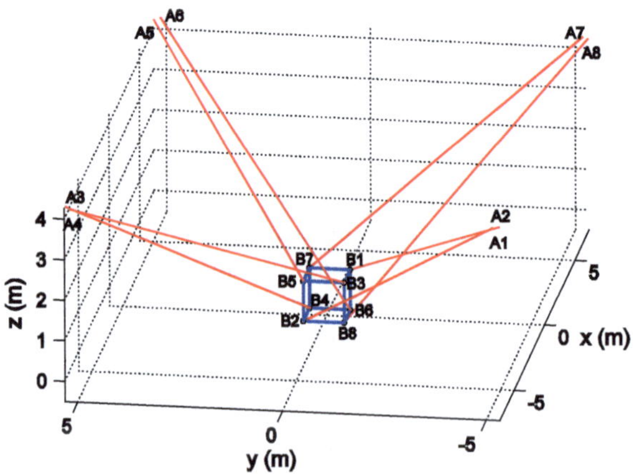

Fig. 7 Example of 8-cable CDPR

Table 1 Collision detection computation time

Destination pose X_d (m, rad)	Cables-cables (ms)	Cables-platform (method 1) (ms)	Cables-platform (method 2) (ms)
$(0\,0\,0,\ 0\,0-\pi/4)$	4.63 (no collision)	4.61 (collision)	0.74 (collision)
$(2\,0\,1,\ \pi/3-\pi/3\,0)$	4.57 (collision)	4.79 (collision)	0.92 (collision)
$(0\,0\,0,\ 0-\pi/3-\pi/4)$	4.54 (collision)	4.50 (collision)	0.39 (collision)
$(0\,0\,1,\ 0\,0\,\pi/4)$	4.65 (no collision)	5.28 (no collision)	1.47 (no collision)
$(0\,2\,0,\ 0\,0\,0)$	4.61 (no collision)	5.30 (no collision)	1.46 (no collision)
$(0\,0\,2,\ \pi/4\,0\,0)$	4.63 (no collision)	5.27 (no collision)	1.44 (no collision)

between a line segment and a triangle is $6 \times m = 48$. The maximum computation time is around $t_{cp1} \approx 5.3$ ms.

On the other hand, in the second method, the mobile platform has a convex shape. For each vertex B_i, there are a minimum of 3 neighboring vertices. The computation time in this case is significantly reduced compared to the first method. The maximum computation time is around $t_{cp2} \approx 1.47$ ms

Assume that one want to verify the CDPR capability over a given workspace where the Cartesian workspace is discretized into $N_p = 20$ points and the orientation workspace is discretized into the minimum number $N_q = 8$ points (taking only the extreme values of each angle into account). The number of considered home poses is $N_c = 10$ points. If we choose to use the second method to check the collision between the cables and the CDPR mobile platform then the maximum computation time to verify the collision free collision condition for the given workspace is around:

$$t_{\max} = N_c \times N_p \times N_q \times (t_{cc} + t_{cp2}) = 10 \times 20 \times 8 \times (4.65 + 1.47) \text{ ms} = 9.792 \text{ s}$$

Currently, it is up to the user to choose appropriate values of N_c, N_p and N_q, considering the trade off between reliability of the result and computation time. It is worth noting that, by using parallel computing (taking the advantages of both powerful CPU and GPU), one can also greatly reduce the computation times of the presented methods.

6 Conclusion

Several algorithms to detect the cable interferences of a CDPR have been discussed in this paper. The presented heuristic approaches improve the usual methods of detecting cable collisions in term of efficiency. Two types of cable interferences have been considered: collisions between cables and cables as well as collisions between cables and the CDPR mobile platform. The application of these tools was illustrated by an example of checking the collision free condition of a CDPR with respect to given Cartesian workspace and orientation workspace. The proposed approach offers the user a fast and reliable method to quantify the CDPR capability in term of position and orientation in the design or path planning phases.

In our future work, we will aim to explicitly "prove" the presented algorithms (to verify the collision free conditions of a CDPR with respect to a given range of orientations and given Cartesian workspace) as well as improve the method of simplifying the CDPR platform shape.

Acknowledgments The research leading to these results has received funding from the European Community's Seventh Framework Programme under grant agreement No. NMP2-SL-2011-285404 (CABLEBOT).

References

1. Albus J, Bostelman R, Dagalakis N (1993) The NIST robocrane. J Robot Syst 10(5):709–724
2. Merlet J-P (2008) Kinematics of the wire-driven parallel robot marionet using linear actuators. Proceedings of IEEE international conference on robotics and automation. pp 3857–3862
3. Merlet J-P, Daney D (2010) A portable, modular parallel wire crane for rescue operations. In: Proceedings of IEEE international conference on robotics and automation (ICRA) , May 2010. pp 2834–2839
4. Lamaury J, Gouttefarde M (2013) A tension distribution method with improved computational efficiency. Cable-driven parallel robots, mechanisms and machine science, vol 12. Springer, Heidelberg, pp 71–85
5. Merlet J-P (2004) Analysis of the influence of wires interference on the workspace of wire robots. In: Advances in Robot Kinematics. Kluwer Academic Publishers, Dordrecht, pp 211–218
6. Maeda K, Tadokoro S, Takamori T, Hiller M, Verhoeven R (1999) On design of a redundant wire-driven parallel robot WARP manipulator. In: Proceedings of IEEE international conference on robotics and automation. pp 895–900
7. Otis M, Perreault S, Nguyen-Dang T-L, Lambert P, Gouttefarde M, Laurendeau D, Gosselin C (2009) Determination and management of cable interferences between two 6-DOF foot platforms in a cable-driven locomotion interface. IEEE Trans Syst Man Cybern Part A Syst Hum 39(3):528–544
8. Perreault S, Gosselin C (2008) Cable-driven parallel mechanisms: application to a locomotion interface. J Mech Des 130:10
9. Perreault S, Cardou P, Gosselin C, Otis MJ-D (2010) Geometric determination of the interference-free constant-orientation workspace of parallel cable-driven mechanisms. J Mech Robot 2:011012
10. Wischnitzer Y, Shvalb N, Shoham M (2008) Wire-driven parallel robot: permitting collisions between wires. Int J Robot Res 27(9):1007–1026
11. Ericson C (2005) Real-time collision detection. Morgan Kaufmann, San Fransisco
12. Meshlab. http://meshlab.sourceforge.net/

Workspace Analysis of Redundant Cable-Suspended Parallel Robots

Alessandro Berti, Jean-Pierre Merlet and Marco Carricato

Abstract This paper focuses on cable-suspended parallel robots (CSPRs), a subclass of cable-driven parallel robots, and particularly on the analysis of their workspace. CSPRs present, among other interesting characteristics, large workspaces and high reconfigurability, which make them attractive for a large variety of applications, especially for pick and place operations over wide spaces. This paper is based on the assumption that the safest (and cheapest) control scheme for a redundant CSPR consists, at the current state of development, in actuating only 6 cables at a time. This paper shows how, under this assumption, it is still possible to take advantage of redundancy to enhance the workspace and eventually reduce the maximal tension among cables. A simple interval-analysis routine is presented as a tool for the workspace and trajectory analysis of a redundant CSPR, and the results of a case study on an existing prototype are discussed.

1 Introduction

Cable-driven parallel robots (CDPRs) employ cables in place of rigid-body extensible legs in order to control the end-effector pose. CDPRs strengthen classic advantages characterizing closed-chain architectures versus serial ones, like reduced mass and inertia, a larger payload to robot weight ratio, high dynamic performances, etc., while providing peculiar advantages, such as a larger workspace, reduced manufacturing

A. Berti · M. Carricato (✉)
Department of Industrial Engineering and Interdepartmental Center for Health Sciences and Technologies, University of Bologna, Bologna, Italy
e-mail: marco.carricato@unibo.it

A. Berti
e-mail: alessandro.berti10@unibo.it

J.-P. Merlet
HEPHAISTOS Project, French National Institute for Research in Computer Science and Control (INRIA), Sophia-Antipolis, France
e-mail: jean-pierre.merlet@inria.fr

A. Pott and T. Bruckmann (eds.), *Cable-Driven Parallel Robots*,
Mechanisms and Machine Science 32, DOI 10.1007/978-3-319-09489-2_4

and maintenance costs, ease of assembly and disassembly, high transportability, and superior modularity and reconfigurability.

A CDPR is *fully-constrained* if the pose of the end-effector is completely determined when actuators are locked and, thus, all cable lengths are assigned. A CDPR is, instead, *under-constrained* if the end-effector preserves some degrees of freedom (dofs) once actuators are locked [5, 14]. Accordingly, to completely restrain the n dofs of the moving platform, the number of cables employed by a fully-constrained CDPR is usually greater or equal to $n+1$, but it can be lowered to n when the robot is in crane configuration and gravity acts like an additional pulling wire. In this case, no cable can exert a downward tension on the platform. This latter kind of manipulators are also called *cable-suspended parallel robots* (CSPRs), and this paper is focused on them.

Fully-constrained CDPRs and CSPRs may operate in non negligible parts of their workspace as under-constrained ones, as mechanical equilibrium may require some cables to be slack. Accordingly, a rigorous investigation from a kinematic point of view must involve all the configurations in which one or more cables may get slack. Even though significant progresses have been made on the study of under-constrained CDPRs [1, 3–6, 13, 15], their control and kinematics still present difficult open issues that are not discussed in this paper.

In this contribution, cables are assumed to be perfectly stiff and massless. Under these assumptions, the maximum number of cables that can be under tension at the same time on a generic n–n CDPR with $n \geq 6$ is 6. This assumption is motivated by the results reported in [11] for the $N-1$ robots (N cables attached at the same point on the platform), in which it is also shown that, if elasticity in cables is taken into account, trying to obtain optimal distribution among the cable tensions through redundancy may lead to moderate positioning errors but, most important, to large error on cable tension. Although not yet formally proved, the authors believe that this result could be extended to any CSPR.

Let an *m-cable configuration* be the set of the m cables (each one identified by a digit h, $h = 1 \ldots m$) which are under tension at a given pose. This paper shows that, for a given pose, there may be several 6-cable configurations, which are not equivalent as far as the cable tensions and other kineto-static performance idexes are concerned. As the current cable configuration cannot be estimated using only the cable length measurements, a control scheme based on the simultaneous control of *all* cables may lead the robot to be, in a given pose, in a 6-cable configuration that is not the one exhibiting the better performance among all possible ones.

Hence, a safer strategy for the control of a redundant n–n CSPR may consist in using only 6 cables to govern the pose of the platform, and forcing the remaining $n-6$ to be slack. This seemingly counterintuitive approach seems not only the safest way to control a redundant CSPR, but also the only choice if the robot is not equipped with additional sensors. Moreover, as shown in [12], even if force sensors are available, they seem to be helpful to correctly distribute tension among more than 6 cables only if cable mechanical properties are perfectly known. However, this is a strong assumption, considering that, for example, characteristics like cable stiffness and density are extremely variable with humidity, temperature or even with time, when creeping occurs [10].

For a robot with $n \geq 6$ cables, assuming that only 6 cables are taut at the same time leads to C_6^n possible cable configurations in which the platform pose is fully controllable. As previously mentioned, in some regions of the workspace one or more cables may get slack and this fact leads to a risky and uncontrolled behavior. This implies that, in order to achieve a desired task, a preliminary analysis should be made to establish if the whole trajectory may be executed with a single set of 6 taut cables or if it is necessary to switch to a better cable configuration in some parts of the movement.

In this paper, an interval-analysis tool is presented to perform this kind of analysis on a generic CSPR with $n \geq 6$ cables. The following section provides some basic notion on interval analysis and describes the algorithm for the workspace analysis. In Sect. 3 the results obtained by applying the developed method on the geometry of an existing prototype are reported, and their consequences on trajectory planning are discussed. In Sect. 4, some conclusions are drawn.

2 The Interval-Analysis Procedure

This section describes the interval-analysis method used to determine the different wrench-feasible workspaces (WFWs [2]) for all sets of 6 cables in tension. In the current version, the algorithm does not take into account operating configurations with less than 6 taut cables. Accordingly, the core of the algorithm consists in solving the static equilibrium for a 6-cable suspended robot.

2.1 Introduction to Interval Analysis

A short introduction to interval analysis is presented in the following. More informations may be found in [7].

The *real interval* $X = \left[\underline{x}, \overline{x}\right]$ is defined as the set of real numbers y such that $\underline{x} \leq y \leq \overline{x}$. The *width* of the interval is $\mathrm{w}(X) = \overline{x} - \underline{x}$ and its *mid-point* is $\mathrm{mid}(X) = \left(\overline{x} + \underline{x}\right)/2$. An *interval vector* $\mathbf{X}$, also called a *box*, is a list of intervals. The mid-point of a box is the vector whose components are the mid-points of its interval components and the width of a box is the width of its largest element.

If $f(\mathbf{x})$ is a function in n unknowns, with $\mathbf{x} = [x_1, x_2, \ldots, x_n]$, and $\mathbf{B} = [X_1, X_2, \ldots, X_n]$ is a box comprising an interval for each unknown, an interval evaluation $F(\mathbf{B})$ of f over $\mathbf{B}$ is an interval $\left[\underline{F}, \overline{F}\right]$ such that, for any $\mathbf{x} \in \mathbf{B}$, $\underline{F} \leq f(\mathbf{x}) \leq \overline{F}$. There are many ways to implement an interval evaluation of a function but the simplest one is the *natural evaluation*, in which each arithmetic operation and elementary mathematical function is substituted by an interval equivalent. For example, if $f(x) = x^2 - 2x + 1$ and $X = [4, 5]$, the natural evaluation of f over X is:

$$f([4, 5]) = [4, 5]^2 - 2[4, 5] + 1 = [16, 25] - [8, 10] + [1, 1] = [7, 18] \quad (1)$$

It is worth emphasizing that the bounds provided by the natural evaluation of f are not exact: the upper (lower) bound may be larger (lower) than the actual maximum (minimum) of the function image, namely $f(\mathbf{B}) = \{f(\mathbf{x}) | x \in \mathbf{B}\} \subseteq F(\mathbf{B})$. Ordinarily, the overestimation decreases with the width of the box over which f is evaluated, and there are cases and methods that allow one to get bounds as tight as possible.

2.2 Geometrico-Static Model for a CSPR with 6 Cables

A mobile platform is connected to a fixed base by 6 cables. The ith cable ($i = 1, \ldots, 6$) exits from the base at point A_i and it is connected to the platform at point B_i. $Oxyz$ is a Cartesian coordinate frame attached to the base in O, whereas $Gx'y'z'$ is a Cartesian frame appended to the moving platform in G. In the following, $\mathbf{x} = G - O$, $\mathbf{a}_i = A_i - O$, $\mathbf{r}_i = B_i - G$, $\mathbf{s}_i = B_i - A_i = \mathbf{x} + \mathbf{r}_i - \mathbf{a}_i$, and ρ_i is the cable length, with $i = 1, \ldots, n$. Furthermore, if $\mathbf{b}_i = B_i - G$ in $Gx'y'z'$, Φ denotes the vector containing the variables parameterizing the orientation of the moving platform with respect to $Oxyz$ and $\mathbf{R}(\Phi)$ is the corresponding rotation matrix, then $\mathbf{r}_i = \mathbf{R}(\Phi)\mathbf{b}_i$. The platform pose may be described by the array $\mathbf{X} = [\mathbf{x}, \Phi]$, with x, y and z be the components of $\mathbf{x}$ in $Oxyz$ and α, β and γ be the rotation angles around the fixed axes x, y and z, respectively.

The normalized Plücker vector of the line associated with the ith cable is $\mathscr{L}_i/\rho_i$, where, in axis coordinates, $\mathscr{L}_i = \left[\mathbf{s}_i;\ \mathbf{a}_i \times \mathbf{s}_i\right]$. Accordingly, the wrench exerted by the ith cable on the platform is $\tau_i(\mathscr{L}_i/\rho_i)$, with τ_i being a positive scalar representing the intensity of the cable tensile force. The platform is acted upon also by a force of constant magnitude Q applied at point G, e.g. the platform weight acting through its center of mass. This force is described as a 0-pitch wrench $Q\mathscr{L}_e$, where $\mathscr{L}_e$ is the normalized Plücker vector of its line of action. Static equilibrium may then be expressed as:

$$\sum_{i=1}^{6} \frac{\tau_i}{\rho_i} \mathscr{L}_i = \underbrace{\left[\mathscr{L}_1/\rho_1 \ldots \mathscr{L}_6/\rho_6\right]}_{\mathbf{M}} \begin{bmatrix} \tau_1 \\ \vdots \\ \tau_6 \end{bmatrix} = -Q\mathscr{L}_e\,, \quad (2)$$

with $\tau_i \geq 0$, $i = 1, \ldots, 6$.

2.3 Description of the Algorithm

The aim of the procedure described hereafter is to determine the different WFWs for all the 6-cable configurations of a CSPR with $n \geq 6$ cables.

First, the program reads the geometry of the n–n redundant CSPR at hand from a convenient input file and generates the corresponding C_6^n configurations of anchor points on the base and the platform. For each configuration, a basic interval-analysis-based algorithm is executed. $\mathbf{B}_1$ is the box whose elements are the starting intervals for the components of $\mathbf{x}$ and Φ, and $\mathcal{L}$ is a list of boxes, initially set as $\mathcal{L} = \{\mathbf{B}_1\}$. An index i, initialized to 1, indicates which box $\mathbf{B}_i$ in $\mathcal{L}$ is currently being processed, while N denotes the number of boxes in $\mathcal{L}$. $\mathcal{S}_r$ and S_n are other two lists, initially empty, storing, respectively, the boxes thar are deemed to be in the WFW and the boxes for which the algorithm is not capable to assess if they belong or not to the WFW. The key component of the algorithm is the inclusion module $\mathcal{I}(\mathbf{B}_i)$, which takes a box $\mathbf{B}_i$ as input, solves the interval linear system (2) and returns the following output:

- 1: if all cable tensions are certainly positive;
- -1: if at least one cable tension is certainly negative;
- 0: otherwise;

The overall algorithm proceeds along the following steps:

1: $i = 1,\ \mathcal{L} = \{\mathbf{B}_1\},\ \mathcal{S}_r = \{\},\ \mathcal{S}_n = \{\},\ N = 1;$
2: if $i > N$, then return $\mathcal{S}_r$ and $\mathcal{S}_n$;
3: if $\mathrm{w}(\mathbf{B}_i) < \varepsilon$, then add $\mathbf{B}_i$ to $\mathcal{S}_n$, $i = i + 1$, go to 2;
4: compute $\mathcal{I}(\mathbf{B}_i)$

(a) if $\mathcal{I}(\mathbf{B}_i) = -1$, then $i = i + 1$, go to 2;
(b) if $\mathcal{I}(\mathbf{B}_i) = 1$, then add $\mathbf{B}_i$ to $\mathcal{S}_r$, $i = i + 1$, go to 2;
(c) if $\mathcal{I}(\mathbf{B}_i) = 0$, select a variable x_k and bisect X_k in the middle point, create two new boxes $\mathbf{B}_i'$ and $\mathbf{B}_i''$ from $\mathbf{B}_i$, replace $\mathbf{B}_i$ with $\{\mathbf{B}_i', \mathbf{B}_i''\}$ in $\mathcal{L}$, $N = N+1$, $i = i + 1$, go to 2.

Provided that ε is an user-input value which prevents the current box to be bisected if its width is below this threshold, the above algorithm always terminates, since the size of a box always decreases after a bisection.

2.3.1 The Inclusion Module $\mathcal{I}$

Hereafter, the function governing the inclusion module is described. This function first computes the interval evaluation of the matrix $\mathbf{M}$ in (2) with the elements of the input box $\mathbf{B}_i$. Then, it executes the interval version of the Gauss elimination scheme provided by the C++ library ALIAS [9] to solve the interval linear system, which leads to three possible outputs:

- if the Gauss elimination succeeds and the lower bounds of the components of resulting interval vector are strictly positive, it returns 1.
- if the Gauss elimination succeeds and at least one upper bound of the elements of the resulting interval vector is negative, it returns -1.
- otherwise, it returns 0;

However, this implementation often gives unsatisfying results, mainly because **M** is often ill conditioned and the Gauss-elimination leads to an overestimation of the solution, except for extremely small boxes. This obviously means a large number of bisections and long computation times. In the following, some procedures to improve the performances of the Gauss-elimination algorithm are proposed. One method relies on *pre-conditioning* the initial system as:

$$\mathbf{HMt} = \mathbf{HQ} \tag{3}$$

where **H** is an arbitrary 6×6 matrix, **t** is the interval vector of the cable tensions and $\mathbf{Q} = -Q\mathscr{L}_e$. A usually good candidate for matrix **H** is the inverse of the mid-point evaluation of **M**. This choice implies that, as the box size decreases, the product between **H** and **M** tends to be the identity matrix which is the best input for Gauss-elimination scheme. It must be remarked that the elements of **M** are functions of the pose variable **X**. Thus, in order to reduce the overestimation introduced by the matrix product, each element of the preconditioned matrix $\mathbf{C} = \mathbf{HM}$ can be rewritten trying to obtain the lowest number of multiple occurrences of the pose parameters. A further refinement comes from considering that each element of the interval matrix **C** has the following form

$$C_{ij} = \sum_{k=1}^{6} H_{ik} M_{kj} \tag{4}$$

and that each column of matrix **M** is a Plücker vector and consequently the following relations must hold:

$$M_{1j}^2 + M_{2j}^2 + M_{3j}^2 - 1 = 0 \tag{5}$$

$$M_{1j}M_{4j} + M_{2j}M_{5j} + M_{3j}M_{6j} = 0 \tag{6}$$

for $j = 1, \ldots, 6$. Thus, the bounds for each element of **C** may be improved by searching the minimum and the maximum of the function

$$C'_{ij}(\mathbf{q}) = \sum_{k=1}^{6} H_{ik} q_k \tag{7}$$

where **q** must satisfy the following constraints:

$$q_1^2 + q_2^2 + q_3^2 - 1 = 0$$

Table 1 Position of the anchor points on the base and the platform for the CoGiRo

Coordinates of the base exit points A_i, $1 = 1 \ldots 8$, in $Oxyz$ [m]							
−7.175	−7.316	−7.303	−7.161	7.182	7.323	7.302	7.161
−5.244	−5.103	5.236	5.373	5.348	5.206	−5.133	−5.270
5.462	5.472	5.476	5.485	5.488	5.499	5.489	5.497
Coordinates of the platform anchor points B_i, $1 = 1 \ldots 8$, in $Gx'y'z'$ [m]							
0.503	−0.510	−0.503	0.496	−0.503	0.500	0.502	−0.505
−0.493	0.351	−0.270	0.356	0.493	−0.340	0.275	−0.346
0.000	0.998	0.000	1.000	0.000	0.999	0.000	0.998

$$
\begin{aligned}
q_1 q_4 + q_2 q_5 + q_3 q_6 &= 0 \\
\underline{M_{kj}} \leq q_k \leq \overline{M_{kj}} \quad k &= 1, \ldots, 6
\end{aligned}
\tag{8}
$$

The optimization problem emerging from relations (7) and (8) can be analytically solved with the generalized Lagrange multiplier method and thus implemented without using numeric optimization algorithms.

3 Examples and Discussion of Results

This section presents the workspace analysis of all 6-cable configurations for the 8-8 cable robot CoGiRo [8] (whose geometry data are reported in Table 1), in relation to different trajectories. The algorithm described in Sect. 2 may be employed both to determine the WFW for each 6-cable configuration and to assess which configurations are active along a desired trajectory. Moreover, if multiple configurations are possible for a certain part of the trajectory, it is possible to rank them by computing for each one of them:

- the maximal tension $T_{\max}$ reached by the cables during the movement on the examined section of the trajectory;
- the maximal positioning error $E_{G\max} = 1/2\sqrt{\mathrm{w}(\Delta x)^2 + \mathrm{w}(\Delta y)^2 + \mathrm{w}(\Delta z)^2}$ and the maximal orientation error $E_{\Phi\max} = \mathrm{Max}[\mathrm{w}(\Delta\alpha), \mathrm{w}(\Delta\beta), \mathrm{w}(\Delta\gamma)]$ when errors on the cable-length measures are considered.

$E_{G\max}$ and $E_{\Phi\max}$ are computed by solving (with the interval Gauss elimination algorithm) the interval linear system

$$
\mathbf{J}^{-1}\Delta\mathbf{X} = \Delta\rho \tag{9}
$$

where the interval matrix $\mathbf{J}^{-1}$ is the inverse kinematic Jacobian of the examined 6-cable configuration, $\Delta\mathbf{X} = [\Delta\mathbf{x}^T, \Delta\Phi^T]^T$ is the interval vector containing the errors

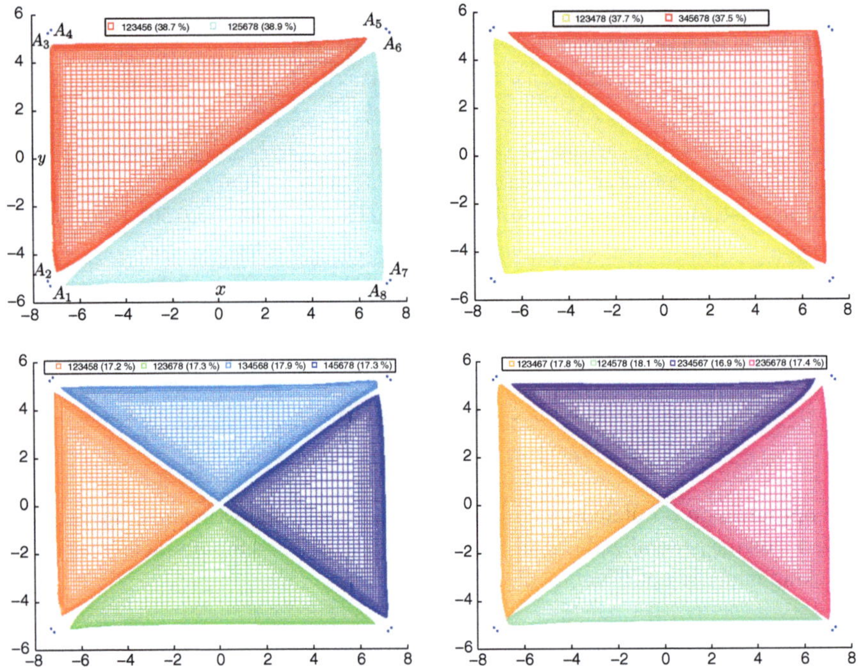

Fig. 1 The WFWs for the 6-cable configurations at $z = 3$ and $\Phi = [0\ 0\ 0]$, obtained with $\varepsilon = 0.05$ (Sect. 2). In the legends, the percentage of the total volume that each configuration occupies with respect to the initial search domain (i.e. with $z \in [0.0, 4.0]$) is shown between parenthesis

on the platform pose parameters, and $\Delta\rho$ is a user-input interval vector containing the bounds on the error in cable lengths.

All the above indexes are determined using the interval optimization procedure contained in ALIAS [9].

The algorithm was executed for $z \in [0.0, 4.0]$, $\Phi = [0\ 0\ 0]$ and $Q = 1000\,\text{N}$. Figure 1 represents sections, for $z = 3\,\text{m}$, of the WFWs for all 6-cable configurations. Some configurations are not represented in the figure because the algorithm, with a precision $\varepsilon = 0.05$, was not able to find any feasible box. It emerges from from Fig. 1 that different configurations must be used to fully explore the area bounded by the hull of the base exit points. In addition to that there is a small area in the middle of the figure where the algorithm did not succeed in finding a feasible 6-cable configuration. It is likely that, due to symmetries in the structure of the robot, this small area contains some points in witch only configurations with less than 6 cables in tension are feasible (from 1 to 5). These cases are out of the scope of this paper and will be addressed in future works. In the following, a 6-cable configuration is denoted by the symbol C_{ijklmn} where $ijklmn$ are the digits denoting the active cables (e.g. C_{123456} is the configurations with the first 6 cables under tension). Unless otherwise specified, the measures of lengths and distances are always expressed in meters and the angles in radians.

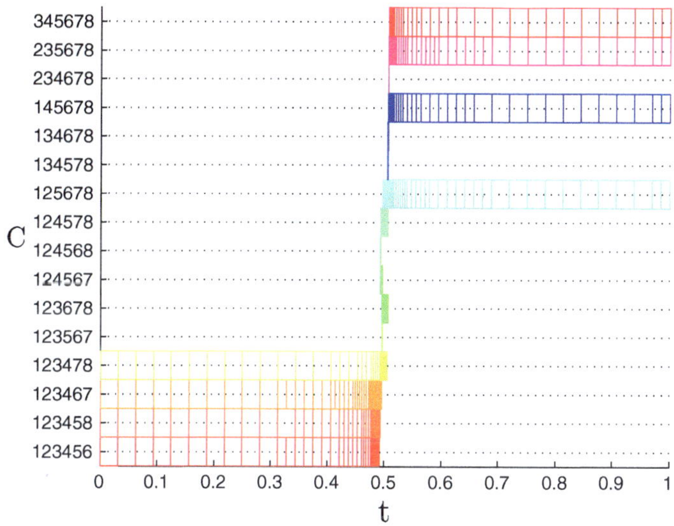

Fig. 2 The different cable configurations available along the trajectory $\overline{P_{0a}P_{1a}}$

3.1 Straight-Line Trajectories

Figure 1 shows a small area around $x = 0$ and $y = 0$ in which the algorithm can not assess (for $\varepsilon = 0.03$) if some 6-cable configuration is feasible. Thus a straight-line trajectory crossing this area is examined, from point $P_{0a} = [-6.0 \;\; 0.0 \;\; 3.0]$ to point $P_{1a} = [6.0 \;\; 0.0 \;\; 3.0]$. The platform has a constant orientation $\Phi = [0 \;\; 0 \;\; 0]$ and the trajectory is described by the parameter $t \in [0, 1]$. The results reported in Fig. 2 show that the minimum number of configurations necessary to perform the whole trajectory is 2, namely C_{123478} and C_{125678}. The intersection between these configurations is extremely small ($t \in [0.4968, 0.5045]$) and the other sets of active cables covering the same section have similar ranges (C_{123678}, C_{124578}). Considering that the computed intersection interval corresponds to roughly 9.2 cm on the trajectory line, a configuration switch seems practically unmanageable, even in quasi-static conditions. It is worth observing that, if the switch between two configurations is done outside this range, the platform may go to a configuration with less than 6 cables in tension and thus out of control.

Now, a linear trajectory from point $P_{0b} = [-6.0, 4.5, 4.0]$ to $P_{1b} = [6.0, -4.0, 0.0]$ is examined. The results in Fig. 3 show that the whole movement may be managed by the configuration C_{345678}. However, it may be interesting to investigate if there are some parts of the trajectory in which another configuration is more convenient. Then, the indexes $T_{\max}$, $E_{G\max}$ and $E_{\Phi\max}$ are computed (with an error on all cable lengths equal to $\Delta\rho_i = [-0.05, 0.05]$) for all feasible configurations in the trajectory sections $t_1 \in [0.0, 0.4]$ and $t_2 \in [0.6, 1.0]$. The results are reported in Table 2.

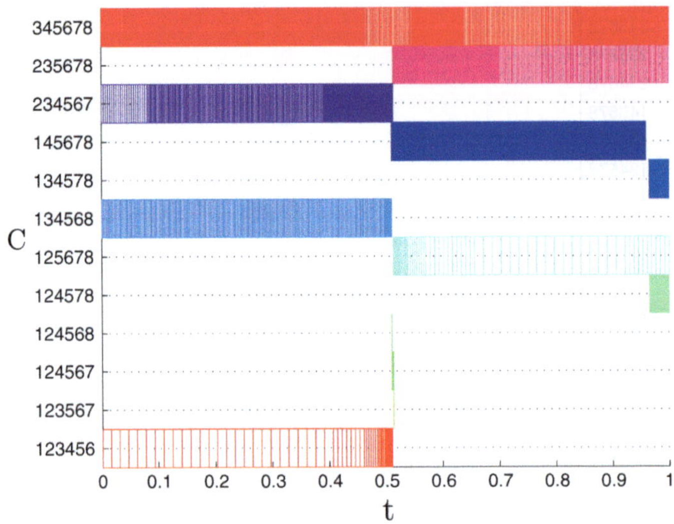

Fig. 3 The different cable configurations available along the trajectory $\overline{P_{0b}P_{1b}}$

Table 2 Comparison between different configurations for two sections of the trajectory $\overline{P_{0b}P_{1b}}$

	$t_1 \in [0.0, 0.4]$				$t_2 \in [0.6, 1.0]$		
	C_{123456}	C_{134568}	C_{234567}	C_{345678}	C_{125678}	C_{235678}	C_{345678}
$T_{\max}$[N]	874	1,069	1,033	1,034	664	724	726
$E_{G\max}$[m]	0.244	0.237	0.254	0.247	0.196	0.204	0.196
$E_{\Phi\max}$[rad]	0.381	0.519	0.496	0.644	0.284	0.431	0.542

While $E_{G\max}$ is similar for all configurations in each section, $T_{\max}$ and $E_{\Phi\max}$ are appreciably different. This shows that, even with a perfect robot model, tension is very sensitive to the configuration, whereas positioning accuracy remains almost constant. Accordingly, the first part of the trajectory could be more efficiently managed by C_{123456}, and the second one by C_{125678}.

3.2 Circular Trajectory

Here, a plane circular trajectory with constant orientation $\Phi = [0\ 0\ 0]$, centered in $P_c = [0.0\ \ 0.0\ \ 3.0]$ and with radius $r_c = 4.0$, is tested. Figure 4 shows that this trajectory requires at least 4 configurations, namely C_{345678}, C_{123456}, C_{123478} and C_{125678}. Table 3 compares the available 6-cable configurations for the trajectory sections $t_3 \in [0.15, 0.35]$ and $t_4 \in [0.65, 0.85]$. The results suggest that for these

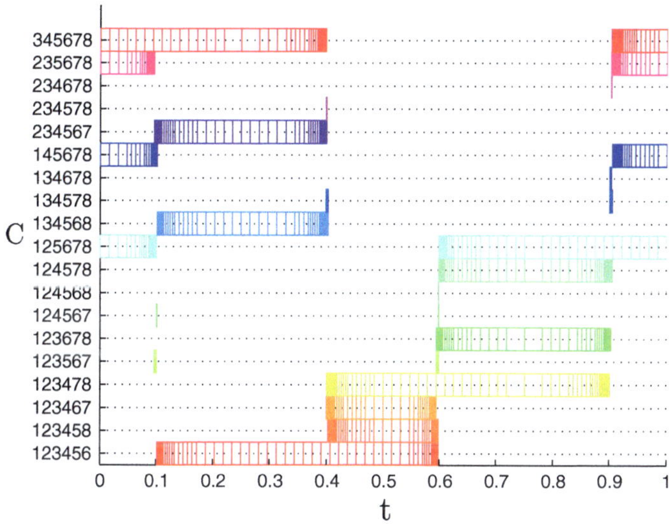

Fig. 4 The different cable configurations available along the circumference centered in $P_c = [0, 0, 4]$ and having radius $r = 4.0$

Table 3 Comparison between different configurations for two sections of the plane circular trajectory with center in $P_c = [0.0\ \ 0.0\ \ 3.0]$ and radius $r_c = 4.0$

	$t_3 \in [0.15, 0.35]$				$t_4 \in [0.65, 0.85]$			
	C_{123456}	C_{134568}	C_{234567}	C_{345678}	C_{123478}	C_{123678}	C_{124578}	C_{125678}
$T_{\max}[N]$	1,002	1,065	940	1,007	1,012	932	1,062	999
$E_{G\max}[m]$	0.258	0.256	0.248	0.244	0.244	0.248	0.256	0.258
$E_{\Phi\max}[rad]$	0.424	0.422	0.402	0.445	0.446	0.406	0.427	0.429

two parts of the path, a reduction of the maximal tension among cables is achievable with the configurations C_{234567} and C_{123678}, respectively.

3.3 Algorithm Performances

The computation time of the WFWs for all the 28 6-cable configuration with $\varepsilon = 0.05$ is roughly 245 min on a Intel® M620, 2.67 GHz CPU. This algorithm is indeed conceived to run off-line and eventually implement the obtained results in the control routine. On the other hand, the time required to analyze a single trajectory and ranking the configurations that are common for certain parts of the path usually varies from 3 to 10 s. For both cases, computation time can be significantly reduced with a distributed version of the proposed interval algorithm in order to take advantage of computer clusters and modern multi-core CPUs.

4 Conclusions

This paper studied the WFW of redundant cable-suspended parallel robots. The adopted approach is based on the assumption that the safest control strategy, at the current state of development of the control schemes for these family of manipulators, consists in using only 6 active cable at a time. However, the advantages of redundancy, such as increased workspace dimensions and reduction of maximal tension in cables, may be preserved by conveniently switching among the possible 6-cable configurations.

In order to explore this strategy, an interval-analysis-based method was developed in order to analyze, for all possible configurations with 6 cables in tension, both the WFW and a user-defined trajectory. Since some parts of a given trajectory may often be managed by different configurations, the possibility to individuate the best set of cables according to different performance indexes was examined and implemented in the algorithm. The results obtained from the conducted tests showed that switching between different configurations may be helpful to reduce the maximal tension in cables and the errors on the positioning of the platform, especially as far as the orientation is concerned. A similar analysis may be useful also at the design stage of a robot, in order to evaluate how many degrees of redundancy are required to cover a certain workspace or to accomplish a given set of tasks.

The approach presented in this paper rises some questions about manage the transition between two working configurations. The paper is based on the assumption that cables are massless and perfectly stiff so it is possible to show that the robot cannot have more than 6 cables under tension. So, at the transition point, coming say from cable configuration 123456, at least one of the tensions must be equal to 0. If the next cable configuration is 123457, while the robot is moving toward the transition point (with cable 7 slack) the controller starts coiling cable 7 so that at the transition point it has the exact length needed to be taut. Then cable 6 is quickly uncoiled and the control continues the movement using the 123457 configuration.

The presented interval-analysis procedure has wide margins of improvement. It would be interesting to develop it and see if it may be compatible with on-line applications. Moreover, future works will consider the extension of the analysis to include configurations with less than 6 cables in tension, as well as uncertainties in the location of the anchor points A_i and B_i.

Acknowledgments The financial support of the Italian Ministry of Education, Universities and Research through the PRIN grant No. 20124SMZ88_002 is gratefully acknowledged.

References

1. Abbasnejad G, Carricato M (2013) Direct geometrico-static problem of underconstrained cable-driven parallel robots with 5 cables. In: Thomas, F, Perez Gracia A (eds) Proceedings of the 6th international workshop on computational kinematics, pp 49–56. Springer, Dordrecht

2. Bouchard S, Gosselin C, Moore B (2010) On the ability of a cable-driven robot to generate a prescribed set of wrenches. ASME J Mech Robot 2(011):010/1–10
3. Carricato M, Abbasnejad G (2013) Direct geometrico-static analysis of under-constrained cable-driven parallel robots with 4 cables. In: Bruckmann T, Pott A (eds) Cable-driven parallel robots. Springer, Berlin, pp 269–285
4. Carricato M, Merlet J (2011) Direct geometrico-static problem of under-constrained cable-driven parallel robots with three cables. In: 2011 IEEE international conference on robotics and automation, pp 3011–3017. Shanghai, China
5. Carricato M, Merlet J (2013) Stability analysis of underconstrained cable-driven parallel robots. IEEE Trans Robot 29(1):288 296
6. Collard J, Cardou P (2013) Computing the lowest equilibrium pose of a cable-suspended rigid body. Optim Eng 14(3):457–476
7. Jaulin L (2001) Applied interval analysis: with examples in parameter and state estimation, robust control and robotics. Springer, London
8. Lamaury J, Gouttefarde M (2013) Control of a large redundantly actuated cable-suspended parallel robot. In: Proceedings of the 2013 IEEE international conference on robotics and automation, pp 4659–4664. Karlsruhe, Germany
9. Merlet JP (2000) Alias: an algorithms library for interval analysis for equation systems. COPRIN Project, INRIA. Available http://wwwsop.inria.fr/coprin/logiciels/ALIAS
10. Merlet JP (2008) Analysis of wire elasticity for wire-driven parallel robots. In: Proceedings of the 2nd European conference on mechanism science, pp 471–478. Cassino, Italy
11. Merlet JP (2012) The kinematics of the redundant n-1 wire driven parallel robot. In: 2012 IEEE international conference on robotics and automation, pp 2313–2318
12. Merlet JP (2013) On the accuracy of n-1 wire-driven parallel robots. In: Padois V, Bidaud P, Khatib O (eds) Romansy 19: robot design, dynamics and control, CISM international centre for mechanical sciences, vol 544. Springer, Vienna, pp 19–26
13. Michael N, Kim S, Fink J, Kumar V (2009) Kinematics and statics of cooperative multi-robot aerial manipulation with cables. In: Proceedings of the ASME 2009 international design engineering technical conferences, vol 7A, pp 83–91, paper no. DETC2009-87677. San Diego, USA (2009)
14. Ming A, Higuchi T (1994) Study on multiple degree-of-freedom positioning mechanism using wires–part 1: concept, design and control. Int J Japan Soc Precis Eng 28(2):131–138
15. Yamamoto M, Yanai N, Mohri A (2004) Trajectory control of incompletely restrained parallel-wire-suspended mechanism based on inverse dynamics. IEEE Trans Robot 20(5):840–850

On the Static Stiffness of Incompletely Restrained Cable-Driven Robot

Hui Li

Abstract The static stiffness of incompletely restrained cable-driven robot (IRCR) is studied and its theoretical formulation of the calculation model is derived. The analysis shows that the static stiffness matrix is a sum of two components. The first part $\mathbf{K}_1$ results from the change of position and orientation of end-effector which is proportional to cable tensions, while the second $\mathbf{K}_2$ is caused by the deformation and deflection of cables. The study shows that $\mathbf{K}_2$ is not only affected by material performances of all cables but by cable tensions. For the IRCR, however, the static stiffness correlates both parts because of the existence of cable catenaries. Two examples concerning the static stiffness are given for verification. Finally a few improvements are proposed for robot design based on stiffness analysis.

1 Introduction

Static stiffness of cable-driven parallel robots has been a topic of interest in that it reflects the control performance of cable robots. The purposes of stiffness study are clear: ① to prepare for the first design of the control and mechanical architecture; ② to know the tendency of stiffness variation in the whole workspace and then the possible payload applied on end-effector. Although finite element analysis makes the study feasible, the repeated mesh generations and calculations according to each possible pose of workspace may make the work tedious. A simple analytical model is then necessary to the conceptual design of cable robots. Cables are tension-only materials which is the first attention to be paid in the controllability design. That is why the stiffness of cable robot is generally lower than that of similar rod-driven robot.

H. Li (✉)
National Astronomical Observatories, Chinese Academy of Sciences, 20A Datun Road, Chaoyang District 100012, Beijing, China
e-mail: lihui@nao.cas.cn

H. Li
Key Laboratory of Radio Astronomy, Chinese Academy of Sciences, 20A Datun Road, Chaoyang District 100012, Beijing, China

A. Pott and T. Bruckmann (eds.), *Cable-Driven Parallel Robots*,
Mechanisms and Machine Science 32, DOI 10.1007/978-3-319-09489-2_5

There are some studies on stiffness analysis of cable robot. Verhoeven et al. [1, 2] deduce the stiffness matrix of cable robot. Lafoucade et al. [3, 4] and Zheng [5] put up with some theoretical tools in studying the static stiffness distribution cable robot. However the current researches are nearly focusing on completely or redundantly restrained cable robots in which cable tensions are much larger than cable weights and cable curves are therefore simplified as straight lines. The simplification makes study easy but can sometimes hardly be applied in IRCRs, especially when the huge cable robot of 500 m aperture spherical radio telescope (FAST) is considered. Kozak et al. [6] began to study the effects of sagging on under-restrained cable robot and the related inverse and forward kinematics on cable robot. Later Riehl et al. [7] also used a model of cable sagging to study numerical computations of the inverse and forward kinematics.

This paper addresses first the general stiffness formulation of cable robot, analyzing the respective physical meaning of its two components. Second a cable curve model is given for a further formulation. Two examples are then given for verifications. Finally the author further gives a few comments on how static stiffness contributes to robot design.

2 Formulation of Static Stiffness

2.1 Configuration

Figure 1 shows a general spatial cable robot with six degree-of-freedoms (Dofs). Let's assume that the end-effector itself is simplified as a 6-Dof rigid body. A frame **PXYZ** parallel is built with the center of gravity (COG) **P** of the end-effector set as the origin. The pose wrench of an arbitrary point on the end-effector is set as $\mathbf{S} = \left\{\mathbf{x}^{\mathrm{T}}, \boldsymbol{\psi}^{\mathrm{T}}\right\}^{\mathrm{T}}$, where is the coordinate vector in the frame **PXYZ** and $\boldsymbol{\psi} = \left\{\psi_x, \psi_y, \psi_z\right\}^{\mathrm{T}}$ the orientation vector of the end-effector making up of three Euler angles.

Let the matrix **K** represent the stiffness of cable robot on a set end-effector pose, provided that a static equilibrium is held under all cable tensions and the gravity (taken as a special cable without change in both direction and amplitude) acting on the end-effector. The force set is $\mathbf{T} = \left\{\mathrm{t}_1, \mathrm{t}_2, \ldots, \mathrm{t}_n\right\}^{\mathrm{T}}$ where the superscript T means transpose, so the induced force wrench should be zero as follows

$$\mathbf{F} = \mathbf{J} \cdot \mathbf{T} = \begin{bmatrix} \mathbf{u}_1, & \mathbf{u}_2, & \ldots, & \mathbf{u}_n \\ \mathbf{r}_1 \times \mathbf{u}_1, & \mathbf{r}_2 \times \mathbf{u}_2, & \ldots, & \mathbf{r}_n \times \mathbf{u}_n \end{bmatrix} \left\{\mathrm{t}_1, \mathrm{t}_2, \ldots, \mathrm{t}_n\right\}^{\mathrm{T}} = 0. \quad (1)$$

Here $\mathbf{F} = \left\{\mathrm{F}_x, \mathrm{F}_y, \mathrm{F}, \mathrm{M}_x, \mathrm{M}_y, \mathrm{M}_z\right\}^{\mathrm{T}}$ is the force wrench induced by cable tensions with all moments corresponding to the origin **P**; **J** represents the Jacobi matrix relating with the pose of end-effector and tension directions; $\mathbf{u}_i$ is unit vector with $\mathbf{u}_i = \left\{\mathrm{u}_x, \mathrm{u}_y, \mathrm{u}_z\right\}_i^{\mathrm{T}}$ $(i = 1, \ldots, n)$ and represents the tangential direction at the ith

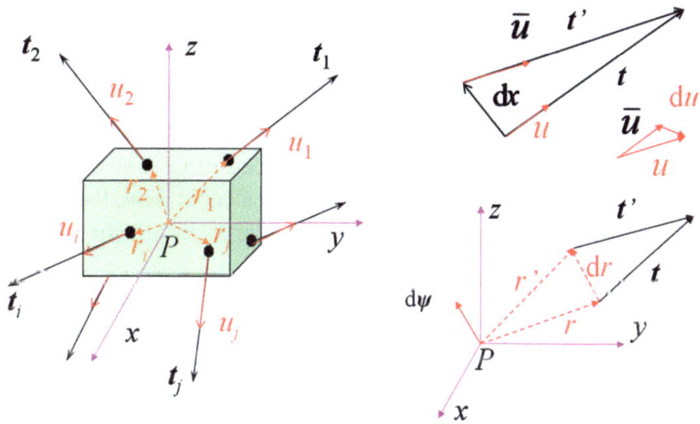

Fig. 1 Configuration of end-effector of cable robot under static equilibrium

cable joint; $\mathbf{r}_i$ is the position vector of the ith cable joint. The static stiffness of cable robot can then come out via differentiation of Eq. (1) as follows

$$\mathbf{K} = -\frac{\mathrm{d}\mathbf{F}}{\mathrm{d}\mathbf{S}} = -\frac{\mathrm{d}\mathbf{J}}{\mathrm{d}\mathbf{S}} \cdot \mathbf{T} - \mathbf{J} \cdot \frac{\mathrm{d}\mathbf{T}}{\mathrm{d}\mathbf{S}} = \mathbf{K}_1 + \mathbf{K}_2. \tag{2}$$

Here the minus means the restoring forces are always contrary in direction to the incremental displacements of the end-effector. Equation (2) shows that the static stiffness matrix is a sum of two components. The first part $\mathbf{K}_1$ results from the change of position and orientation of end-effector which correlates with cable tensions (so called configuration stiffness), while the second $\mathbf{K}_2$ is caused by the deformation and deflection of cables (so called eslatic or deformation stiffness).

2.2 Configuration Stiffness

By further expansion of Eq. (2), the configuration stiffness can be written in matrix form as:

$$\mathbf{K}_1 = -\frac{\mathrm{d}\mathbf{J}}{\mathrm{d}\mathbf{S}} \cdot \mathbf{T} = -\frac{\mathrm{d}}{\mathrm{d}\mathbf{S}} \begin{bmatrix} \mathbf{u}_1, & \mathbf{u}_2, & \ldots, & \mathbf{u}_n \\ \mathbf{r}_1 \times \mathbf{u}_1, & \mathbf{r}_2 \times \mathbf{u}_2, & \ldots, & \mathbf{r}_n \times \mathbf{u}_n \end{bmatrix}$$
$$\cdot \mathbf{T} = \sum_{i=1}^{n} \left(-\frac{\mathrm{d}}{\mathrm{d}\mathbf{S}} \begin{bmatrix} \mathbf{u} \\ \mathbf{r} \times \mathbf{u} \end{bmatrix}_i \cdot t_i \right), \tag{3}$$

where $-\frac{\mathrm{d}}{\mathrm{d}\mathbf{S}} \begin{bmatrix} \mathbf{u} \\ \mathbf{r} \times \mathbf{u} \end{bmatrix}_i$ is a 6×6 matrix and can be further expressed as follows:

$$-\frac{d}{d\mathbf{S}}\begin{bmatrix}\mathbf{u}\\ \mathbf{r}\times\mathbf{u}\end{bmatrix}_i=\begin{bmatrix}-\frac{d\mathbf{u}}{d\mathbf{S}}\\ -\mathbf{r}\times\frac{d\mathbf{u}}{d\mathbf{S}}-\frac{d\mathbf{r}}{d\mathbf{S}}\times\mathbf{u}\end{bmatrix}=\begin{bmatrix}-\frac{d\mathbf{u}}{d\mathbf{x}}, & -\frac{d\mathbf{u}}{d\psi}\\ -\mathbf{r}\times\frac{d\mathbf{u}}{d\mathbf{x}}-\frac{d\mathbf{r}}{d\mathbf{x}}\times\mathbf{u}, & -\mathbf{r}\times\frac{d\mathbf{u}}{d\psi}-\frac{d\mathbf{r}}{d\psi}\times\mathbf{u}\end{bmatrix}_i. \tag{4}$$

The child matrices in the right part of Eq. (4) can be further deduced and expanded. Provided that a translational increment happens on a cable joint, the tension vector $\mathbf{t}$ changes as $\mathbf{t}'$ while the unit vector $\mathbf{u}$ changes as $\bar{\mathbf{u}}$. The increment $d\mathbf{u}$ can be written as:

$$d\mathbf{u}=\bar{\mathbf{u}}-\mathbf{u}=\frac{\mathbf{t}'}{|\mathbf{t}'|}-\frac{\mathbf{t}}{|\mathbf{t}'|}=\frac{1}{L_C}\begin{bmatrix}u_x^2-1, & u_xu_y, & u_xu_z\\ & u_y^2-1, & u_yu_z\\ & & u_z^2-1\end{bmatrix}d\mathbf{x}. \tag{4a}$$

Here L_C can be approximated as the cord length of the related cable curve if the cable sag is relatively much smaller than its span. As the end-effector rotate an angular increment $d\boldsymbol{\psi}$ around the origin P, the related displacement increment is $d\mathbf{r}=d\mathbf{x}=d\boldsymbol{\psi}\times\mathbf{r}$. After a few vector and matrix calculations, a further deduction can be gain from equation Eq. (4a) as follows

$$-\frac{d\mathbf{u}}{d\boldsymbol{\psi}}=-\frac{d\mathbf{u}}{d\boldsymbol{\psi}\times\mathbf{r}}\cdot\frac{d\boldsymbol{\psi}\times\mathbf{r}}{d\boldsymbol{\psi}}=-\frac{d\mathbf{u}}{d\mathbf{x}}\cdot(\mathbf{I}\times\mathbf{r})=-\left(\mathbf{r}^T\times\frac{d\mathbf{u}}{d\mathbf{x}}\right)\cdot\mathbf{I}=-\left(\mathbf{r}\times\frac{d\mathbf{u}}{d\mathbf{x}}\right)^T$$
$$=\frac{1}{L_C}\begin{bmatrix}r_zu_xu_y-r_yu_xu_z, & r_z\left(1-u_x^2\right)+r_xu_xu_z, & -r_y\left(1-u_x^2\right)-r_xu_xu_y\\ -r_z\left(1-u_y^2\right)-r_yu_yu_z, & r_xu_yu_z-r_zu_xu_y, & r_x\left(1-u_y^2\right)+r_yu_xu_y\\ r_y\left(1-u_z^2\right)+r_zu_yu_z, & -r_x\left(1-u_z^2\right)-r_zu_xu_z, & r_yu_xu_z-r_xu_yu_z\end{bmatrix}. \tag{4b}$$

Here $\mathbf{I}$ is the unit matrix. Furthermore the other parts of the matrix in Eq. (4) can be expanded as follows:

$$-\frac{d\mathbf{r}}{d\mathbf{x}}\times\mathbf{u}=-\mathbf{I}\times\mathbf{u}=\begin{bmatrix}0 & -u_z & u_y\\ u_z & 0 & -u_x\\ -u_y & u_x & 0\end{bmatrix}; \tag{4c}$$

$$-\frac{d\mathbf{r}}{d\boldsymbol{\psi}}\times\mathbf{u}=-\frac{d\boldsymbol{\psi}\times\mathbf{r}}{d\boldsymbol{\psi}}\times\mathbf{u}=-(\mathbf{I}\times\mathbf{r})\times\mathbf{u}=-[(\mathbf{ru})-\mathbf{I}(\mathbf{r}\cdot\mathbf{u})]$$
$$=\begin{bmatrix}r_yu_y+r_zu_z & -r_xu_y & -r_xu_z\\ -r_yu_x & r_xu_x+r_zu_z & -r_yu_z\\ -r_zu_x & -r_zu_y & r_xu_x+r_yu_y\end{bmatrix}; \tag{4d}$$

$$-\mathbf{r}\times\frac{d\mathbf{u}}{d\psi}=\frac{1}{L_C}\begin{bmatrix}r_y^2+r_z^2-(r_yu_z-r_zu_y)^2, & -r_xr_y-(r_yu_z-r_zu_y)(r_zu_x-r_xu_z), & -r_zr_x-(r_xu_y-r_yu_x)(r_yu_z-r_zu_y)\\ & r_z^2+r_x^2-(r_zu_x-r_xu_z)^2, & -r_zr_y-(r_xu_y-r_yu_x)(r_zu_x-r_xu_z)\\ & & r_x^2+r_y^2-(r_xu_y-r_yu_x)^2\end{bmatrix}. \tag{4e}$$

Here in Eq. (4d) the expressions $(\mathbf{ru})$ and $(\mathbf{r}\cdot\mathbf{u})$ represent the dyad and dot product between $\mathbf{r}$ and $\mathbf{u}$ respectively. Based on Eq. (1) we get $\sum_{i=1}^{n}\mathbf{u}_i t_i = 0$ and $\sum_{i=1}^{n}(\mathbf{r}_i\times\mathbf{u}_i)\,t_i = 0$, so the following two equations are further derived:

$$\sum_{i=1}^{n}\left[-\frac{d\mathbf{r}}{d\mathbf{x}}\times\mathbf{u}\right]_i t_i = \sum_{i=1}^{n}\begin{bmatrix}0 & -u_z & u_y\\ u_z & 0 & -u_x\\ -u_y & u_x & 0\end{bmatrix}_i t_i = 0 \tag{5a}$$

$$\begin{cases}\sum\limits_{i=1}^{n}(r_x u_y)_i t_i = \sum\limits_{i=1}^{n}(r_y u_x)_i t_i\\ \sum\limits_{i=1}^{n}(r_y u_z)_i t_i = \sum\limits_{i=1}^{n}(r_z u_y)_i t_i\\ \sum\limits_{i=1}^{n}(r_x u_z)_i t_i = \sum\limits_{i=1}^{n}(r_z u_x)_i t_i\end{cases} \tag{5b}$$

Since the Eqs. (4d), (5a) and (5b) all give symmetric matrices, the configuration stiffness $\mathbf{K}_1$ is in total symmetric as well. Considering the Eqs. (5a) and (5b), it can be further written as follows:

$$\mathbf{K}_1 = -\sum_{i=1}^{n}\frac{d}{d\mathbf{S}}\begin{bmatrix}\mathbf{u}\\ \mathbf{r}\times\mathbf{u}\end{bmatrix}_i\cdot t_i = \sum_{i=1}^{n}\begin{bmatrix}-\frac{d\mathbf{u}}{d\mathbf{x}}, & -\frac{d\mathbf{u}}{d\psi}\\ -\mathbf{r}\times\frac{d\mathbf{u}}{d\mathbf{x}}, & -\mathbf{r}\times\frac{d\mathbf{u}}{d\psi}-\frac{d\mathbf{r}}{d\psi}\times\mathbf{u}\end{bmatrix}_i t_i. \tag{6}$$

Here its four child matrices are expanded in the Eqs. (4b), (4d) and (4e) respectively.

2.3 Deformation stiffness

By further expansion of Eq. (2), the deformation stiffness can be written in matrix form as:

$$\begin{aligned}\mathbf{K}_2 = -\mathbf{J}\cdot\frac{d\mathbf{T}}{d\mathbf{S}} &= -\mathbf{J}\cdot\begin{bmatrix}\frac{dT_1}{d\mathbf{x}}, & \frac{dT_2}{d\mathbf{x}}, & \ldots, & \frac{dT_n}{d\mathbf{x}}\\ \frac{dT_1}{d\boldsymbol{\psi}}, & \frac{dT_2}{d\boldsymbol{\psi}}, & \ldots, & \frac{dT_n}{d\boldsymbol{\psi}}\end{bmatrix}^T\\ &= -\mathbf{J}\cdot\begin{bmatrix}\nabla T_1, & \nabla T_2, & \ldots, & \nabla T_n\\ \mathbf{r}_1\times\nabla T_1, & \mathbf{r}_2\times\nabla T_2, & \ldots, & \mathbf{r}_n\times\nabla T_n\end{bmatrix}^T,\end{aligned} \tag{7}$$

where the symbol ∇ means the gradient calculator.

In the case of completely or redundantly restrained cable robot that tensile cables are simplified as straight lines, the gradients of cable tensions is in accord with their related cable lines and their values are determined by the cable materials, i.e., the modulus, sectional area, length, and so on. Therefore the gradient matrix in Eq. (7) is further expanded as follows:

$$\begin{bmatrix} \nabla T_1, & \nabla T_2, & \ldots, & \nabla T_n \\ \mathbf{r}_1 \times \nabla T_1, & \mathbf{r}_2 \times \nabla T_2, & \ldots, & \mathbf{r}_n \times \nabla T_n \end{bmatrix}^T$$

$$= \begin{bmatrix} -\frac{E_1A_1}{L_1}\mathbf{u}_1, & -\frac{E_2A_2}{L_2}\mathbf{u}_2, & \ldots, & -\frac{E_nA_n}{L_n}\mathbf{u}_n \\ -\frac{E_1A_1}{L_1}\mathbf{r}_1 \times \mathbf{u}_1, & -\frac{E_2A_2}{L_2}\mathbf{r}_2 \times \mathbf{u}_2, & \ldots, & -\frac{E_nA_n}{L_n}\mathbf{r}_n \times \mathbf{u}_n \end{bmatrix}^T$$

$$= -\mathrm{diag}\left(\frac{E_1A_1}{L_1}, \frac{E_2A_2}{L_2}, \ldots, \frac{E_nA_n}{L_n}\right) \cdot \left[\begin{Bmatrix} \mathbf{u}_1 \\ \mathbf{r}_1 \times \mathbf{u}_1 \end{Bmatrix}, \begin{Bmatrix} \mathbf{u}_2 \\ \mathbf{r}_2 \times \mathbf{u}_2 \end{Bmatrix}, \ldots, \begin{Bmatrix} \mathbf{u}_n \\ \mathbf{r}_n \times \mathbf{u}_n \end{Bmatrix}\right]^T$$

$$= -\mathrm{diag}\left(\frac{E_1A_1}{L_1}, \frac{E_2A_2}{L_2}, \ldots, \frac{E_nA_n}{L_n}\right) \cdot \mathbf{J}^T \tag{8}$$

Here similar to Eq. (2), the minus means the gradient is always contrary in direction to the incremental cable force. So the deformation stiffness of completely or redundantly restrained cable robot can be written as:

$$\mathbf{K}_2 = -\mathbf{J} \cdot \frac{d\mathbf{T}}{d\mathbf{x}} = \mathbf{J} \cdot \mathrm{diag}\left(\frac{E_1A_1}{L_1}, \frac{E_2A_2}{L_2}, \ldots, \frac{E_nA_n}{L_n}\right) \cdot \mathbf{J}^T \tag{9}$$

In the case of IRCR that on many occasions tensile cable has to be processed by cable catenary, however, the tension gradient is relatively much difficult to expect. It is determined by the shape change of cable catenary which is in turn caused via the position increment of cable joint. Before the derivation of the deformation stiffness, it is necessary to analyze at first the model of suspended cable as shown in Fig. 2.

Let's set **q** as the lineal weight of suspended cable, **l** as the span, **h** as the height, **f** as the cable sag in the middle span, **H** and **V** as the respective horizontal and vertical components of the cable tension. Here **H** is a constant along the whole cable. Then the curve function of cable catenary can be written as follows:

$$y(x) = \frac{H}{q}\cosh\left(\frac{qx}{H} + C\right) + D, \tag{10}$$

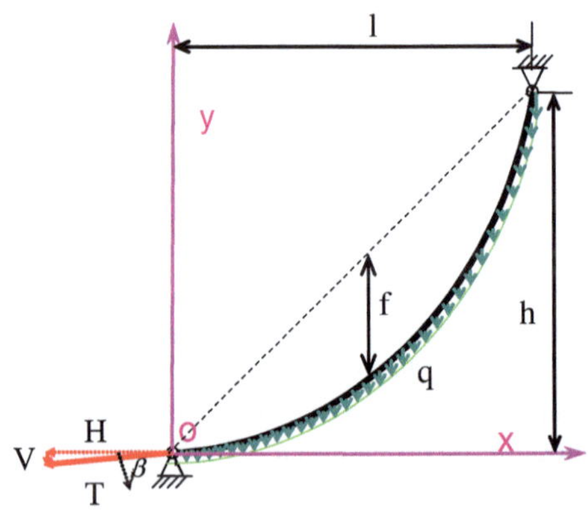

Fig. 2 Analytical model of cable catenary

where C and D are constants and can be expressed by aforementioned parameters as:

$$\begin{cases} C = \log\left(\frac{qh/H+\sqrt{(qh/H)^2+4\sinh^2\left(\frac{ql}{2H}\right)}}{\exp(qh/H)-1}\right) \\ D = -\frac{H}{q}\cosh(C) \end{cases} \tag{10a}$$

Considering the change of cable length before and after gravity bearing, we get:

$$\int_0^l \sqrt{1+\left(\frac{dy}{dx}\right)^2}dx - L_0 = \frac{H}{EA}\int_0^l \left(1+\left(\frac{dy}{dx}\right)^2\right)dx. \tag{11}$$

Here L_0 is the original length of cable before deformation; the left and right of Eq. (11) represent the total length change and elastic extension of cable respectively. Substituting Eq. (10) into Eq. (11) produces the following equation:

$$L - L_0 = \frac{H}{2EA}\left(l + \frac{q}{2H}\left(L^2+h^2\right)\frac{\exp(ql/H)+1}{\exp(ql/H)-1}\right), \tag{12}$$

where L the cable length after deformation. As the function of l and h, it satisfies:

$$L = \frac{H}{q}\sqrt{\left(\frac{qh}{H}\right)^2 + 4\sinh^2\left(\frac{ql}{2H}\right)}. \tag{13}$$

Generally cable tension of IRCR is not suggested too small or the controllability becomes worse. if $ql/H \leq 0.5$ can be satisfied, Eq. (13) can be further simplified with neglecting higher order terms, and its partial derivatives according to l and h have more simple forms:

$$L \approx \sqrt{l^2+h^2}\left(1+\frac{l^2}{24\left(l^2+h^2\right)}\left(\frac{ql}{H}\right)^2\right) \Rightarrow \begin{cases} \frac{\partial L}{\partial l} \approx \frac{l}{\sqrt{l^2+h^2}} - \frac{q^2l^4}{12H^3\sqrt{l^2+h^2}}\frac{\partial H}{\partial l} \\ \frac{\partial L}{\partial h} \approx \frac{h}{\sqrt{l^2+h^2}} - \frac{q^2l^4}{12H^3\sqrt{l^2+h^2}}\frac{\partial H}{\partial h}. \end{cases} \tag{14}$$

Equation (12) shows the relation among the variables like **H**, **l** and **h**. Taking **l** and **h** as the independent variables of function **H**, differentiating Eq. (12) according to **l** and **h** respectively, using the relation in Eq. (14) and neglecting higher order terms, we get the derivatives $\partial H/\partial l$ and $\partial H/\partial h$ as follows:

$$\begin{cases} \frac{\partial H}{\partial l} = \frac{2EA}{\sqrt{l^2+h^2}} \Big/ \left(1+\frac{q^2(L^2+h^2)}{H^2}\cdot\frac{\exp(ql/H)}{(\exp(ql/H)-1)^2}+\frac{q^2l^3EA}{6H^3\sqrt{l^2+h^2}}\right) \\ \frac{\partial H}{\partial h} = \frac{2EAh}{l\sqrt{l^2+h^2}} \Big/ \left(1+\frac{q^2(L^2+h^2)}{H^2}\cdot\frac{\exp(ql/H)}{(\exp(ql/H)-1)^2}+\frac{q^2l^3EA}{6H^3\sqrt{l^2+h^2}}\right) \end{cases} \tag{15}$$

At cable joint we get the relation between cable tension t and its horizontal component H as follows:

$$T = H\sqrt{1 + (y'(0))^2} = \frac{1}{2}\left(\frac{q\,(L+h)}{\exp(ql/H) - 1} + \frac{H^2}{q} \cdot \frac{\exp(ql/H) - 1}{L + h}\right). \tag{16}$$

Differentiating Eq. (16) according to **l** and **h** respectively, using the relation in Eq. (15) and neglecting higher order terms, we get the derivatives $\partial t/\partial l$ and $\partial t/\partial h$ as follows:

$$\begin{cases} \frac{\partial T}{\partial l} = \frac{\partial H}{\partial l} \cdot \left(\frac{q^2 l(L+h)\exp(ql/H)}{2H^2(\exp(ql/H)-1)^2} + \frac{H(\exp(ql/H)-1)}{q(L+h)} - \frac{\exp(ql/H)\cdot l}{2(L+h)}\right) \approx \frac{\partial H}{\partial l} \cdot \frac{\sqrt{l^2+h^2}}{l} \\ \frac{\partial T}{\partial h} = \frac{\partial H}{\partial h} \cdot \left(\frac{q^2 l(L+h)\exp(ql/H)}{2H^2(\exp(ql/H)-1)^2} + \frac{H(\exp(ql/H)-1)}{q(L+h)} - \frac{\exp(ql/H)\cdot l}{2(L+h)}\right) \approx \frac{\partial H}{\partial h} \cdot \frac{\sqrt{l^2+h^2}}{l} \end{cases}. \tag{17}$$

It indicates from Eqs. (15) and (17) that $|\nabla T| = \sqrt{(\partial H/\partial l)^2 + (\partial H/\partial h)^2} \rightarrow EAl/\left(l^2 + h^2\right)$ is sure if $ql/H \rightarrow 0$, therefore we get the relation: $|\nabla T| \rightarrow EA/\sqrt{l^2 + h^2} = EA/L$ that is just the tension gradient of completely or redundantly restrained cable robot.

Equation (17) further shows that the tangent of tension gradient is equal to $\frac{\partial T}{\partial h}/\frac{\partial T}{\partial l} = \frac{\partial H}{\partial h}/\frac{\partial H}{\partial l} = \frac{h}{l}$, so its direction is parallel to the cord line of cable curve (defined as vector $\bar{\mathbf{u}}$). Similarly, the direction of the gradient is opposite to $\bar{\mathbf{u}}$, namely $\nabla T_i = -\left(\sqrt{(\partial T/\partial l)^2 + (\partial T/\partial h)^2}\right)_i \bar{\mathbf{u}}_i$. Substituting it into Eq. (7) produces the following:

$$\mathbf{K}_2 = \mathbf{J} \cdot \mathrm{diag}\left(\ldots, \left(\sqrt{\left(\frac{\partial T}{\partial l}\right)^2 + \left(\frac{\partial T}{\partial h}\right)^2}\right)_i, \cdots\right) \cdot \begin{bmatrix} \bar{\mathbf{u}}_1, & \bar{\mathbf{u}}_2, & \cdots, & \bar{\mathbf{u}}_n \\ \mathbf{r}_1 \times \bar{\mathbf{u}}_1, & \mathbf{r}_2 \times \bar{\mathbf{u}}_2, & \cdots, & \mathbf{r}_n \times \bar{\mathbf{u}}_n \end{bmatrix}^T. \tag{18}$$

Equation (18) indicates that in the case of cable catenary usually we get $\bar{\mathbf{u}} \neq \mathbf{u}$, so K_2 is asymmetric matrix. The phenomenon comes mainly from the geometrically nonlinear performance of suspended cable. Only if ql/H is quite small, i.e., less than 0.2, is $\bar{\mathbf{u}} \approx \mathbf{u}$ reasonable.

In summary there are two components that make up of the static stiffness of cable robot. The first is defined configuration stiffness which can be calculated by Eqs. (3)–(6). The second is defined deformation stiffness and its estimation is different according to different type of cable robots. In the case of completely or redundantly restrained cable robot, it can be calculated by Eq. (9) while in the case of IRCR it should be calculated based on Eq. (18).

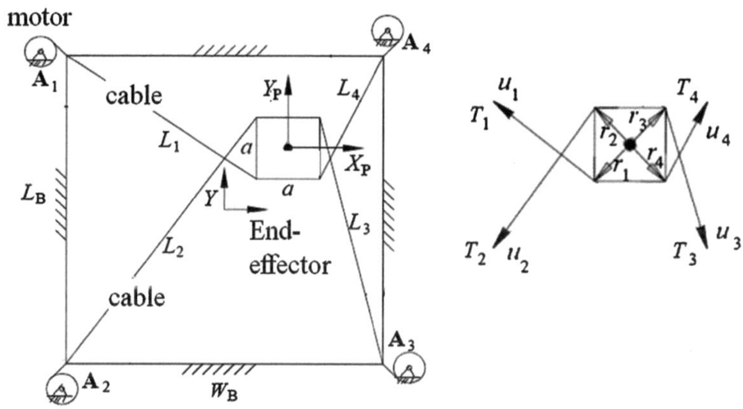

Fig. 3 A planar 3-DoF cable robot with 4 parallel cables

3 Numerical Examples

3.1 Redundantly Restrained Cable Robot

Figure 3 shows a planar 3-DoF cable robot which has 4 cables with each end of a cable connected to a fixed motor on the base frame or a joint on the end-effector. The square end-effector has in plane two independent translational DoFs, x and y, and a rotational DoF around z axis. Since the number of cables is larger than that of DOFs, the robot is redundantly restrained. However, the design requires that the rotation of the end-effector should be forbidden to keep it always horizontal. Some basic parameters are given, such as the dimension of the base frame: $L_B = W_B = 1.4\,m$; the dimension of the end-effector: $a = 0.2\,m$; the cable: Φ 8 mm steel rope with modulus equal to 1.5 × 105 MPa and sectional area equal to $30\,mm^2$.

Let's take the COG of the end-effector (origin of the local frame $O_pX_pY_p$) as the moment center. The robot,s static stiffness at any point of its workspace is then calculated based on Eqs. (4) and (9). When the end-effector moves to the center of base frame where the global coordinate is equal to (0,0), let's assume that the tension of each cable share the same value: 4,500 N; the length is 1 m; their direction vectors are in sequence: $u_1 = (-0.6, 0.8)$; $u_2 = (-0.6, -0.8)$; $u_3 = (0.6, -0.8)$; $u_4 = (0.6, 0.8)$. The estimation based on aforementioned equations gives:

$$K_1 + K_2 = \begin{bmatrix} 1.1520 & 0 & 0 \\ 0 & 0.6480 & 0 \\ 0 & 0 & -0.0353 \end{bmatrix} \times 10^4 + \begin{bmatrix} 0.6480 & 0 & 0 \\ 0 & 1.1520 & 0 \\ 0 & 0 & 0.0353 \end{bmatrix} \times 10^7$$

$$= \begin{bmatrix} 0.6492 & 0 & 0 \\ 0 & 1.1526 & 0 \\ 0 & 0 & 0.0353 \end{bmatrix} \times 10^7. \quad (19)$$

The above calculation can be verified via finite element analysis according to the definition of static stiffness which gives nearly the same result as diag(0.6491, 1.1526, 0.3525) $\times 10^7$.

When the end-effector moves to the point with the global coordinate equal to (0.5,0.5), let's assume that the cable tensions are in sequence: $t_1 = 1{,}898$N, $t_2 = 620$ N, $t_3 = 5{,}560$ N and $t_4 = 5{,}397$ N; the length in sequence: $L_1 = 1.14$ m, $L_2 = 1.703$ m, $L_3 = 0.316$ m and $L_4 = 1.304$ m; their direction vectors are in sequence: $u_1 = (-0.965, 0.263)$; $u_2 = (-0.646, -0.763)$; $u_3 = (0.077, -0.997)$; $u_4 = (0.316, 0.949)$. The estimation based on aforementioned equations gives:

$$K_1 + K_2 = \begin{bmatrix} 1.6982 & -0.2666 & 0.0016 \\ & 0.9552 & -0.0057 \\ & & -0.0520 \end{bmatrix} \times 10^4 + \begin{bmatrix} 0.6219 & 0.4306 & 0.0768 \\ & 1.8051 & 0.1665 \\ & & 0.0379 \end{bmatrix} \times 10^7$$
$$= \begin{bmatrix} 0.6239 & 0.4302 & 0.0768 \\ & 1.8055 & 0.1665 \\ & & 0.0379 \end{bmatrix} \times 10^7. \tag{20}$$

Similarly the finite element analysis gives nearly the same result. The above two equations indicates that $\mathbf{K}_1$ contributes little to the total stiffness compared with $\mathbf{K}_2$. It is because $\mathbf{K}_1$ is determined by cable tension while $\mathbf{K}_2$ by the material stiffness of cable which is nearly three order higher than the former. Considering only the torsional component (the last diagonal element) of the stiffness matrix, according to Eqs. (4) and (6) it can be written as follows:

$$K_T = K_{T1} + K_{T2} = \sum_{i=1}^{4} \left\{ (r_x u_x + r_y u_y)_i + \frac{\left[r_x^2 + r_y^2 - (r_x u_y - r_y u_x)^2 \right]_i}{L_i} \right\} t_i$$
$$+ \sum_{i=1}^{4} \frac{\left[(r_x u_y - r_y u_x)^2 \right]_i}{L_i} E_i A_i. \tag{21}$$

The equation indicates that it is true: $K_{T1} \ll K_{T2}$ if r and L has the similar order. But if $r \ll L$, the situation may change. Let's take the case of above planar 3-DoF cable robot for example. when the end-effector is at the center of base frame, keeping all parameters unchanged except that the dimension of the base frame increases as: $L_B = 800$ m , $W_B = 600$ m and cable length increasing correspondingly, we get the torsional stiffness as $K_{T1} + K_{T2} = -360.69 + 705.52 = 344.83$ N $\cdot$ m/rad. The finite element analysis gives similar value: 344.95 N $\cdot$ m/rad. Obviously the contribution of K_{T1} is quite a lot.

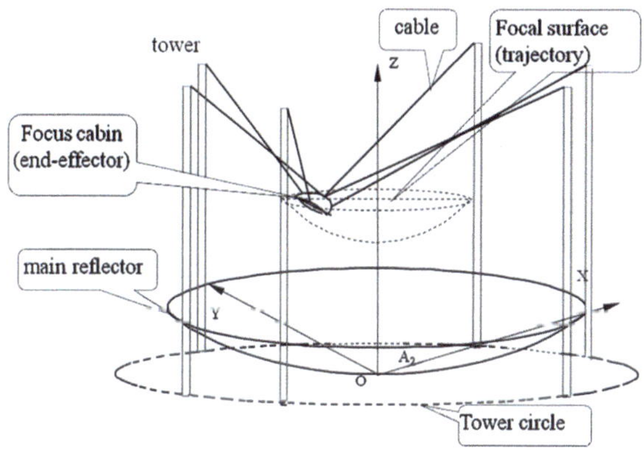

Fig. 4 Sketch of FAST cable robot

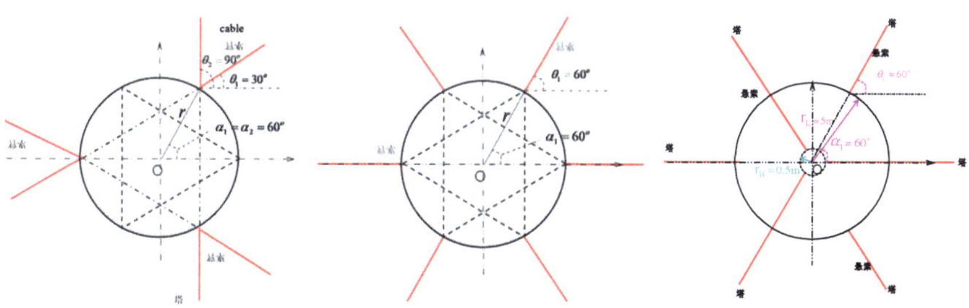

Fig. 5 Design concept of cabin-cable connection

3.2 Incompletely Restrained Cable Robot

A typical example is the cable robot of the FAST telescope whose static equilibrium must depend on the gravity. Figure 4 shows one of its conceptual designs. Its end-effector, the airborne 30-ton focus cabin with the diameter of 10 m, is driven by six parallel steel cables. Each cable has a sectional area of about 531 mm^2 and a lineal weight of about 71 N/m. Its trajectory curve, the focal surface, is a part of spherical surface with the curvature diameter equal to about 160 m, the aperture equal to about 203 m, and it shares the same curvature center with the main reflector of the telescope.

The connection between the cabin and cables is an interesting research topic. There are a few design concepts as shown in Fig. 5. Here both in the concept 1 and concept 3 there are six cable joints and each neighbor joints have an equal angle of 60° in the horizontal plane while only three cable joints exist with angle of 120° in concept 2. On the other side, in both concept 1 and concept 2 all the cable joints share the same circle plane with a diameter of 10 m. In concept 3, however, the six cable joints are distributed on two planes: three inner top joints on a circle of 1 m in diameter and three outer bottom joints on a circle of 10 m in diameter.

For comparison and searching for an optimal design, an effective way is to study the torsion stiffness of the cable robot as the cabin stays at the center of the focal surface. At that time both the torsion stiffness and the damping of the cable system are very low. The fact in turn greatly reduces the robot's controllability. In such case the calculation of cable tension is relatively simple because of the symmetry. According the reference [8, 9], the results are given as 1.220×10^5 N for the cable tension t and 1.090×10^5 N for its horizontal component $\mathbf{H}$.

For each cable the condition $ql/\mathbf{H} \leq 0.2$ is satisfied, so it is reasonable to get $\mathbf{u} \approx \bar{\mathbf{u}}$. Using Eqs. (4), (6) and (18), the torsion stiffness is obtained as follows:

$$\mathrm{K_T} = \mathrm{K_{T1}} + \mathrm{K_{T2}} = \sum_{i=1}^{6} \left\{ (r_x u_x + r_y u_y)_i + \frac{\left[r_x^2 + r_y^2 - (r_x u_y - r_y u_x)^2\right]_i}{L_i} \right\} T_i + \sum_{i=1}^{6} \left[(r_x u_y - r_y u_x)^2 \sqrt{\left(\frac{\partial T}{\partial l}\right)^2 + \left(\frac{\partial T}{\partial h}\right)^2} \right]_i \tag{22}$$

According to Fig. 5, the ith tension vector at its cable joint can be written as $\mathbf{u}_i = \{\cos\beta_i \cos\theta_i,\ \cos\beta_i \sin\theta_i,\ \sin\beta_i\}^T$ and the corresponding position vector is given as $\mathbf{r}_i = \{r_i \cos\alpha_i,\ r_i \sin\alpha_i,\ 0\}^T$, where β_i represents the angle between the *i*th tension vector and the horizontal plane. Substituting these expressions of vectors into Eq. (22) produces the following:

$$\mathrm{K_T} = \sum_{i=1}^{6} \left[H_i r_i \cos(\alpha_i - \theta_i) + \frac{T_i r_i^2 \left(1 - \cos^2\beta_i \sin^2(\alpha_i - \theta_i)\right)}{L_i} \right] + \sum_{i=1}^{6} \left[\left(\sqrt{\left(\frac{\partial T}{\partial l}\right)^2 + \left(\frac{\partial T}{\partial h}\right)^2} \right)_i r_i^2 \cos^2\beta_i \sin^2(\alpha_i - \theta_i) \right]. \tag{23}$$

In the case of concept 1, we get $|\alpha_i - \theta_i| = 30°$, the gradient of cable tension as 9.870×10^4 N/m and $\cos\beta = 0.8944$. So the torsion stiffness is calculated as:

$$K_T = 3\sqrt{3}\mathbf{H}r + \frac{6Tr^2\left(1 - 0.25\cos^2\beta\right)}{L} + 6 * 9.8704e4 \cdot r^2 \cos^2\beta * 0.25$$
$$= 6.189 \times 10^6\ \mathrm{N \cdot m/rad}.$$

In the case of concept 2, we get $|\alpha_i - \theta_i| = 0°$, so $\mathrm{K_{T2}}$ is zero. So the torsion stiffness is calculated as:

$$\mathrm{K_T} = \mathrm{K_{T1}} = 6Hr + \frac{6Tr^2}{L} = 3.330 \times 10^6\ \mathrm{N \cdot m/rad}.$$

The case of concept 3 is somehow similar to that of concept 2 as $|\alpha_i - \theta_i| = 0°$ is obtained, so K_{T2} is also zero. Since it has 2 joint circles at different height, the torsion stiffness is calculated as:

$$K_T = K_{T1} = 3Hr_L + 3Hr_H + T \cdot \frac{3r_L^2 + 3r_H^2}{L} = 1.828 \times 10^6 N \cdot m/rad.$$

In summary concept 1 gives the highest torsion stiffness, nearly twice as the value of concept 2 and four times as that of concept 3, so it is singled out as the optimal design. Both in concept 2 and 3 the tension vectors intersect with **z** axis, so the deformation stiffness has no contribution. Furthermore it is worse that the configuration of concept 3 makes the moment arm of three top tensions smaller, in turn reduces its configuration stiffness.

4 Static Stiffness and First Design

Generally the first design of cable robot prefers to higher stiffness for good stability and controllability of its end-effector. The above-mentioned stiffness formulation helps to quantify the necessary stiffness estimation and therefore to draw a preliminary conclusion whether a design is good or not. The formulation as well as above numerical examples further indicate three effective ways for the improvement of such stiffness. These key factors may be related to cable tension vectors, namely their magnitudes and moments, and cable length as well.

The first is obviously observed in the formulation to increase the total level of cable tensions. In the case of completely or redundantly restrained cable robot, the cable force system is self-equilibrating. A given combination of position and pose of the end-effector may give rise to infinite tension solutions of a cable system. That means cable tension can theoretically go up without bound. It is not easy in the case of IRCR, however, for its tension level is limited by external force or even unique according to a given position and pose as analyzed in reference [9]. That is why IRCR usually seems much more flexible.

The second is focused on cable length as well as the ratio of such tension arm to it. It is easily concluded from both Eqs. (9) and (15) that the shorter the cable length, the higher the deformation stiffness. If cable length has to be designed very long in the case of giant FAST cable robot, another feasible way may be proposed to increase tension arm and improve its configuration stiffness as possible. Usually configuration stiffness plays a more important role in large IRCRs.

The third pays more attention to the details of cable robot like the relation between cable-platform joint and tension direction, as shown in the configuration of both numerical examples. The main goal is to keep the original moment of each tension as large as possible wherever end-effector stays within its work space. These detailed designs aim not to lift the global level of robot stiffness, but they surely give rise to local improvements in rotation which is ultimate to vibration control of end-effector, as proved in the second numerical example.

5 Conclusions

The static stiffness of cable robot can be decomposed as two parts. One is configuration stiffness proportional to cable tensions and can be calculated with Eq. (6); the other is deformation stiffness. For completely or redundantly restrained cable robot, it is only related to cable material and can be calculated with Eq. (9). For IRCR, however, it is related to both cable material and cable tension and can be calculated with Eq. (18).

In the case of completely or redundantly restrained cable robot, the deformation stiffness generally contributes almost the total. In the case of IRCR, however, on many occasions both parts are important.

Static stiffness is an important factor to be considered in the first design. Stiffness formulation indicates three effective ways to improve such stiffness. In the case of usually more flexible IRCR, keeping end-effector balanced via non-zero moments may obtain promising stiffness improvement.

Acknowledgments The author would like to acknowledge the financial support of the National Natural Science Foundation (NNSF) under grant No. 10973023, 11203048 and 11103046.

References

1. Verhoeven R, Hiller M, Tadoroko S (1998) Workspace, stiffness, singularities and classification of tendon driven Stewart platforms [C]. In: Proceedings of the 6th international symposium on advances in robot kinematics, June/July, 1998, Strobl/Salzburg, Austria. Salzburg: Kluwer Academic Publishers, pp 105–114.
2. Verhoeven R (2004) Analysis of the workspace of tendon-based Stewart platforms. Ph.D. Dissertation, Gerhard Mercator University, Duisburg
3. Lafourcade P, Libre M, Rboulet C (2002) Design of a parallel wire-driven manipulator for wind tunnels. In: Proceedings of the workshop on fundamental issues and future directions for parallel mechanisms and manipulators, Kluwer Academic Publishers, Quebec City, Canada, pp 187–194, 3–4 Oct 2002
4. Lafourcade P, Zheng YQ, Liu XW (2004) Stiffness analysis of wire-driven parallel kinematic manipulators. In: Proceedings of the 11th world congress in mechanism and machine science, China Machine Press, Tianjin, China, pp 1878–1882, 18–21 Aug 2003
5. Zheng YQ (2004) Research on key theoretical issues of wire-driven parallel kinematic manipulators and the application to wind tunnel support system [D]. Huaqiao University, Quanzhou
6. Kozak K, Zhou Q, Wang JS (2006) Static analysis of cable-driven manipulators with non-negligible cable mass. IEEE Trans Rob 22(3): 425–433
7. Riehl N, Gouttefarde M, et al. (2009) Effects of non-negligible cable mass on the static behavior of large workspace cable-driven parallel mechanisms. In: Proceedings of the IEEE international conference on robotics and automation, pp 2193–2198

8. Li HUI, Rendong NAN, Kaercher H, et al (2008) Working space analysis and optimization of the main positioning system of FAST cabin suspension. In: Proceedings of SPIE, astronomical instrumentation, ground-based and airborne telescopes II, SPIE-International Society for Optical Engine, Marseille, France. Sandieg, pp 70120T–1–70120T-11, June, 2008
9. Li H, Zhang X, Yao R, et al. (2013) Optimal force distribution based on slack rope model in the incompletely constrained cable-driven parallel mechanism of FAST telescope. In: Cable-driven parallel robots. Springer, Heidelberg, 87–102

Simulation and Control with XDE and Matlab/Simulink of a Cable-Driven Parallel Robot (CoGiRo)

Micaël Michelin, Cédric Baradat, Dinh Quan Nguyen and Marc Gouttefarde

Abstract We present in this paper the process allowing to create a cable-driven parallel robot (CDPR) simulation within the XDE software environment in C++ language. The elementary classes constituting a CDPR are shown with their constructor specificities. The winches, the pulleys, the cable fastenings and the platform are presented. The parameterization of elements such as the cable material characteristics, structure and size are detailed. An interface between the XDE cable-driven parallel robot simulator and a Matlab/Simulink controller have been developed. Inputs and outputs are exchanged between the controller and the simulated cable-driven robot, exactly as it is done with a physical robot.

1 Introduction

The simulation of systems allows engineers to test concepts, improve and optimize the design allowing substantial economy in the development process by avoiding the manufacturing of expensive real prototypes and reducing the time of development. This is particularly true for cable-driven parallel robot (CDPR) design where the size of robot and then the cost can grow very fast. Classical rigid robots are well simulated in environments such as Simulia [1], Microsoft Robotics Studio [2], Marilou [3], SpaceClaim [4] or Solidworks [5]. In the case of cable-driven parallel robot simulation, the highly flexible nature of the cables is extremely challenging to simulate.

A CDPR consists of several key elements such as the cables, the winches, the pulleys and the mobile platform. It can be seen that, to be able to successfully simulate

M. Michelin (✉) · C. Baradat
Tecnalia France, MIBI, 672 rue du Mas de Verchant, 34000 Montpellier, France
e-mail: micael.michelin@tecnalia.com

D.Q. Nguyen · M. Gouttefarde
LIRMM-CNRS—UM2, 161 Rue Ada, 34095 Montpellier Cedex 5, France

A. Pott and T. Bruckmann (eds.), *Cable-Driven Parallel Robots*,
Mechanisms and Machine Science 32, DOI 10.1007/978-3-319-09489-2_6

a CDPR, a simulation tool should be capable of at least: "correctly" simulating the nonlinear behavior of the cables, the process of winding or un-winding cables on drums and effectively managing different collision types (e.g. between cables-cables, cables-mobile platform, cable-self-collision) as well as important effects such as friction contacts (e.g. between cables-pulleys, cables-drums).

Few simulation tools have been developed with specific components to achieve such goals. Some of these tools can be listed. Algoryx [6] or Vortex [7] are used to create training simulators for crane or construction equipment and have to be real time capable. Others focus on the precise and quantitative simulation of cables for design process such as Dymola [8] or Adams [9]. The multi-physics software Sofa [10] has been developed to provide deformable objects simulation for medical applications, and can be used for cable modelling. However, from our experience, obtaining stable simulations in this open-source framework is a critical issue for a complicated CDPR scene including a lot of finite elements and friction contacts.

In the present work, we focus on the XDE [11] multi-physics software which is more intended to research developments. We aim to develop a cable-driven robot simulation framework based on XDE which offers the user two main features: simulation and control of CDPR with XDE and MATLAB/Simulink.

The paper is organized as follows. Section 2 introduces the concepts of simulating a CDPR in XDE, including the key features of XDE and the method to construct different parts of a CDPR (e.g. winch system, cables, pulleys, mobile platform). An illustration is made by creating the CoGiRo prototype [12] in XDE. The interfacing between XDE and MATLAB/Simulink is presented in Sect. 3. Finally, Sect. 4 gives some discussions on the simulation features, current issues and future work.

2 Principle of Modeling in XDE Simulation Environment

The use of XDE for CDPR simulation consists in a first step to create classes of the different sub structures needed to model the whole robot. Indeed a eight cables robot uses eight winches, eight cable fastenings on the platform, etc. Classes have been developed to create the cables, winches, cable fastenings, pulleys and the mobile platform. These classes integrate the conceptual elements needed to simulate the sub systems. Each sub system is constituted by the assembly of the following main elements:

- the rigid body in which we define the inertial parameters (mass, inertia matrix)
- the 3D mesh used for the collision model
- the 3D mesh used for the visual model
- the material parameters involved in the contact model
- the material parameters involved in the deformable model
- the joints between the rigid bodies to create mechanical assembly with some degrees of freedom (DOF) taking into account dry and viscous frictions.

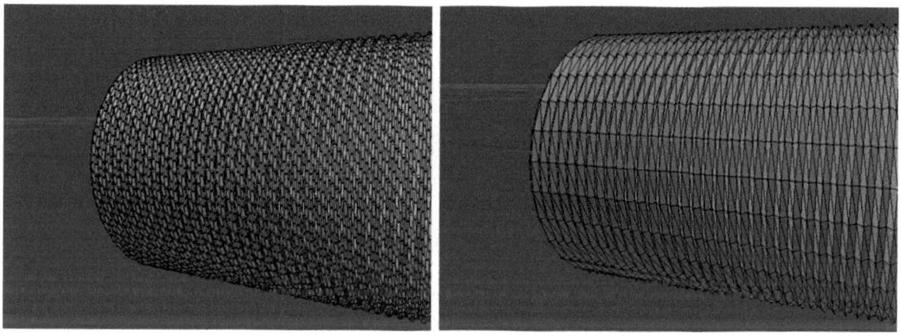

Fig. 1 Original 3D mesh of the drumfrom CAD software (*left*) and simplified 3D mesh used in simulation (*right*)

2.1 Winch Modeling

The class "Winch" is composed of two rigid bodies, the motor and the drum, linked together by a pivot joint on which we can apply a torque via a "setWrench" method. The 3D mesh for the motor and the drumare imported from a CAD software. These 3D meshes have to be drastically simplified to reduce the loading time of the scene and the computation time during the simulation (Fig. 1). For example, from the original CAD file with 150,000 faces drum's mesh, we have reduced by 10 the number of faces to be able to run the simulation in good conditions. This simplified mesh is used to compute the collision interactions with the cable via a Signorini contact law. The position and the orientation of a winch are arguments of the constructor of the "Winch" class which allow to get a complete winch (Fig. 2).

The moment of inertia of the drum has been computed from the CoGiRo robot data. A gear reduction equal to three is used in the CoGiRo robot winches. Is has not been simulated but the inertia effects have been included in the drum one as described in the following:

- Drum inertia vs rotation axis: $0.0285\,\mathrm{Kg\,m^2}$
- Motor rotor inertia vs rotation axis: $0.015\,\mathrm{Kg\,m^2}$
- Simulated drum inertia vs rotation axis: $0.0285 + 9 \times 0.015 = 0.1635\,\mathrm{Kg\,m^2}$

2.2 Pulley Modeling

The pulleys are modeled with three rigid bodies: a base to be fixed to the robot frame, a bracket linked to the base with a pivot joint and the wheel of the pulley linked to the bracket with another pivot joint (Fig. 3).

The 3d mesh of the wheel is used for the collision interaction with the cable. The collision interaction for the base and the bracket are deactivated.

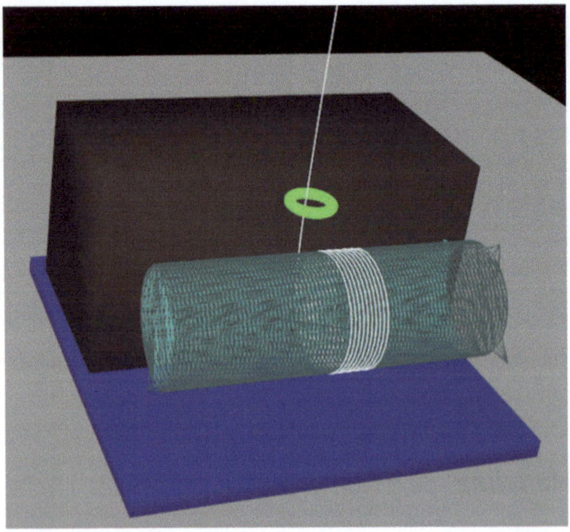

Fig. 2 Complete winch assembly in XDE

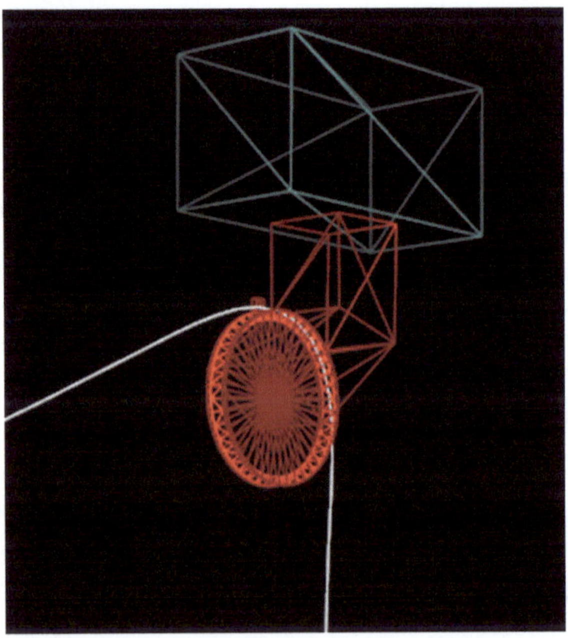

Fig. 3 Pulley modelling

Dry and viscous friction parameters of the pivot joints can be tuned. The position and the orientation of a pulley are arguments of the constructor of the "Pulley" class.

The mass parameters are:

- Bracket mass: 3.456 Kg
- Wheel mass: 1 Kg

2.3 Cable Fastening Modeling

The cable interfaces with the platform are implemented in the "Cable_Ball_Joint" class. They are made of two rigid bodies, the base and the fastening part, linked together by two pivot joints forming a U-joint (Fig. 4).

The base has to be welded tothe object we want to manipulate with cable and the fastening part has to be weldedto a cable. In addition, the base integrates three prismatic joints which are "weld", this means that these joints are blocked but are necessary to allow the measure of the cable tensions at the fastening location, without producing any motion. Indeed these joints have an orthogonal disposition and give an access to the force applied on it via "getWeldReactionForce" method. The position of a cable interface is given versus the object linked with and which has to be given as

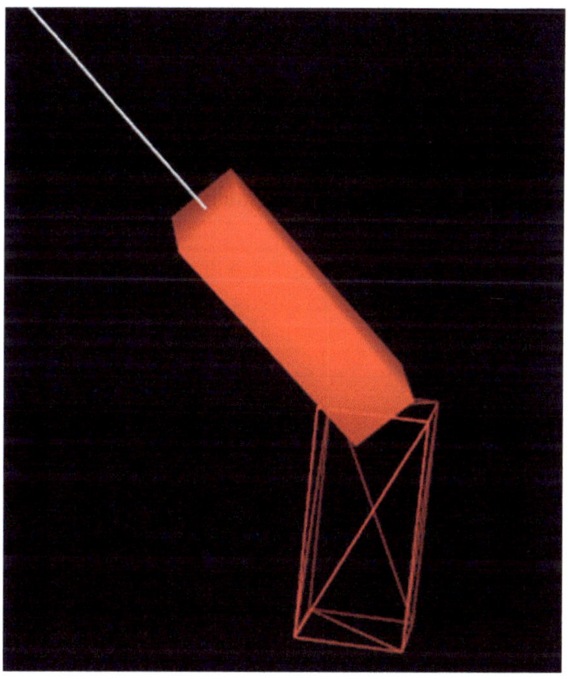

Fig. 4 Cable interface with platform

Fig. 5 CoGiRo platform in XDE simulator

argument of the “Cable_Ball_Joint” constructor. The orientation of a cable interface is an argument of the constructor of the “Cable_Ball_Joint” class.

The mass parameters are the following:

- Base mass: 0.23 Kg
- Fastening mass: 0.122 Kg

The inertia matrices have been computed considering these two parts as homogenous cuboid.

2.4 Platform Modeling

The CoGiRo platform is made of two major element, an aluminum alloy cubic frame and a steel fork allowing to lift up loads (Fig. 5). Two rigid bodies have been implemented and connected together with their own 3D mesh. The number of element faces in the fork mesh has been reduced from 13,000 faces in the original to 400 faces in the simplified one (Fig. 6).

Fig. 6 Fork original 3D mesh (*left*) and simplified 3D mesh used in simulation (*right*)

Fig. 7 CoGiRo cable structure

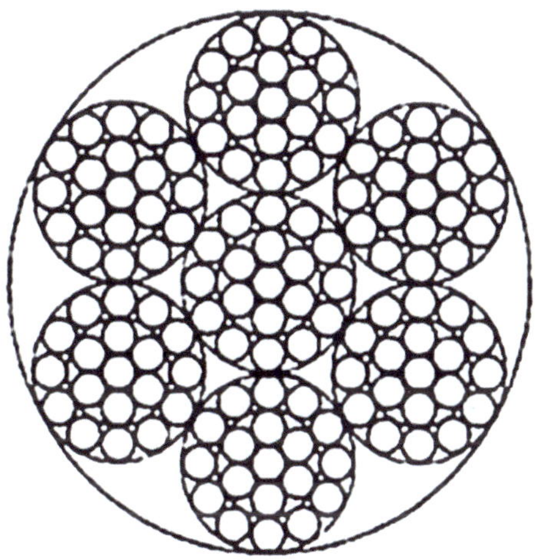

2.5 Cable Modeling

The cable modeling in XDE consists in using Reissner beam made of multiple nodes. The number of nodes and the position of each one at the initial state have to be defined. A 0.02 m node resolution has been chosen.

The radius of the Reissner beam which compose the cable has to be defined.

As a cable is not really a beam, we have computed the equivalent metallic section corresponding to the cable. The CoGiRo cable (Fig. 7) is 2 mm radius and has a filling rate of 0.5911. So the equivalent metallic section has a radius of 1.537 mm.

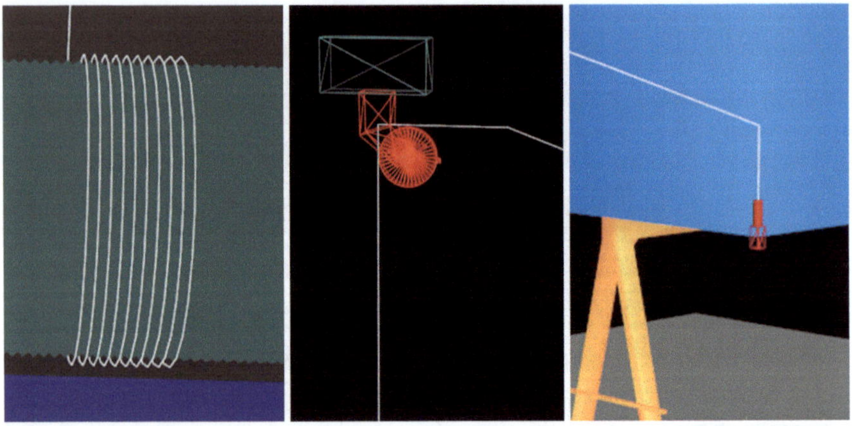

Fig. 8 Initial cable routing

The "Cable" class constructor needs as arguments: the winch, on which the cable is attached, the pulley for guiding the cable and the cable interface on which the end of cable is connected.

In a first step, the constructor computes the cable nodes position allowing the first meters of cable to be winded on the winch (Fig. 8, left). Then the cable is routed via two points around the pulley (Fig. 8, center). Finally, a third routing point is used before cable connection to the cable attachment (Fig. 8, right) which is fixed to the platform.

XDE gives the opportunity to define cables made of several coaxial materials but in our case only one material was used. The material parameters such the Young modulus, the shear modulus, the linear mass, the damping coefficient, the torsional ratio and the flexural ratio have to be defined in the simulation environment. The CoGiRo cable Young modulus has been evaluated at 100e9 Pa. The shear modulus is difficult to define and has been evaluate around 10e9 Pa. The linear mass is 0.064 Kg/m.

Except the linear mass, all the other parameters are difficult to tune within the simulation. They may have a significant influence on the stability and the increasing of cable tensions needs an increase of the damping coefficients to avoid instabilities. But if we increase the damping, then the dynamic behavior is not realistic. We didn't reach a good tradeoff between stability and high cable tension. High loaded cable stability by damping increase is contradictory with a realistic dynamic behavior.

To avoid cable instabilities when subjected to high tension, the XDE software provider advices to use sampling frequencies ten times higher than the phenomenon frequency we want to simulate.

Typically if we want to simulate a 100 Hz oscillating cable (a guitar cord for example), we have to use a sampling frequency of 1,000 Hz, which leads to long computation time as described in the following section.

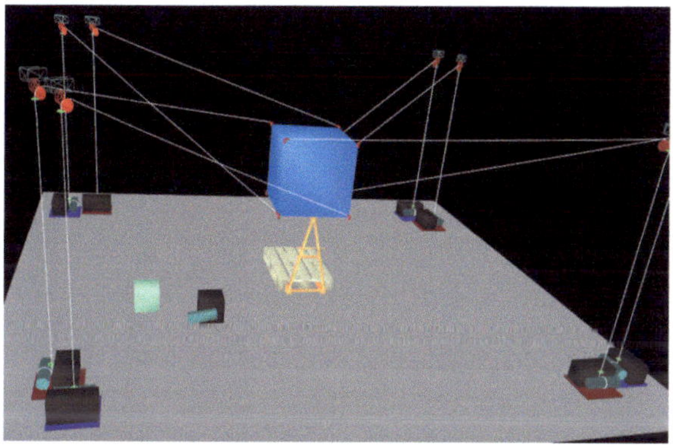

Fig. 9 Complete CoGiRo robot scene in XDE

In order to perform a preliminary simulation without stability problems, it has been decided to use the following parameter values, despite the fact that they are not realistic:

- Linear mass: 0.064 kg/m
- Young modulus: 10e9 Pa (ten times lower than the evaluated one)
- Shear modulus: 10e8 Pa
- Damping coefficient: 0.1
- Torsional ratio: 0.5
- Flexural ratio: 0.5

2.6 Complete CoGiRo Robot Scene

All the classes described before allow the user to create as much objects as needed to create the cable-driven robot. The constructors arguments allow specifying the position and orientation of each object easily. In Fig. 9, winches, pulleys, cables and fastenings have been added in the scene in order to model the CoGiRo robot with eight cables. The objective of this simulation is to pick and place a pallet. The pulleys orientation is computed from the cable fastening position versus the winch position.

The loading and execution times of the scene depend on the number of cable nodes, and thus on the cable length. Thus, in order to reduce the loading and execution time, the footprint of the CoGiRo robot has been reduced. Only the relative positions of the winches and pulleys have been modified, not the size of the different modules of the robot. The real size of the CoGiRo robot is 16 m × 12 m × 6 m, whereas in the simulation it is 6 m × 6 m × 3 m. With these dimensions and the node resolution, each cable has between 400 and 600 nodes.

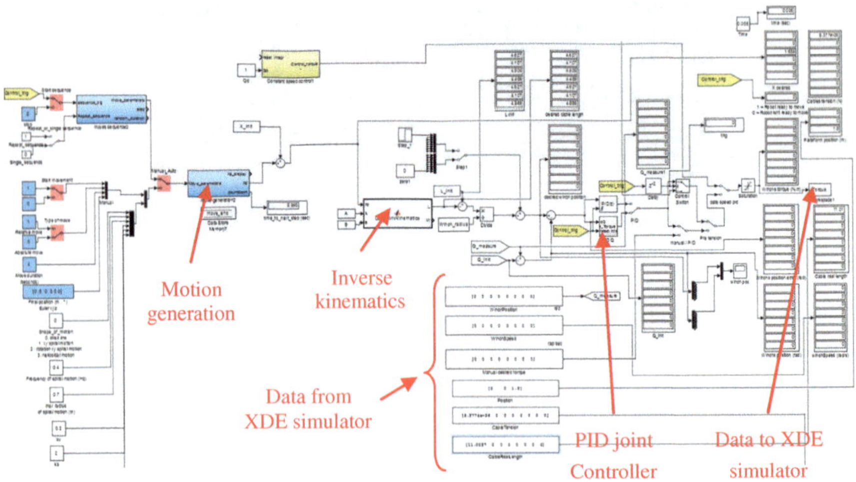

Fig. 10 Simulink cable-driven robot controller

3 Interface with Matlab/Simulink Controller

With the aim of controlling the simulated CoGiRo robot, an interface with Matlab/Simulink has been developed, which uses Matlab Engine C++ library within the simulator project. Matlab is configured to be used as a server and the simulat orcan use the functions available in the Matlab Engine to interact with the Matlab Workspace and the Simulink model. With this interface, we can control the execution of a Simulink model via the XDE simulator.

A Simulink robot controller including the motion generation, the inverse kinematics and a PID controller for motors position has been developed (Fig. 10). For each step time, XDE simulator sends data to the Simulink controller:

- Time (simulated real time)
- Winches joint position
- Winches joint speed
- Platform position and orientation
- Cables tension at fastening locations
- Cables length

Then XDE simulator reads the data coming from the Simulinkcontroller, in our case the desired torques for the motors. Thenthe XDE simulator executes one step time of the Simulink controller.

Thus, the XDE simulator execution is synchronized with the Simulink controller execution, under the condition that sampling time are set the same in XDE and Simulink environments.

The control scheme used here is the same we use for a real prototype, only the input/output exchanges mechanisms are adapted to suit the XDE simulator.

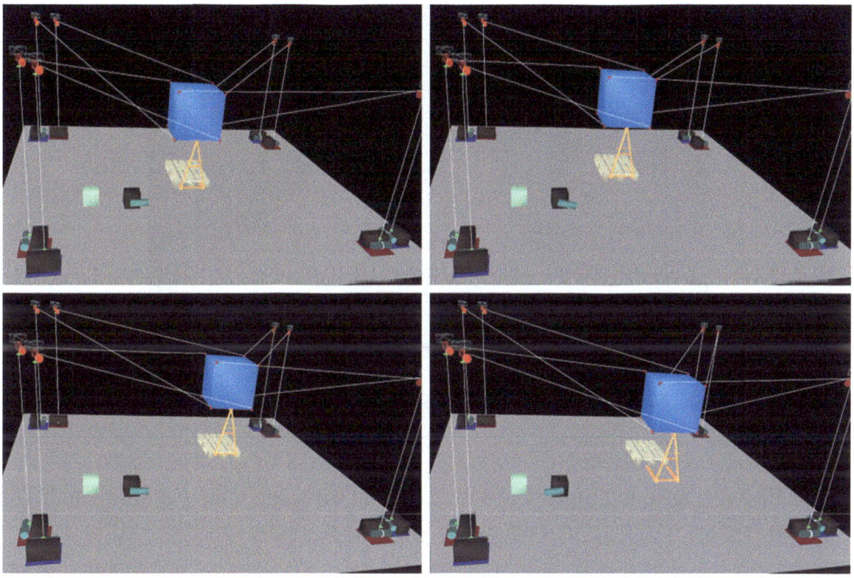

Fig. 11 Motion sequence of a pallet displacement with the CoGiRo simulated robot controlled by the Matlab/Simulink controller

With this Simulink controller allowing to control the platform position, we have implemented a motion sequence to show the validity of this dual system XDE simulator/Simulink controller.

Figure 11 shows the displacement of a pallet from a central position to a new decentralized position.

4 Simulation Discussion

The simulators developed in this work presents interesting features but also drawbacks which have to be improve in future milestones.

4.1 Simulation Interesting Features

The simulation allows validation of the application process: simulation of the motions of the robot and execution of the tasks, validating workspace, detecting collisions, accessibility, maneuverability etc.

As explained in the dedicated section, the fastening of the cable are made of two parts that are fixed together. It is possible within XDE environment to provoque the separation of these two parts, such that we simulate a cable rupture.

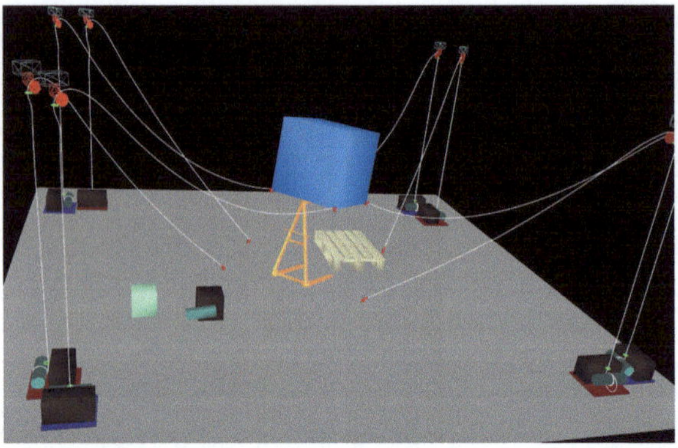

Fig. 12 Simulation of cable failure on the CoGiRo robot

In a first scenario, the rupture of four cables at the same time was simulated. The platform falls on to the ground and cable winding is totally disordered (Fig. 12).

4.2 Simulation Issues and Alternatives

The sampling frequency was set to 200 Hz, and in order to simulate 10 s, the computing time is about 10 h.

One of the major issue comes from the collision computation between the cable and the winch. Indeed the winding of the cable generates a great number of collisions which have to be computed and these collisions could be unnecessary in some use cases.

4.3 Future Work

XDE developers are working on new cable modelling and dynamic node resolution allowing one to decrease complexity and improve the computation time, unfortunately this feature is not ready to use yet.

Developers are working on a winding cable disappearing after one or two turn around the winch to avoid too much collisions computation, it's also an unavailable work in progress. When those new features will be implemented, XDE will offer a very promising tool to simulate cable robots.

5 Conclusions

We have developped a XDE framework with C++ classes allowing to create a simulation of a cable robot composed of elementary components such as winches, pulleys, cable attachments and a mobile platform. We can create as much cables as needed and attach these cables to any object, this object can be a mobile platform or any object in the scene such as a piece to assemble with another one. The winches are implemented in a such way that it's possible to apply a torque on the rotating drum. The cable fastenings allow the feedback of the cable tensions. The platform position, winches rotating position and speed are also available.

We have developped an interface between Matlab/Simulink and the XDE simulator which allows a Simulink control model to interact with the simulated robot. Sequence of motion, motion generation, inverse kinematics and PID control have been implemented to control the cartesian platform position. This combinaison of XDE simulation and Matlab/Simulink controller is able to simulate the CoGiRo robot displacing a pallet.

The cable attachment can be unwelded from the platform, this allowing one to show the behaviour of the system in case of a cable breakdown.

Acknowledgments The research leading to these results has received funding from the European Community's Seventh Framework Program under grant agreement No. NMP2-SL-2011-285404 (CABLEBOT). The authors would like to thank the CEA for their support.

References

1. http://www.3ds.com/products-services/simulia/
2. http://www.microsoft.com/en-us/download/details.aspx?id=29081
3. http://www.anykode.fr/index.php
4. http://www.spaceclaim.com/en/default.aspx
5. http://www.solidworks.com/
6. https://www.algoryx.se/
7. http://www.cm-labs.com/robotics
8. http://www.3ds.com/fr/produits-et-services/catia/fonctionnalites/ingenierie-systeme/simulation-de-systemes-avec-modelica/dymola/
9. http://www.mscsoftware.com/fr/product/adams
10. http://www.sofa-framework.org/
11. http://www.kalisteo.fr/lsi/en/aucune/a-propos-de-xde
12. CoGiRo Project. [Online] http://www2.lirmm.fr/cogiro/

Part II
Accuracy

Presentation of Experimental Results on Stability of a 3 DOF 4-Cable-Driven Parallel Robot Without Constraints

Valentin Schmidt, Werner Kraus and Andreas Pott

Abstract This paper presents the findings of several experiments done on the stability of a planar three degree-of-freedom (DOF) cable-driven parallel manipulator. The major interest is the deflection caused by a wrench or force applied normal to the plane. Here the planar cable robot has no direct constraint. A simple mathematical model based on stiffness is presented to predict the outcomes. It was found that a planar cable robot of dimensions four by 10 m deflects between 20 and 400 mm under forces up to 50 N applied through the normal. Deflections under movement were found to be small in comparison. Oscillations frequencies are in the range of 0.5–1 Hz. The results are also compared to a big mock up robot with dimensions eight by twenty meters. The deflections on this mock-up were found to be very similar to those found on the robot after considering scale.

Keywords Stiffness · Parallel cable-driven robot · Wire robot · Parallel kinematic

1 Introduction

The first cable robots appeared around thirty years ago, an aerial camera device built around 1985 [6]. Many prototypes have been implemented since then, which make use of the cable-driven principle [7, 8, 11].

There are a unique set of challenges for this technology. Cables are non-rigid links which need to be kept under tension at all times. This has significant impact on workspace and stiffness [12].

V. Schmidt (✉) · W. Kraus · A. Pott
Fraunhofer Institute for Manufacturing Engineering and Automation IPA, Stuttgart, Germany
e-mail: Valentin.Schmidt@ipa.Fraunhofer.de

W. Kraus
e-mail: Werner.Kraus@ipa.Fraunhofer.de

A. Pott
e-mail: Andreas.Pott@ipa.Fraunhofer.de

A. Pott and T. Bruckmann (eds.), *Cable-Driven Parallel Robots*,
Mechanisms and Machine Science 32, DOI 10.1007/978-3-319-09489-2_7

It is well known that to achieve n DOF, a cable robot needs $n + 1$ cables. Then a robot can be considered *fully constrained*. A construction which does not fulfill this criterion is said to be *underconstrained*, which has important implications for the stability of such a system.

3-DOF cable robot is a particular subcategory of general cable robots, and will have four or more cables. The robot discussed in this paper has all the actuating winches, and the moving platform placed in a single plane. This means that by design the robot is only capable of moving in 3-DOF. Two linear x and y and one rotary θ. The additional three degrees of freedom that exist for a rigid body in three dimensional space still need to be constrained, thus it is technically underconstrained. Hence, stability is very important. Especially in the normal to the plane, where the cables will provide no restraint. An example is shown in Fig. 1.

2 Related Work

There are many analytical tools which have been developed for parallel robots, and indeed for cable robots.

Especially workspace [5, 13], kinematics and control [9] have been consistent topics of research also for planar robots [3]. Stability and accuracy of planar parallel robot as investigated by [2] mainly focuses on the degrees of freedom on the plane itself. While some analyses go even further to investigate changes in cable forces and the effects on vibrations in the plane [4].

Such investigations are valid in cases where the additional DOFs of a rigid body are somehow constrained. Many constructs of planar robots constrain movement via a horizontal arrangement where the platform or end-effector are resting on a planar surface. In effect using gravity to constrain the platform degrees of freedom. Another

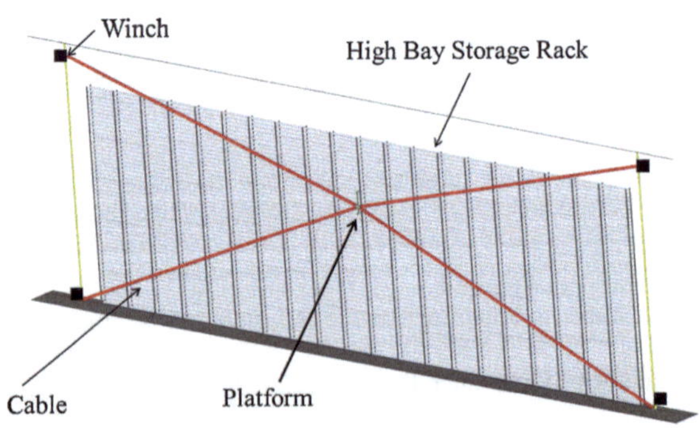

Fig. 1 Concept cable robot for inspecting high bay storage racks

approach could be structural constraints as seen in the form of acrylic glass in [10], effectively building a fully constrained robot.

Instead of providing a physical constraint for the robot, this paper focuses on inherent stability of the system with regards to forces applied along the normal vector of the plane formed by the robot geometry.

Bosscher investigated stability for under-constrained cable robots, but under the influence of a constant external wrench, namely gravity [1]. He even provides a stability criterion which can be applied to different positions. Hence, effects of finite displacement caused by a non-zero wrench are investigated. Changes in cable forces also play a critical role in this investigation.

3 Model of the Cable Robot

Many factors combine to result in the robot stability in regards to displacement and oscillations acting on the normal to the plane.

Figure 2 shows the geometry data used for the model, and the parameters which were investigated. The geometry of each pose is defined as follows. Platform $\mathscr{P}$ moves on the x, y plane with origin $\mathscr{O}$. Each position is described through a position vector $\mathbf{r}$ and an orientation $\mathbf{R}$ which mainly consist of a rotation θ_z. The position of anchor points is described by vector $\mathbf{a}_i$ and $\mathbf{b}_i$ which give the rope vector $\mathbf{l}_i$. Along each wire there acts a force f_i from the winch on the platform. These forces combined represent the force vector $\mathbf{f}$. A unique set of cable forces will be referred to as *tension niveau* for the rest of this paper.

The scalar d is the resulting displacement of force w_z, the component force acting in the z-direction of the wrench $\mathbf{w}$. This displacement is of primary interest in the investigations and experiments. During the execution frequencies of oscillations along the z-axis were also investigated.

We differentiate between two factors which effect stability. The first is the geometric factor. This is effective damping which occurs through the resulting deformation as the platform moves along the z-axis. This will cause the z-components of $\mathbf{f}$ to act

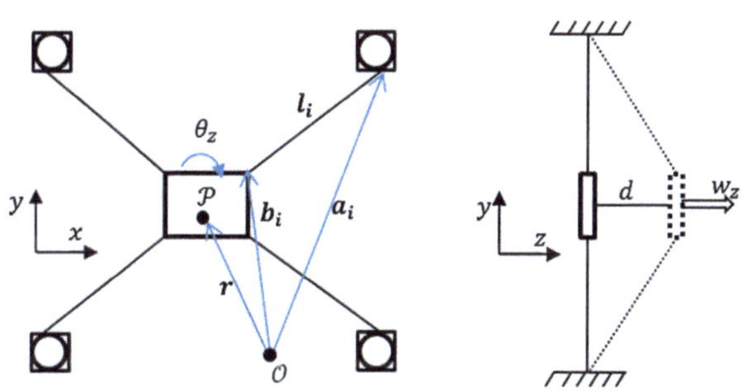

Fig. 2 Diagramatic representation of parameters

in the opposite direction of w_z and thus limit d. Secondly the length of l_i will also increase for each of the cables. This increases the magnitude of each f_i slightly, due to the spring characteristic of each cable. An infinite number of force distributions per pose thus allows increasing the stiffness by increasing the cable forces.

3.1 Stiffness

The concept of stiffness as discussed by [12] is used in order to model the robot's behavior. For the kinematic considerations the robot is modeled with three DOF. This view is reasonable, as due to the planar configuration only these DOFs are controllable. Effects like vibration or stiffness are not limited to the plane but rather are spatial. Therefore, the model of the robot must be extended to be considered as a three dimensional rigid body and all six degrees of freedom. For the further investigation, the platform is described as rigid body with six DOF by its position $\mathbf{r}_{6DOF}$ and orientation $\mathbf{R}_{6DOF}$. The pose is summarized in the generalized coordinate $\mathbf{x} = [r_x, r_y, r_z, \theta_x, \theta_y, \theta_z]^T$. With the cable vector $\mathbf{l}_i$

$$\mathbf{l}_i = \mathbf{a}_i - \mathbf{r}_{6DOF} - \mathbf{R}_{6DOF}\,\mathbf{b}_i . \tag{1}$$

the well known structure matrix yields

$$\mathbf{A}^T(\mathbf{r}_{6DOF}, \mathbf{R}_{6DOF}) = \begin{bmatrix} \mathbf{u}_1 & \cdots & \mathbf{u}_m \\ \mathbf{b}_1 \times \mathbf{u}_1 & \cdots & \mathbf{b}_m \times \mathbf{u}_m \end{bmatrix} \text{where } \mathbf{u}_i = \mathbf{l}_i \, \|\mathbf{l}_i\|^{-1}. \tag{2}$$

The Cartesian stiffness matrix $\mathbf{K}_x$ describes the linear relation

$$\delta\mathbf{w} = \mathbf{K}_x \delta\mathbf{x} \tag{3}$$

between an infinitesimal wrench $\delta\mathbf{w}$ and deflection $\delta\mathbf{x}$. For the analytic expression for $\mathbf{K}_x$ the structure equation

$$\mathbf{w} = \mathbf{A}^T\mathbf{f}, \tag{4}$$

which describes the static equilibrium for the endeffector between the external wrench $\mathbf{w}$ and cable forces $\mathbf{f}$, is necessary. Differentiating the structure equation yields to

$$\delta\mathbf{w} = \delta\mathbf{A}^T\mathbf{f} + \mathbf{A}^T\delta\mathbf{f} \tag{5}$$

where $\delta\mathbf{A}^T$ is the derivation of the structure matrix by $\mathbf{x}$ and $\delta\mathbf{f}$ is a differential change in the cable forces.

The relation between the incremental cable forces and an incremental change in cable length can be expressed by

$$\delta\mathbf{f} = \mathbf{K}_l \delta\mathbf{l} \tag{6}$$

where $\mathbf{K}_l$ is the joint space stiffness matrix, where the stiffness of each cable is on the diagonal. Incremental changes in the platform pose $\delta\mathbf{x}$ are transformed to the endeffector space with

$$\delta\mathbf{l} = \mathbf{A}\delta\mathbf{x}. \tag{7}$$

With Eqs. 3, 6 and 7 one receives from Eq. 5

$$\mathbf{K}_x d\mathbf{x} = \frac{\partial \mathbf{A}^T}{\partial \mathbf{x}} \mathbf{f} d\mathbf{x} + \mathbf{A}^T \mathbf{K}_l . \mathbf{A} d\mathbf{x} \tag{8}$$

It is worth mentioning that the first term does not depend on the parameters of the cables, but is mainly proportional to the height of the tension niveau. The stiffness of the cables only has an influence on the second term $\mathbf{K}_c$. For illustration, Eq. 8 is now applied on the robot geometry defined in Table 2b for a deflection in the z-direction. This is performed for two different tension niveaus with a mean cable force of 200 and 400 N respectively. The resulting stiffness for a deflection in z are shown in Fig. 3. The stiffness in the undeflected state is only provided by the first term, whereas the term $\mathbf{K}_c$ is zero. Only under deflection the second term becomes important and rises overproportionally. This means, the cable stiffness has no influence on the total stiffness of the endeffector in undeflected state. As the diagram reveals, the stiffness

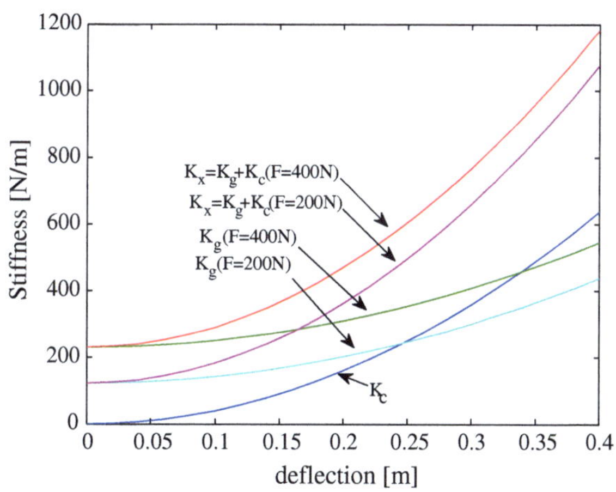

Fig. 3 Composition of the Cartesian stiffness in z-direction for two force niveaus over the deflection of the endeffector

Table 1 Model results

	f_{mid}			
k'	200 N	300 N	400 N	500 N
(*a*) Deflection [mm] at different tension niveaus and specific spring constants k'				
80000 N/m	0.275	0.242	0.213	0.188
120000 N/m	0.251	0.225	0.201	0.180
160000 N/m	0.234	0.212	0.19	0.174
(*b*) Calculated natural frequencies				
Natural frequency (Hz)	0.65	0.78	0.89	0.99

in undeflected state depends mainly on the tension niveau. Although a linear spring model is applied, the resulting spring characteristic is strongly progressive.

3.2 Deflection Under Transverse Force

With knowledge of the stiffness of the endeffector, now the deflection for a given wrench can be calculated. Integrating numerically to determine the deflection takes the progressive stiffness characteristic into account. Relevant influence parameters are the initial force distribution and the specific stiffness of the cables. In Table 1 the resulting deflection for a traverse force of 50 N is shown for the combinations of the relevant parameters.

3.3 Natural Frequency

One characteristic value for evaluating the vibrations of the endeffector is the natural frequency f_0. It can be directly estimated with the results from the stiffness analysis. The natural frequency of a pendulum can be computed with

$$f_0 = 2\pi\sqrt{\frac{c}{m}} \tag{9}$$

with the stiffness c and the mass m. As the planar cable robot is mainly susceptible for vibration in the z-direction, Eq. 9 can be used to calculate the natural frequecy. As stiffness c the z-component of the Cartesian stiffness matrix $\mathbf{K}_x$ is used and for the mass m the mass of the endeffector.

Table 2 Robot geometrical parameters: base vector $\mathbf{a}_i$ and platform vector $\mathbf{b}_i$

Cable i	Base vector $\mathbf{a}_i$ in [m]					Platform vector $\mathbf{b}_i$ in [m]				
(a) Robot geometrical parameters										
		x	y	z			x	y	z	
1	[	5.319	−0.050	−0.017	$]^T$	[	0.459	−0.450	0.0	$]^T$
2	[	5.320	3.615	0.028	$]^T$	[	0.452	0.442	0.0	$]^T$
3	[	−4.838	3.676	0.042	$]^T$	[	−0.446	0.437	0.0	$]^T$
4	[	−4.829	−0.049	−0.018	$]^T$	[	−0.446	−0.450	0.0	$]^T$
(b) Mock-up parameters										
		x	y	z[a]			x	y	z	
1	[	10.85	−2.771	0.0	$]^T$	[	0.459	−0.450	0.0	$]^T$
2	[	10.85	5.3595	0.0	$]^T$	[	0.452	0.442	0.0	$]^T$
3	[	−10.05	−2.773	0.0	$]^T$	[	−0.446	0.437	0.0	$]^T$
4	[	−10.05	5.397	0.0	$]^T$	[	−0.446	−0.450	0.0	$]^T$

[a] Due to the mock-up size, imperfections in z of the winches were not measurable

4 Experimental Results

4.1 Experiment Setup

The experimental set up was simple. A weight attached to a pulley was used to exert a defined force w_z upon the platform. This was done on several poses in the workspace of the planar cable robot with the dimensions shown in Table 2b. For measuring the geometric parameters of the kinematic chain, deflection, and oscillations a Leica LaserTracker was used. With this optical device the absolute position of a reflector can be determined with an accuracy of $\pm 15\,\mu\text{m} + 6\,\mu\text{m/m}$.

The cable material used in this experiment was LIROS D-Pro 01505-0250 based on Dyneema SK 75 fibre (Polyethylene). This has a diameter of 2.5 mm and a specific weight w of 0.0035 kg/m.

In addition to different poses, different force distributions were also considered and matched. Vibrations after the force w_z was reduced to zero were also recorded. Finally deflection induced by the robot's linear movement when actuated by the joints for different trajectories were also compared.

The experiments for deflection and frequency were also repeated on a much larger scale mock-up of the planar cable robot. This was done to test the scalability of the results gained in the previous experiments. Here the same cable was used, but at much larger dimensions shown in Table 2 these were found with a laser range finder as the winches of the mock-up span distances beyond the range of the Leica LaserTracker.

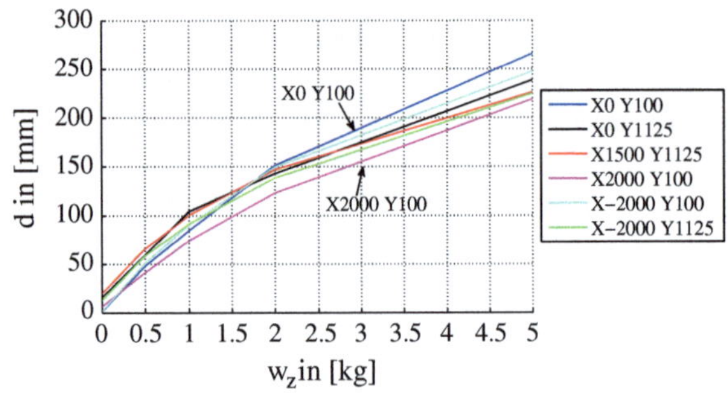

Fig. 4 Deflection d for different applied forces w_z at different poses

4.2 Results

A few selected results are given here to show the characteristics of the phenomena, and give a comparison to the modeled results. Poses for the robot will be expressed in the following form: [x in (mm), y in (mm)]

Figure 4 shows the deflection experienced at different poses in the workspace. Pose [0, 1125] which is very close to the center of the base frame shows the most deflection and thus has the least stiffness. Pose [2000, 100] towards the edge of the workspace shows much more stiffness.

Table 3 shows results varying the tension niveau at different poses. Cable Force indicates a the maximum force in a particular tension niveau. Figure 5 shows the frequency spectrum of oscillations in d these oscillations under different tension niveaus.

Figure 6 shows the deflection that occurs when the robot moves along a trajectory. A given path results in a displacement in all axes of the coordinate system. This is due to geometrical deficiencies and the fact that the measurements were produced in a different coordinate system than the one programmed by the kinematics. Hence a corrected d is also displayed which gives the displacement from the trajectory between the two measured point on the z-axis. In fact, all stationary measurements are normalized in the same way.

Table 4a and Fig. 4b show the results for different tension niveaus for the big mock-up, similar to the information contained in Table 3.

4.3 Discussion

It can be seen that the impact on stability is primarily result of the geometric configuration. The stability is mainly governed by $\mathbf{K}_g$ rather than $\mathbf{K}_l$ which coincides with the first quarter of the graph at Fig. 3. From Table 3b stiffness at a tension

Table 3 Results for different cable forces at different positions

(a) Pose: [0, 1125]			(b) Pose: [0, 100]			(c) Pose: [0, 750]		
w_z in (kg)	Cable Force (N)	Deflection d in (mm)	w_z in (kg)	Cable Force (N)	Deflection d in (mm)	w_z in (kg)	Cable Force (N)	Deflection d in (mm)
0.5	200	53.9109	0.5	200	40.7645	0.5	200	38.7358
	300	34.0864		300	30.9638		300	27.6953
	400	26.6526		400	24.7105		400	23.0817
	500	22.8604		500	21.7184		500	20.6585
1	200	98.4713	1	200	77.5834	1	200	67.5565
	300	59.5159		300	59.2566		300	48.3678
	400	43.7081		400	47.7208		400	40.0389
	500	35.5547		500	42.3703		500	34.7599
2	200	185.946	2	200	165.431	2	200	127.972
	300	105.98		300	116.353		300	92.376
	400	74.0245		400	93.3925		400	74.4883
	500	62.1933		500	80.8146		500	62.6368
5	200	416.169	5	200	386.278	5	200	297.214
	300	242.596		300	292.786		300	217.305
	400	181.99		400	235.568		400	177.82
	500	139.261		500	192.454		500	145.233
(d) Pose: [1500, 1125]			(e) Pose: [1500, 100]			(f) Pose: [1500, 750]		
w_z in (kg)	Cable Force (N)	Deflection d in (mm)	w_z in (kg)	Cable Force (N)	Deflection d in (mm)	w_z in (kg)	Cable Force (N)	Deflection d in (mm)
0.5	200	49.5651	0.5	200	41.2883	0.5	200	39.1305
	300	33.9899		300	31.8738		300	29.6688
	400	27.9604		400	24.8672		400	24.9784
	500	24.8945		500	21.1144		500	22.055
1	200	88.3799	1	200	83.0119	1	200	67.9625
	300	59.6591		300	59.0051		300	50.1867
	400	43.8635		400	47.2317		400	42.0422
	500	36.5778		500	39.259		500	34.9966
2	200	173.225	2	200	161.976	2	200	135.944
	300	99.7939		300	119.072		300	89.6771
	400	72.0893		400	90.9264		400	72.2515
	500	59.0626		500	76.2177		500	61.638
5	200	375.615	5	200	368.67	5	200	312.465
	300	176.364		300	278.397		300	207.995
	400	166.744		400	221.483		400	165.031
	500	134.912		500	180.714		500	137.186

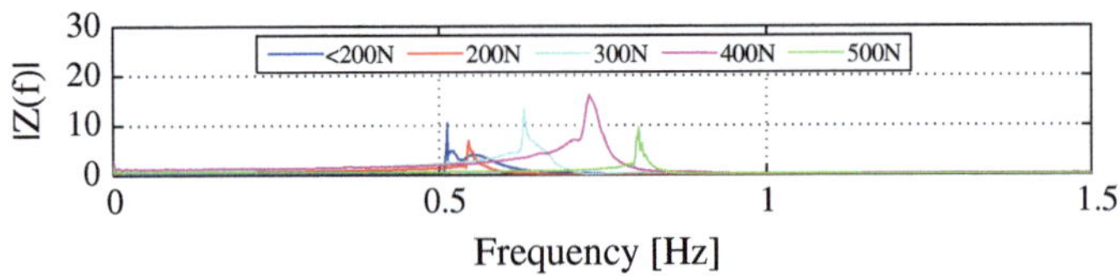

Fig. 5 Amplitude spectrum $z(t)$ of oscillations at Pose [0, 100]

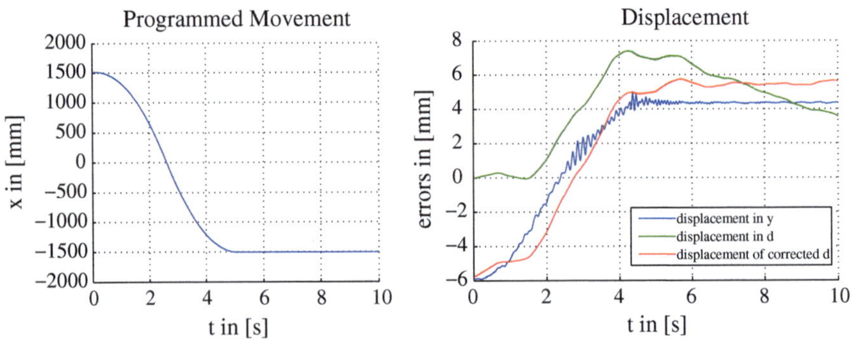

Fig. 6 Deflection on horizontal path

niveau of 200 N can be estimated and gives an average stiffness of $\approx$ 125 N/m for deflections < 0.1 m which coincides with the model.

It can also be seen from the data that different positions contribute to differences in deflection of 50–100 mm. Poses at the center of the workspace having more deflection that poses closer to the winches.

As expected the stability can be increased drastically by increasing the tension niveau, which can be seen in Table 3. Increasing the tension niveau by slightly more than a factor of two reduced the deflection by about the same amount. The Results of this vary between different poses, which is predicted by modeling the stiffness.

The model also predicted the expected result on the frequency of the oscillations. These coincide for lower tension niveaus, but become increasingly innaccurate as the tension niveau is increased. It was also found that positions with different absolute cable lengths have a broader frequency spectrum. This is to be expected, as each joint will contribute its own oscillating frequency, which is directly correlated to the cable length. The model for frequency is also very simple in comparison to the geometric model, only acting in one dimension.

The deflection when the robot follows a trajectory is fairly small. An ideal robot would have no deflection, but uncontrollable external forces, and geometrical imperfections cause deflections to occur. It can be seen that the geometric imperfections have a big impact. Oscillations in d are also caused and in the order of magnitude of 1 mm while the maximum absolute corrected deflection was 14 mm. This is highly variable for different trajectories. For the current application concept the scale of

Table 4 Mock-Up Deflection for Tension Niveaus at Pose [0, 0]

(a)

Pose [0,0]		
w_z in [kg]	Cable Force [N]	Deflection d in [mm]
0.5	80-100	157.949
	200-230	75.7707
	250-280	51.2554
1	80-100	258.299
	200-230	143.589
	250-280	105.709
2	80-100	387.406
	200-230	247.662
	250-280	198.006
5	80-100	629.655
	200-230	461.833
	250-280	408.997

(b)

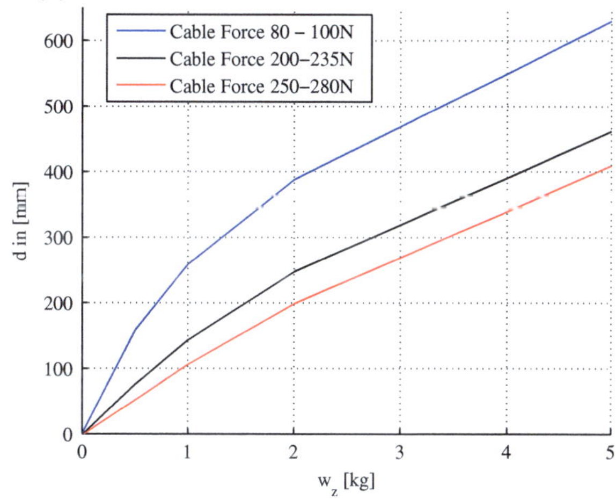

the results is highly promising. A deflection of 20 mm is two thousandths of the length of the robot, and is much less than when a force w_z is applied. This makes many applications for planar cable robots feasible, without the need of additional constraints. In the case of high bay inspection, the expected w_z is much lower than the values which were tested for in this set of experiments, and all the results were within acceptable limits, even when scaled to larger dimensions.

At a larger scale, the results are very similar. The big mock up was about double in size, with a length of $\approx$20 m and a height of $\approx$8 m compared to $\approx$10, 4 m. Deflection increased by a very similar factor (keeping in mind position [0, 1125] on the robot is more similar to position [0, 0] on the mock-up than [0, 100]). Unfortunately the importance of geometric factors in the stability make comparing the results from the actual robot and the big mock-up difficult. It is important to note that even the results for the mock-up were at acceptable levels.

5 Conclusion

This paper presents experimental results regarding the stability of a 3-DOF planar cable robot. In light of certain applications, this stability concern may render applications of such a robot design useless. So far very few such robots have been built and investigated.

Results show that while a deflection from the plane is inevitable, the magnitude is manageable. Through increasing tension niveau, one can control the magnitude of d in a very predictable manner. This can be estimated through a simple stiffness model. Such model can also predict frequencies of oscillation. Unfortunately, due to many factors contributing to imprecision, the model is too simplified to give very exact estimates for all tension niveaus, and poses.

One very important factor which is not modeled so far, is the deflection caused by the robot moving along a trajectory. While this is highly unpredictable and variable between different trajectories, the deflection was found to be small and therefore manageable. This makes the application of such a robot plausible, from a stability point of view. Especially considering that very little w_z is actually expected in this application of non-contact inspection. The robot is expected to move undisturbed along its path.

A robot using this concept is possible, but knowledge of the horizontal stiffness is very important. The results presented here have shown that a simple stiffness model can approximate a general estimate, for different geometries and positions but is not very precise due to numerous factors.

This information can be used when considering building a robot which relies only on the tension niveau in the cable for stability but is technically underconstrained. An overview of the magnitudes in deflection was given, and a method for predicting these magnitudes through the stiffness model. Further work would include making this model more precise, and more detailed investigations of the deflections caused through moving along a trajectory. The latter being very unpredictable and essential for the operation of such a planar cable robot.

Acknowledgments This work was supported by the Fraunhofer-Gesellschaft Internal Programs under Grant No. WISA 823 244.

References

1. Bosscher P, Ebert-Uphoff I (2004) A stability measure for underconstrained cable-driven robots. IEEE Int Conf Robot Autom 5:4943–4949
2. Briot S, Bonev IA (2008) Accuracy analysis of 3-dof planar parallel robots. Mech Mach Theory 43:445–458
3. Briot S, Bonev IA, Chablat D, Wenger P, Arakelian V (2008) Self-motions of general 3-rpr planar parallel robots. Int J Robot Res 27:855–866
4. Diao X, Ma O (2009) Vibration analysis of cable-driven parallel manipulators. In: Multibody system dynamics, vol. 21. Springer, Netherlands, pp 347–360
5. Fattah A, Agrawal SK (2005) Design and simulation of a class of spatial reactionless manipulators. IEEE Int Conf Robot Autom 23:75–81
6. Lawerence LC (1985) Skycam: an aerial robotic camera system. Byte Mag 10:122–132
7. Merlet JP (2010) Marionet, a family of modular wire-driven parallel robots. In: International symposium on advances in robot kinematics. Springer-Verlag, pp 53–61
8. Ottaviano E, Ceccarelli M, Paone A, Carbone G (2005) A low-cost easy operation 4-cable driven parallel manipulator. In: IEEE international conference on robotics and automation, pp 4008–4013
9. Pott A (2010) An algorithm for real-time forward kinematics of cable-driven parallel robots. In: International symposium on advances in robot kinematics. Springer-Verlag
10. So-Ryeok Oh, Agrawal S (2003) Cable-suspended planar parallel robots with redundant cables: controllers with positive cable tensions, pp 3023–3028

11. Trevisani A, Gallina P, Williams RL II (2006) Cable-direct-driven robot (cddr) with passive scara support. J Int Robot Syst: Theory Appl 46:73–94
12. Verhoeven R (2004) Analysis of the workspace of tendon-based stewart platforms. Ph.D. thesis, University of Duisburg-Essen, Duisburg
13. Verhoeven R, Hiller M, Tadokoro S (1998) Workspace, stiffness, singularities and classification of tendon-driven stewart platforms. In: International symposium on advances in robot kinematics, pp 1–10

Experimental Determination of the Accuracy of a Three-Dof Cable-Suspended Parallel Robot Performing Dynamic Trajectories

Clément Gosselin and Simon Foucault

Abstract The experimental determination of the accuracy of a three-degree-of-freedom (three-dof) spatial cable-suspended parallel robot is addressed in this paper. The concept of the dynamic trajectory planning of a three-dof spatial cable-suspended parallel robot is first briefly recalled. Then, periodic trajectories are planned and an external three-dimensional measurement system is used to determine the actual trajectory of the end-effector. Linear regression is used to fit the measured trajectory with the planned trajectory and eliminate the bias error. The accuracy of the trajectories is then assessed.

1 Introduction

Cable-suspended parallel robots are cable-driven mechanisms in which gravity is used to maintain the cables in tension. They include as many actuators (or sometimes fewer actuators) as degrees of freedom. Cable-suspended parallel robots have the potential to provide very large workspaces with very effective payload to mass ratios. Examples of cable-suspended robots can be found in [1, 5, 9] and in several other works. In the latter references, cable-suspended parallel mechanisms are treated as quasi-static devices which move slowly and which are assumed to always remain within the bounds of their static workspace. Techniques to determine the static workspace of cable-suspended robots were proposed in the literature, for instance in [10]. Also, their dynamics were studied in [8, 12] for control purposes.

On the other hand, if the dynamics of cable-suspended robots is considered at the trajectory planning stage, it is possible to extend their workspace beyond the static workspace and exploit their so-called dynamic workspace [2]. Indeed, by including

C. Gosselin (✉) · S. Foucault
Département de génie mécanique, Université Laval, 1065 Avenue de la médecine, Québec, QC G1V 0A6, Canada
e-mail: gosselin@gmc.ulaval.ca

S. Foucault
e-mail: foucault@gmc.ulaval.ca

A. Pott and T. Bruckmann (eds.), *Cable-Driven Parallel Robots*,
Mechanisms and Machine Science 32, DOI 10.1007/978-3-319-09489-2_8

cable tension constraints in the trajectory planning algorithms, it is possible to design trajectories that satisfy these constraints by construction.

Dynamically controlled (pendulum-like) cable-suspended robots were proposed in [3, 7, 11] and [13]. The robots proposed in the latter references are underactuated and techniques are developed to control the actuator inputs in order to produce point-to-point motion between prescribed poses. Such techniques require the on-line numerical integration of the differential equations.

In [6], a fully actuated planar 2-dof robot suspended on two cables was considered. By expressing the tensions in terms of the kinematic variables, it is possible to design trajectories and to obtain conditions on the trajectory parameters that can guarantee that the tensions remain positive. This approach greatly simplifies the trajectory planning and opens the avenue for the global planning of families of feasible trajectories. The approach was extended to spatial three-dof parallel cable-suspended robots with a point-mass end-effector in [4].

In this paper, the accuracy of the three-dof parallel cable-suspended robot described in [4] is investigated. First, the mechanism architecture and the dynamic trajectory planning technique are recalled. Examples of parametric trajectories that ensure that the cables are maintained under tension are given. Then, an external measurement system is used to determine the position of the end-effector. Based on the experimental results, the assumed fixed reference coordinate system is adjusted and the positioning error of the mechanism is determined.

2 Kinematic and Dynamic Modelling

The spatial three-dof cable-suspended parallel robot studied in [4] is represented schematically in Fig. 1. Three cables are attached to a common end-effector which is considered as a point mass. At their other end, the cables are wound on fixed actuated spools which are used to control the extension of the cables. By controlling the cable extensions, the position of the point mass in the three-dimensional space can be prescribed. The robot includes three actuators and three degrees of freedom. It was shown in [4] that it is possible to perform Cartesian trajectories that extend beyond the static workspace of the robot by imposing that all cable tensions remain positive and by using the mathematical conditions obtained to set the trajectory parameters.

The dynamic model of the cable-driven robot with a point-mass end-effector developed in [4] is now briefly recalled.

Referring to Fig. 1, a fixed reference frame is first defined on the base of the robot with its Z axis pointing downwards, i.e., in the direction of gravity. The points corresponding to the cable outputs of the spools are assumed to be fixed—in practice an eyelet or a pulley can be used—and are noted A_i, with $i = 1, 2, 3$. The vector connecting the origin of the fixed reference frame to point A_i is noted $\mathbf{a}_i$ and the position of the end-effector of mass m with respect to the origin of the fixed reference frame is noted $\mathbf{p} = [x, y, z]^T$. The cable lengths, which are used as joint coordinates, are respectively noted ρ_i, $i = 1, 2, 3$. The inverse kinematic equations can therefore

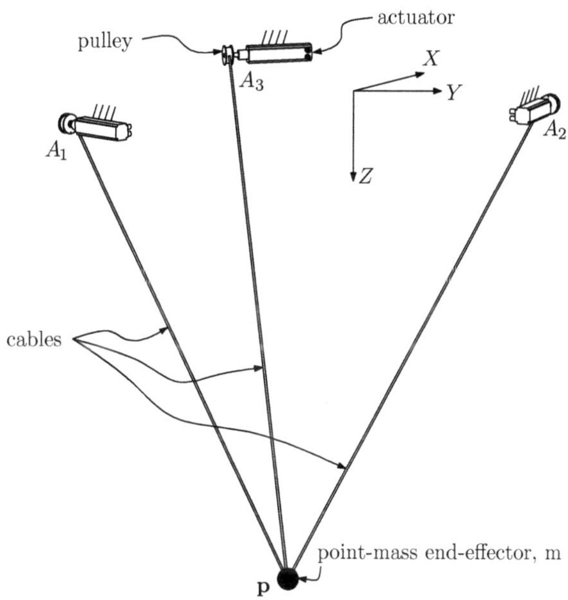

Fig. 1 Spatial three-dof cable-suspended robot with a point-mass end-effector and three cables

be simply written as

$$\rho_i = \sqrt{(\mathbf{p} - \mathbf{a}_i)^T(\mathbf{p} - \mathbf{a}_i)}, \quad i = 1, 2, 3. \tag{1}$$

Since the mass of the cables is neglected, the dynamic model of the robot can be obtained by writing the force balance on the end-effector, which is considered as a point mass. One obtains

$$\sum_{i=1}^{3}\left(\frac{-F_i}{\rho_i}(\mathbf{p} - \mathbf{a}_i)\right) + m\,\mathbf{g} = m\,\ddot{\mathbf{p}} \tag{2}$$

where F_i is the tension in cable i and $\mathbf{g}$ is the vector of gravitational acceleration, namely $\mathbf{g} = [0, 0, g]^T$, in which g is the magnitude of the gravitational acceleration.

Equation (2) constitutes a system of three linear equations in three unknowns (tensions F_1, F_2 and F_3) which can be assembled in a vector defined as $\mathbf{f} = [F_1 \quad F_2 \quad F_3]^T$. Hence, (2) can be explicitly solved for the cable tensions as

$$\tau = \mathbf{M}^{-1}(\mathbf{g} - \ddot{\mathbf{p}}), \text{ with } \mathbf{M} = [\mathbf{e}_1 \quad \mathbf{e}_2 \quad \mathbf{e}_3] \tag{3}$$

where $\tau = \frac{1}{m}\mathbf{f}$ is the vector of cable forces per unit mass of the end-effector and vector $\mathbf{e}_i$ is defined as a unit vector in the direction of the i-th cable and oriented from the spool to the end-effector, namely

$$\mathbf{e}_i = \frac{1}{\rho_i}(\mathbf{p} - \mathbf{a}_i), \quad i = 1, 2, 3. \tag{4}$$

In order to obtain dynamically feasible trajectories, it must be ensured that the above solution yields tensions in the cables, i.e, it must be guaranteed that the components of τ are positive throughout the trajectory. Referring to (3), the inverse of matrix $\mathbf{M}$ can be written as

$$\mathbf{M}^{-1} = \frac{\mathrm{Adj}(\mathbf{M})}{\det(\mathbf{M})} \tag{5}$$

where $\mathrm{Adj}(\mathbf{M})$ is the adjoint matrix of $\mathbf{M}$ and $\det(\mathbf{M})$ is its determinant. It can readily be observed that the determinant of $\mathbf{M}$ can be written as

$$\det(\mathbf{M}) = (\mathbf{e}_1 \times \mathbf{e}_2)^T \mathbf{e}_3 \tag{6}$$

and that this quantity is always strictly negative as long as the end-effector remains below the plane defined by the three attachment points A_1, A_2 and A_3. This assumption is used and it suffices to consider only the adjoint matrix in order to determine whether the tensions in the cables are positive. The condition corresponding to positive tensions is written as

$$\tau \succeq 0 \tag{7}$$

where $\succeq$ stands for the componentwise inequality. Based on (3) and the expression of the adjoint matrix and noting that the determinant of $\mathbf{M}$ is always negative, the constraints then become

$$\left[\mathbf{p} \times (\mathbf{a}_2 - \mathbf{a}_3) + (\mathbf{a}_2 \times \mathbf{a}_3)\right]^T (\ddot{\mathbf{p}} - \mathbf{g}) > 0 \tag{8}$$

$$\left[\mathbf{p} \times (\mathbf{a}_3 - \mathbf{a}_1) + (\mathbf{a}_3 \times \mathbf{a}_1)\right]^T (\ddot{\mathbf{p}} - \mathbf{g}) > 0 \tag{9}$$

$$\left[\mathbf{p} \times (\mathbf{a}_1 - \mathbf{a}_2) + (\mathbf{a}_1 \times \mathbf{a}_2)\right]^T (\ddot{\mathbf{p}} - \mathbf{g}) > 0. \tag{10}$$

The above inequalities represent the constraints to be satisfied in order to ensure that the cables are kept under tension. If these conditions are satisfied at all points of a given trajectory, then it can be guaranteed that the cables will remain under tension throughout the trajectory. These conditions are necessary and sufficient.

3 Trajectory Planning

In [4], it was shown that the above inequality constraints can be used to plan periodic feasible trajectories such as oscillations along a straight line, circles in horizontal or vertical planes or spatial trajectories inscribed on the surface of a sphere or a cylinder. Feasible trajectories are obtained by manipulating the inequality constraints to obtain conditions on the global trajectory parameters that ensure positive cable tensions.

A symmetric architecture in which the three spools are located on the vertices of a horizontal equilateral triangle whose centroid is at the origin of the fixed reference frame is used here. The distance between the fixed attachment points—the length of one side of the equilateral triangle—is noted a. The simple case of a periodic oscillation along a horizontal straight line is now recalled for quick reference.

The trajectory corresponding to a periodic horizontal motion along a straight line intersecting the vertical line passing through the centroid of the base triangle is designed as follows:

$$x = r\cos\alpha\sin(\omega t),\ \ y = r\sin\alpha\sin(\omega t),\ \ z = z_0,\ \ z_0 > 0 \tag{11}$$

where z_0 is the elevation of the horizontal trajectory, r is one half of the total horizontal range of motion, ω is the frequency of the periodic motion, α is the angle corresponding to the direction of the straight line in the horizontal plane and t is the time. Substituting the above parametric equations and their time derivatives into inequalities (8–10) leads to three inequalities that can be written as

$$A_i \sin(\omega t) + B_i > 0, \quad i = 1, 2, 3 \tag{12}$$

where

$$A_1 = r(z_0\omega^2 - g)\cos\alpha \tag{13}$$
$$A_2 = -r(z_0\omega^2 - g)(\cos\alpha + \sqrt{3}\sin\alpha) \tag{14}$$
$$A_3 = -r(z_0\omega^2 - g)(\cos\alpha - \sqrt{3}\sin\alpha) \tag{15}$$
$$2B_1 = B_2 = B_3 = \sqrt{3}ag/3. \tag{16}$$

Given the bounds on the sine function, the conditions given in (12) are satisfied throughout the trajectory if the following conditions are satisfied:

$$|A_i| < B_i, \quad i = 1, 2, 3. \tag{17}$$

When inequalities (17) are satisfied, the horizontal trajectory can be performed while maintaining all cables in tension. Since conditions (17) involve only the geometric parameters of the robot and the global parameters of the trajectory—elevation z_0, amplitude r, horizontal direction α and frequency ω—the parameters can be adjusted to produce feasible trajectories.

It can also be observed, from (13) to (16) that when a frequency of

$$\omega_n = \sqrt{\frac{g}{z_0}} \tag{18}$$

is selected, inequalities (17) are always satisfied and arbitrary amplitudes of motion r can theoretically be produced, irrespectively from the direction of the horizontal line. This frequency, ω_n, can be thought of as a kind of *natural frequency* for the robot performing horizontal oscillations. It is remarkable that this frequency corresponds to the natural frequency of a single cable pendulum of length z_0 and that it does not depend on the direction of the line in the horizontal plane. Similar relationships between the trajectory parameters can be obtained for other periodic trajectories such as circles, hypocycloidal trajectories and others using a similar approach.

4 Experimental Determination of the Accuracy

A prototype of a three-dof spatial cable-suspended robot was built in order to validate the approach and the dynamic trajectories extending beyond the workspace of the mechanism. The successful performance of these trajectories was reported in [4]. However, in the latter reference, no external measurement was provided to assess the actual positioning accuracy of the mechanism. This issue is addressed here using a three-dimensional positioning system (VICON). To this end, a marker is mounted on the end-effector and eight cameras are placed around the workspace of the mechanism. The tracking of the end-effector is performed at 500 Hz using triangulation since the marker is always visible from a sufficient subset of the cameras. The experimental set-up is shown in Figs. 2 and 3. The distance between the cable attachment points on the frame is approximately 9 m and the mass of the end-effector is $m = 0.071$ kg. Three servo-controlled winches are used to control the length of the cables. Vertical, horizontal, circular, hypocycloidal and other trajectories were successfully demonstrated with the prototype. Three trajectories for which external measurements were made with the VICON system are now described. It is pointed out that the objective of this work is to determine the accuracy of the robot and not the precision.

4.1 Horizontal Straight-Line Trajectory

A periodic oscillation along the direction of the X axis of the fixed reference frame is performed. The trajectory is described mathematically in Eq. (11), where the following parameters are used:

$$\alpha = 0, \; r = 2.5\,m, \; z_0 = 4\,m, \; \omega = \sqrt{\frac{g}{z_0}} \simeq 1.566\,s^{-1}. \qquad (19)$$

The above trajectory leads to maximum velocities of approximately 6 m/s. The trajectory is performed and the motion of the end-effector is measured using the VICON system for several back and forth oscillations. A linear regression is then applied to the measured coordinates in order to fit a straight line to the measured trajectory. The

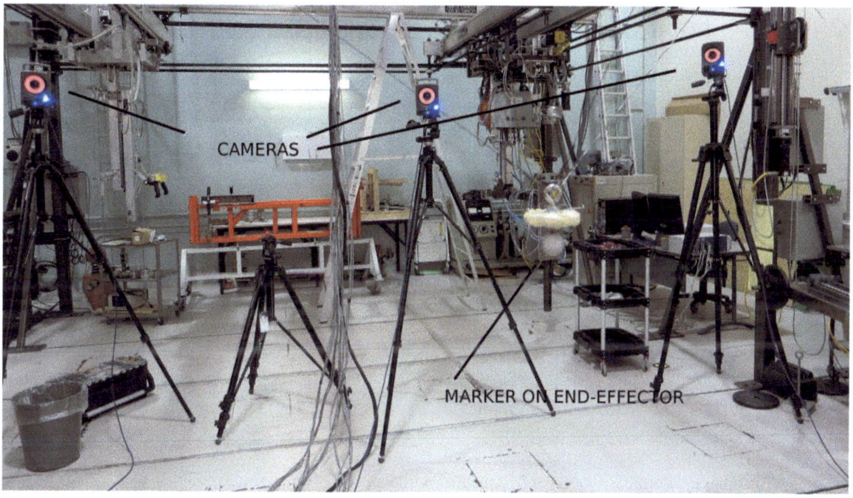

Fig. 2 Prototype of a three-dof spatial cable-suspended robot, cameras and marker used in the experiments

Fig. 3 End-effector of the prototype with the marker

results are shown in Fig. 4a. First, it can be observed that the measured trajectory is not exactly along the X direction. This bias corresponds to the alignment error between the mechanism's reference frame and the measurement system's reference frame. Using the best linear fit, this alignment error is easily determined and compensated for. Then, the accuracy of the trajectory can be determined using the error between the best linear fit and the actual trajectory. The results are shown in Fig. 4b. It can be observed that, after calibration to account for the reference frame alignment, the

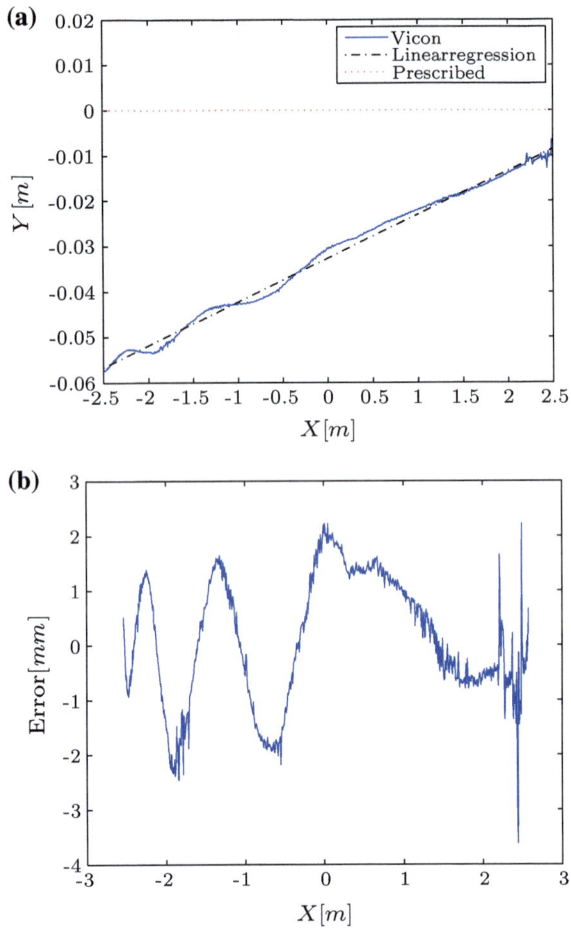

Fig. 4 Measured position for an oscillation along the X axis, best linear fit and error (**a** and **b**)

positioning error for this dynamic trajectory is less than ±4 mm which is relatively small, considering that the cables are several metres long.

4.2 Circular Trajectory in a Vertical Plane

The circular trajectory in the vertical XZ plane can be described mathematically as

$$x = r\sin(\omega t),\ \ y = 0,\ \ z = z_0 + r\cos(\omega t),\ \ z_0 > r \tag{20}$$

with

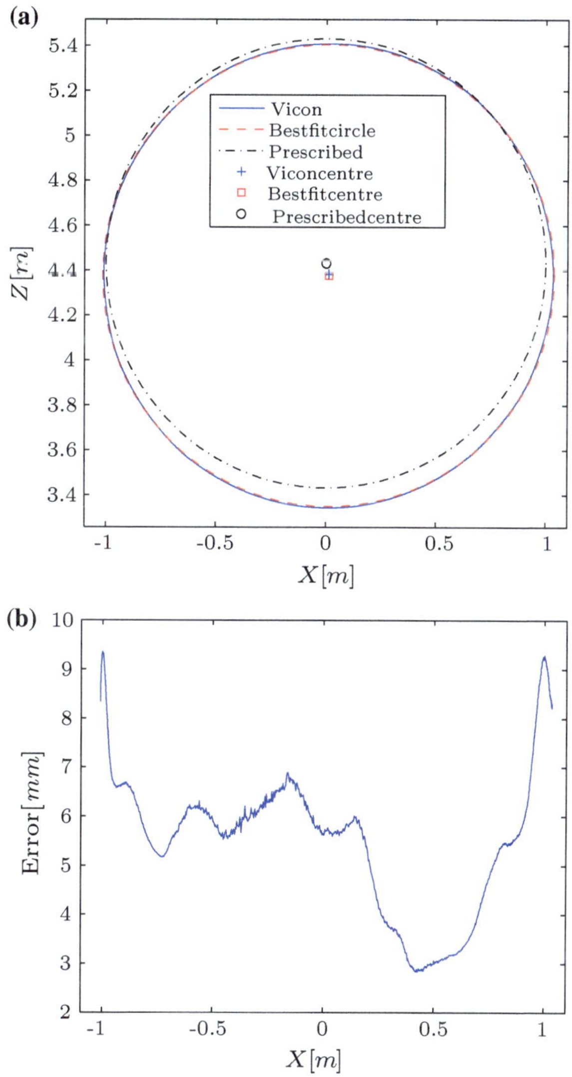

Fig. 5 Measured position for the circular trajectory in a vertical plane, best circular fit and error (**a** and **b**)

$$r = 1\,\text{m},\ z_0 = 4.5\,\text{m},\ \omega = \sqrt{\frac{g}{z_0}} \simeq 1.476\,\text{s}^{-1}. \tag{21}$$

The above trajectory produces a velocity of the end-effector of approximately 1.5 m/s along the circular path. The trajectory is performed with the prototype and the motion of the end-effector is measured. A linear regression is then applied in order to fit a circular path to the measured trajectory and the results are shown in Fig. 5a. Similarly to what was observed with the first trajectory, a bias is noted, which is easily compensated for. Hence, the accuracy of the mechanism is obtained by considering the error between the best circle fit and the measured position. The results are shown in Fig. 5b. It can be observed that the accuracy is similar to that obtained with the straight line motion and that the error can go up to almost 1 cm. It can also be observed that the largest errors are obtained when the end-effector is the furthest away from the centre of the static workspace.

4.3 Circular Trajectory in a Horizontal Plane

Finally, a circular trajectory in the horizontal XY plane is performed. The trajectory is described mathematically as

$$x = r \sin(\omega t), \quad y = r \cos(\omega t), \quad z = z_0 > 0 \tag{22}$$

with

$$r = 1.5\,\text{m}, \quad z_0 = 3\,\text{m}, \quad \omega = \sqrt{\frac{g}{z_0}} \simeq 1.808\,\text{s}^{-1}. \tag{23}$$

The above trajectory produces a velocity of the end-effector of approximately 2.7 m/s along the circular path. Similarly to the previous cases, a linear regression is applied to the experimental data in order to fit a circular path to the measured trajectory and the results are shown in Fig. 6a. The bias observed mainly consists of an offset of the centre of the circle. This offset is compensated for in the controller. Hence, the accuracy of the mechanism is obtained by considering the error between the best circle fit and the measured position. As mentioned above, the main interest is the determination of the accuracy and not the precision. The results are shown in Fig. 6b. It can be observed that the accuracy is not as good as in the previous cases and that the error can go up to almost 4 cm. It can also be observed that the largest errors are obtained when the end-effector is the furthest away from the centre of the static workspace. Since the circle has a radius of 1.5 m, the end-effector is located further from the workspace centre, which may explain the larger error. Also, since the end-effector is moving in a plane that is located only 3 m below the attachment points of the cables to the fixed frame, the tension in the cables is larger which may also increase the error.

5 Conclusion

This paper provided an experimental determination of the accuracy of a three-dof cable-suspended parallel mechanism with a point mass end-effector. The kinematic and dynamic models of the robot were first recalled and an example of a dynamically feasible periodic trajectory was given. Then, three trajectories were presented for

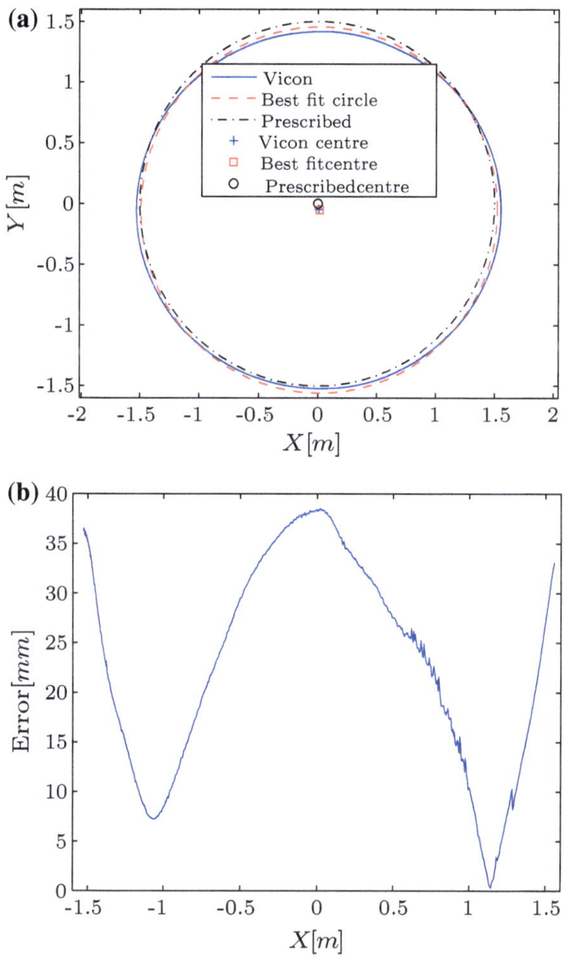

Fig. 6 Measured position for the circular trajectory in a horizontal plane, best circular fit and error (**a** and **b**)

which experimental data was obtained. A VICON system was used to determine the coordinates of the end-effector and compare them with the prescribed trajectory. For two of the trajectories, the positioning error along the path was found to be of the order of one centimetre, which is acceptable considering the length of the cables (several metres) and the dynamic character of the trajectory. It was also observed that points located far away from the central vertical axis tend to produce larger errors. Future work includes the development of compensation schemes to improve the positioning accuracy of the mechanism and the implementation of a prototype of a three-dof translational cable-suspended parallel mechanism with a rigid-body platform.

Acknowledgments This work was supported by The Natural Sciences and Engineering Research Council of Canada (NSERC), by the Fonds de la Recherche du Québec sur la Nature et les Technologies (FRQNT) and by the Canada Research Chair Program.

References

1. Albus J, Bostelman R, Dagalakis N (1993) The NIST Robocrane. J Robot Syst 10(5):709–724
2. Barrette G, Gosselin C (2005) Determination of the dynamic workspace of cable-driven planar parallel mechanisms. ASME J Mech Des 127(2):242–248
3. Cunningham D, Asada H (2009) The Winch-Bot: a cable-suspended, under-actuated robot utilizing parametric self-excitation. In: Proceedings of the IEEE international conference on robotics and automation, pp 1844–1850
4. Gosselin C (2012) Global planning of dynamically feasible trajectories for three-dof spatial cable-suspended parallel robots. In: Bruckmann T, Pott A (eds) Proceedings of the first international conference on cable-driven parallel robots. Stuttgart, Germany, 2–4 Sep 2012, pp 3–22
5. Gosselin C, Bouchard S (2010) A gravity-powered mechanism for extending the workspace of a cable-driven parallel mechanism: application to the appearance modelling of objects. Int J Autom Technol 4(4):372–379
6. Gosselin C, Ren P, Foucault S (2012) Dynamic trajectory planning of a two-dof cable-suspended parallel robot. In: Proceedings of the IEEE international conference on robotics and automation, St-Paul, Minnesota, pp 1476–1481
7. Lefrançois S, Gosselin C (2010) Point-to-point motion control of a pendulum-like 3-dof underactuated cable-driven robot. In: Proceedings of the IEEE international conference on robotics and automation, pp 5187–5193
8. Oh SR, Ryu JC, Agrawal SK (2006) Dynamics and control of a helicopter carrying a payload using a cable-suspended robot. ASME J Mech Des 128:1113–1121
9. Pusey J, Fattah A, Agrawal S, Messina E (2004) Design and workspace analysis of a 6–6 cable-suspended parallel robot. Mech Mach Theory 39:761–778
10. Riechel AT, Ebert-Uphoff I (2004) Force-feasible workspace analysis for underconstrained, point-mass cable robots. In: Proceedings of the IEEE international conference on robotics and automation, pp 4956–4962
11. Zanotto D, Rosati G, Agrawal SK (2011) Modeling and control of a 3-DOF pendulum-like manipulator. In: Proceedings of the IEEE international conference on robotics and automation, pp 3964–3969
12. Zi B, Duan BY, Du JL, Bao H (2008) Dynamic modeling and active control of a cable-suspended parallel robot. Mechatronics 18:1–12
13. Zoso N, Gosselin C (2012) Point-to-point motion planning of a parallel 3-DOF underactuated cable-suspended robot. In: Proceedings of the IEEE international conference on robotics and automation, St-Paul, Minnesota, pp 2325–2330

Efficient Calibration of Cable-Driven Parallel Robots with Variable Structure

Dragoljub Surdilovic, Jelena Radojicic and Nick Bremer

Abstract This paper presents an efficient practical approach for the combined explicit and implicit approximated calibration of cable-driven parallel robots (wire robots, CDPR) mainly developed to tackle the problems with variable system structures, i.e. often reconfigurable common robot platform. Indeed, the developed calibration procedure can also be applied to the systems with stationary (end-effector-i.e. gripper-like) platforms, however the benefits of the new methods are mainly expressed in the variable structure systems. The variable structure CDPR systems cover classes of robots in which the common robot platform represents a working object to be manipulated, itself. Such systems are typical in novel CDPRs referred as *extended-cranes wire robots* or rehabilitation wire robots (e.g. STRINGMAN [1]). An additional specific system case belonging to the considered class is large mobile CDPR developed for applications in agriculture, which with changes of the application fields should be often periodically reconfigured and commissioned (calibrated). The paper provides detailed mathematical modelling of the novel calibration approach based on the parameter sensitivity analysis of the robot kinematic models including wire pulley systems. The implementation of the calibration procedures including required sensory systems and control supports has also been considered. Finally, practical examples illustrate the performance of the developed calibration method.

Keywords Wire-robots calibration · External- and self-calibration · Parameter sensitivity analysis · Sequential linear estimation · Force/impedance control · Wire tension

---

D. Surdilovic (✉) · J. Radojicic · N. Bremer
Department Robotics and Automation, Fraunhofer Institute for Production Systems and Design Technology IPK-Berlin, Pascalstr. 8-9, 10587 Berlin, Germany
e-mail: dragoljub.surdilovic@ipk.fraunhofer.de

A. Pott and T. Bruckmann (eds.), *Cable-Driven Parallel Robots*,
Mechanisms and Machine Science 32, DOI 10.1007/978-3-319-09489-2_9

1 Introduction

The *cable-driven parallel robots* (CDPR) or *wire-robots* have been recently addressed in numerous researches focusing on their advantages for implementing large spans, fast moving, lightweight and heavy-duty active spatial mechanisms. In comparison to a more general class of cable robots (e.g. spatial advanced robotized crane systems), the wire robots, especially the so called over-constrained structures, offer benefits such as ease of reconfiguration and adaptation to specific applications. Thereby the robot structures may be often changed, commonly by exchanging the working platform or varying locations of the robot wires attachments (pulleys, winches) in the space. Moreover, in specific CDPR systems, the change of common working platform or wire locations is closely related to the robot function.

A typical example represents the gait rehabilitation robot STRINGMAN [1]. This wire robot is unique, since its common platform interconnecting all cables represents the upper body of a patient (Fig. 1). The wires are connected to a harness carried by a patient performing gait training. The role of STRINGMAN is to provide weight suspension and gait balancing support (see [1] for more details). The procedures of patient attachments (Fig. 1) and detachments, facilitated by specific wire-tension control algorithms (based on force and impedance control) and patient lift-systems, represent a part of typical robot operations. The dimensions of the established "platform" practically vary for each patient, and efficient robot calibration and commissioning become crucial functions of the control system. Typical for the STRINGMAN are considerable inaccuracies, due to elasticity of the harness (human body attachment), which are tackled by the control system, relying on the interaction (force and impedance) control, rather than on the position control, supported by human tracking sensors (IMU, local wire-position sensors, vision etc.).

Another typical variable CDPR system requiring continuous calibration represents the so called *extended-crane systems* (Fig. 2). An extended-crane represents a combination of a wire robot and a conventional crane. The robot configuration is created to support task decomposition between the overhead crane (mainly performs

Fig. 1 STRINGMAN—patient attachments supported by lifting-system and wire-attach function (position based force/damping control)

Fig. 2 Extended-crane system

the weight balancing and gross motion) and the side wire system (mainly responsible for fine lateral positioning and orientation). The extended crane can considerably improve flexibility and efficiency of assembly of heavy parts with complex irregular geometry. The side wires have been still applied for such operations in industrial practice but in pure manual manipulation form, to support fine-positioning of heavy parts carried by cranes. However, such repetitive operations are due to higher inertia of handling parts (ergonomic safe limit is ca. 20 kg) quite dangerous for human health and may cause serious back-pain problems. In extended crane systems the workpiece itself represents a common platform providing several attachment points. The side-wires (winches) carriers are commonly realized by mobile units with uncertain locations (Fig. 2). Fast calibration after commissioning, as well as following each new part attachment, should improve wire-robot models and model-based control.

A further example of variable structure and reconfigurable wire robot systems represent large CDPRs for agricultural applications (Fig. 3). To meet higher system flexibility and application requirements in various agricultural plants, it is convenient to implement these systems by means of mobile pillars that transport winches and platform, and can be fixed on stand-on legs at some locations to provide stable wire-robot structure [2]. The exact locations of winches thereby should be identified by calibration after system configuration/reconfiguration and commissioning.

Calibration, i.e. estimation of geometric and kinematic parameters of CDPRs, or in more general case of parallel robots, has been recently addressed in several researches [2–8]. Both system classes with characteristic structures, including closed-loops between attached legs/wires and platform, offer a specific possibility for the

Fig. 3 Large wire-robots with mobile pillars for agricultural applications

calibration based only on internal (*proprioceptive*) sensing, referred to as "*implicit loop method*" [3]. As remarked by Wampler et all. [2], closed-loop relations provide additional information for calibrations that are equivalent to the end-point pose measurements in convenient industrial-robots with open kinematic chains. Additionally, redundancy of over-constrained wire-robots provides supplementary information for calibration [4]. The procedures for wire-robot calibration based on "implicit loop method" and proprioceptive sensors applications (e.g. local motor rotation/sliding sensors used for control purposes) are often referred to as *"self-calibration"* [4, 5]. Various specific self-calibration procedures (referred to as *tension-* and *jitter-based* self-calibration) have been developed in [6] for a planar wire-robot structures, in which some wires keep the length constant while remaining wires vary length or tension.

As is well known, the *self-calibration* in complex (e.g. non-negligible pulleys) and large wire-robots structures described by non-linear models, including wire cable elasticity [9, 10] with considerable parameter uncertainties may become also complex and converge to local minima that provide wrong information for identification. Linearized iterative methods that include direct-kinematics in the loop [6], also in general doesn't ensure convergence (contraction processes), producing the steps that deteriorate calibration procedure. Therefore complete identification schemes that tackle both parameter and pose uncertainties have been recently proposed [6].

In large scale robots (Figs. 2 and 3), however, the self-calibration procedure may become time-consumable and information from the external sensors can considerably improve the calibration, especially in non-linear systems with expressed cable sag and elasticity effects [8]. Thereby beside costly laser trackers [8] also relatively cheap sensors, such as cameras [3] or IMU's [1] (with additional external sensors, i.e. closed-loop based data fusion and drift compensation) may be applied.

This paper provides novel calibration methods that combine force/impedance control with external/internal CDPR calibration, which is based on affordable sensors and quite efficient for calibration of frequently reconfigurable wire-robots. This approach is based on an explicit linearized mathematical parameter sensitivity model of complex wires structures including pulleys. To cope with large dimension and non-linearity, an approximation method based on sequential linear programming (SLP) and linear recursive approximations based on smaller calibration sub-problems [11] has been developed. Several simulation examples demonstrate the feasibility and robustness of the calibration performance against higher parametric deviations and measurement disturbances.

2 Kinematic Analysis

Kinematic models of wire-robots provides background for detailed parameter sensitivity analysis In Fig. 4, a general model of wire-robot with n-wires (i = 1,..., n) is given. In a over-constrained wire-robot structure ensuring 6DOF motion plus wires tension is $n \geq 7$. Using the notion from (Fig. 4), the position of the ith wire platform attachment point B_i is defined by

$$\mathbf{p}_i = \mathbf{a}_i + \mathbf{L}_i = \mathbf{p} + \mathbf{b}_i \tag{1}$$

where $\mathbf{a}_i$ and $\mathbf{b}_i$ are position vectors of pulley and platform attachment points A_i and B_i wrt. base and local platform frames respectively, $\mathbf{p}$ is the position vector of the platform reference frame and

$$\mathbf{L}_i = \overrightarrow{A_i C_i} + \overrightarrow{C_i T_i} + \mathbf{l}_i \tag{2}$$

where $\mathbf{l}_i$ is the wire-length vector, while C_i and T_i denote centre of the pulley and wire tangent points (Fig. 4).

During an arbitrary platform displacement, the wire performs a complex composite motion consisting of: *transferred motion*, representing the entire wire

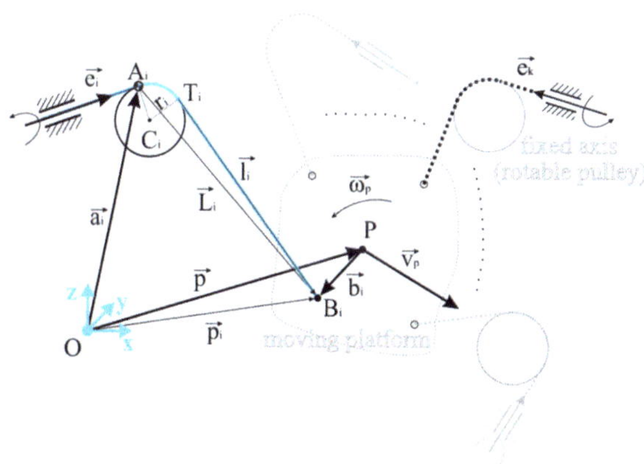

Fig. 4 Wire-robot structure

plane $\{A_i B_i C_i T_i\}$ rotation (rolling motion of the pulley) around the fixed pulley axis $\mathbf{e}_i$, and *relative motion* in the wire plane. The relative motion consists of relative translation i.e. change of the relative length in the actual cable direction (due to cable control, i.e. via a winch or linear slider mechanism (not presented in the Fig. 4), and a relative rotation of the wire around the pulley (i.e. point T_i that represents actual pole of the velocity). Based on this analysis, the expressions for absolute wire end-point velocities and accelerations are obtained

$$\mathbf{v}_i = \dot{\mathbf{p}}_i = \boldsymbol{\omega}_{ei} \times \mathbf{L}_i + \boldsymbol{\omega}_{ri} \times \mathbf{l}_i + \overset{*}{\mathbf{l}}_i = \mathbf{v}_p + \boldsymbol{\omega}_p \times \mathbf{b}_i \tag{3}$$

$$\begin{aligned}\mathbf{a}_i = \ddot{\mathbf{p}}_i &= \boldsymbol{\varepsilon}_{ei} \times \mathbf{L}_i + \boldsymbol{\varepsilon}_{ri} \times \mathbf{l}_i + \boldsymbol{\omega}_{ei} \times (\boldsymbol{\omega}_{ei} \times \mathbf{L}_i) + \boldsymbol{\omega}_{ei} \times (\boldsymbol{\omega}_{ri} \times \mathbf{l}_i) + \boldsymbol{\omega}_{ri} \\ &\times (\boldsymbol{\omega}_{ri} \times \mathbf{l}_i) + 2(\boldsymbol{\omega}_{e}i + \boldsymbol{\omega}_{ri}) \times \overset{*}{\mathbf{l}}_i + \overset{**}{\mathbf{l}}_i = \mathbf{a}_p + \boldsymbol{\varepsilon}_p \times \mathbf{b}_i + \boldsymbol{\omega}_p \times (\boldsymbol{\omega}_p \times \mathbf{b}_i)\end{aligned} \tag{4}$$

where $\boldsymbol{\omega}_{ei}$ and $\boldsymbol{\varepsilon}_{ei}$ denote pulley rotation velocity and acceleration around $\mathbf{e}_i$ (Fig. 5), $\boldsymbol{\omega}_{ri}$ and $\boldsymbol{\varepsilon}_{ri}$ relative wire rotation velocity and acceleration around wire-plane normal $\mathbf{n}_i$, $\overset{*}{\mathbf{l}}_i$ and $\overset{**}{\mathbf{l}}_i$ are linear wire relative velocity and acceleration due to cable length changes, $\mathbf{v}_p$ and $\boldsymbol{\omega}_p$, $\mathbf{a}_p$ and $\boldsymbol{\varepsilon}_p$ are platform linear and angular velocities and acceleration vectors respectively. Relative velocity components and their directions are shown in (Fig. 5). The projections of the velocity and acceleration vectors (3, 4) into wire-length vector direction, defined by unit vector $\mathbf{l}_{i0} = \mathbf{l}_i / l_i$, i.e. scalar multiplication of these equations by $\mathbf{l}_{i0}$ yields the magnitudes of wire linear relative velocity

$$\overset{*}{l}_i = [\mathbf{l}_{i0}{}^T \quad -\mathbf{l}_{i0}{}^T \underline{\underline{\mathbf{b}_i}}]\mathbf{t}_p \tag{5}$$

where $\mathbf{t}_p = [\mathbf{v}_p^T \quad \boldsymbol{\omega}_p^T]^T$ is the platform twist vector, while $\underline{\underline{\mathbf{b}_i}}$ denotes *skew-symmetric* 3×3 matrix formed from the elements of the vector $\mathbf{b}_i$ in order to represent the vector product in the matrix form. Scalar multiplication of (4) by $\mathbf{l}_{i0}$ yields the magnitude

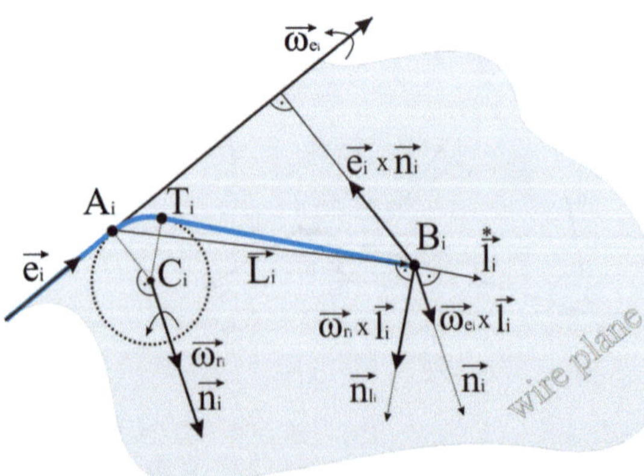

Fig. 5 Velocity vectors components

of wire relative acceleration

$$\overset{**}{l}_i = \omega_{e_i}^2[(\vec{\mathbf{e}}_i \times \vec{\mathbf{L}}_i)\cdot\vec{\mathbf{n}}_i][(\vec{\mathbf{e}}_i \times \vec{\mathbf{n}}_i)\cdot\vec{\mathbf{l}}_{i0}] + \omega_{r_i}^2 l_i + [\mathbf{l}_{i0}{}^T \quad -\mathbf{l}_{i0}{}^T \underline{\underline{\mathbf{b}_i}}]\dot{\mathbf{t}}_p - \mathbf{l}_{i0}{}^T \underline{\underline{\boldsymbol{\omega}_p}}\,\underline{\underline{\mathbf{b}_i}}\boldsymbol{\omega}_p \tag{6}$$

where the first two components represent the projections of centrifugal accelerations components (corresponding to the pulley and relative wire rotations), while the remaining parts define projections of platform tangential and centrifugal accelerations into the wire directions.

The expressions for vectors of angular pulley and wire relative rotations are obtained by scalar multiplication of (3) by vectors $\mathbf{n}_{l_i} = \mathbf{n}_i \times \mathbf{l}_{i0}$ and $\mathbf{n}_i$ respectively in terms of platform twist vector

$$\omega_{e_i} = \frac{1}{(\vec{\mathbf{e}}_i \times \vec{\mathbf{L}}_i)\cdot\vec{\mathbf{n}}_i}[\mathbf{n}_i^T \; -\mathbf{n}_i^T \underline{\underline{\mathbf{b}_i}}]\mathbf{t}_p \tag{7}$$

$$\omega_{r_i} = \frac{1}{l_i}[-\mathbf{l}_{i0}{}^T \underline{\underline{\mathbf{n}_i}} \quad \mathbf{l}_{i0}{}^T \underline{\underline{\mathbf{n}_i \mathbf{b}_i}}]\mathbf{t}_p \tag{8}$$

3 Wire-Robot Jacobian and Its Time-derivative

The relationship between relative wire velocity, defining the cable length variations, and platform twist vector is defined by the wire-robot Jacobian

$$\overset{*}{\mathbf{l}} = \mathbf{J}\mathbf{t}_p \tag{9}$$

where $\overset{*}{\mathbf{l}} = [\overset{*}{l}_1 \ldots \overset{*}{l}_i \ldots \overset{*}{l}_n]^T$ and Jacobian matrix $\mathbf{J} \in \Re^{n\times 6}$ is

$$\mathbf{J}^T = \begin{bmatrix} \mathbf{l}_{10} & \cdots \mathbf{l}_{i0} & \cdots \mathbf{l}_{n0} \\ \underline{\underline{\mathbf{b}_1}}\mathbf{l}_{10} & \cdots \underline{\underline{\mathbf{b}_i}}\mathbf{l}_{i0} & \cdots \underline{\underline{\mathbf{b}_n}}\mathbf{l}_{n0} \end{bmatrix} \tag{10}$$

The time derivative of wire Jacobian is obtained by differentiating (9) (the same result is obtained by substituting (7) and (8) in (6))

$$\overset{**}{\mathbf{l}} = \mathbf{J}\dot{\mathbf{t}}_p + \dot{\mathbf{J}}t_p \tag{11}$$

where based on (10) is

$$\dot{\mathbf{J}}^T = \begin{bmatrix} \dot{\mathbf{l}}_{10} & \cdots & \dot{\mathbf{l}}_{i0} & \cdots & \dot{\mathbf{l}}_{n0} \\ \underline{\underline{\dot{\mathbf{b}}_1}}\mathbf{l}_{10} + \underline{\underline{\mathbf{b}_1}}\dot{\mathbf{l}}_{10} & \cdots & \underline{\underline{\dot{\mathbf{b}}_i}}\mathbf{l}_{i0} + \underline{\underline{\mathbf{b}_i}}\dot{\mathbf{l}}_{i0} & \cdots & \underline{\underline{\dot{\mathbf{b}}_n}}\mathbf{l}_{n0} + \underline{\underline{\mathbf{b}_n}}\dot{\mathbf{l}}_{n0} \end{bmatrix} \tag{12}$$

Taking into account that the time derivatives of the constant intensity vectors $\mathbf{l}_{i0}$ and $\mathbf{b}_i$ (considering an ideal rigid platform) are

$$\begin{aligned} \dot{\mathbf{l}}_{i0} &= \boldsymbol{\omega}_{e_i} \times \mathbf{l}_{i0} + \boldsymbol{\omega}_{r_i} \times \mathbf{l}_{i0} \\ \dot{\mathbf{b}}_i &= \boldsymbol{\omega}_i \times \mathbf{b}_i \end{aligned} \tag{13}$$

and substituting (7) and (8) yields

$$\dot{\mathbf{J}} = \mathbf{t}_p^T \otimes \mathbf{J} = \mathbf{t}_p^T \otimes [\mathbf{J}_1^T \ldots \mathbf{J}_\mathrm{i}^T \ldots \mathbf{J}_\mathrm{n}^T]^T \tag{14}$$

where $\mathbf{J}$ is a $n(6) \times 1(6)$ block matrix (numbers outside parenthesis define block matrix dimension, while within parenthesis the dimension of each block-matrix element has been given), $\otimes$ is the Kronecker's product (each block-element of $\mathbf{J}$ is by $\mathbf{t}_p^T$ multiplied) and the block element $\mathbf{J}_\mathrm{i} \in \Re^{6\times 6}$ has the form

$$\mathbf{J}_\mathrm{i} = \begin{bmatrix} -\frac{1}{l_i}\underline{\underline{\mathbf{n}_i}}\mathbf{l}_{i0}{\mathbf{l}_{i0}}^T\underline{\underline{\mathbf{n}_i}} - \frac{1}{(\vec{\mathbf{e}}_i \times \vec{\mathbf{L}}_i)\cdot\vec{\mathbf{n}}_i}\mathbf{n}_i{\mathbf{l}_{i0}}^T\underline{\underline{\mathbf{e}_i}} & \frac{1}{l_i}\underline{\underline{\mathbf{n}_i}}\mathbf{l}_{i0}{\mathbf{l}_{i0}}^T\underline{\underline{\mathbf{n}_i\mathbf{b}_i}} + \frac{1}{(\vec{\mathbf{e}}_i \times \vec{\mathbf{L}}_i)\cdot\vec{\mathbf{n}}_i}\mathbf{n}_i{\mathbf{l}_{i0}}^T\underline{\underline{\mathbf{e}_i\mathbf{b}_i}} \\ -\frac{1}{l_i}\underline{\underline{\mathbf{b}_i\mathbf{n}_i}}\mathbf{l}_{i0}{\mathbf{l}_{i0}}^T\underline{\underline{\mathbf{n}_i}} - \frac{1}{(\vec{\mathbf{e}}_i \times \vec{\mathbf{L}}_i)\cdot\vec{\mathbf{n}}_i}\underline{\underline{\mathbf{b}_i}}\mathbf{n}_i{\mathbf{l}_{i0}}^T\underline{\underline{\mathbf{e}_i}} & \frac{1}{l_i}\underline{\underline{\mathbf{b}_i\mathbf{n}_i}}\mathbf{l}_{i0}{\mathbf{l}_{i0}}^T\underline{\underline{\mathbf{n}_i\mathbf{b}_i}} + \frac{1}{(\vec{\mathbf{e}}_i \times \vec{\mathbf{L}}_i)\cdot\vec{\mathbf{n}}_i}\underline{\underline{\mathbf{b}_i}}\mathbf{n}_i{\mathbf{l}_{i0}}^T\underline{\underline{\mathbf{e}_i\mathbf{b}_i}} + \underline{\underline{\mathbf{b}_i\mathbf{l}_{i0}}} \end{bmatrix} \tag{15}$$

4 Parameter Sensitivity Model

The kinematic model sensitivity analysis provides a framework for the development of the calibration procedure. The above presented CDPR kinematic model includes several parameters definitions: robot structure (vectors $\vec{\mathbf{e}}_i$ and $\vec{\mathbf{a}}_i$, pulley radius r_i), platform dimension, i.e. wire attachment points (vectors $\tilde{\vec{\mathbf{b}}}_i$, where $\tilde{\cdot}$-denotes vectors in local-platform coordinate systems), and variable platform position and rotation vectors $\mathbf{x}_p = [\mathbf{p}_p^T \quad \mathbf{o}_p^T]^T$. The rotation vector $\mathbf{o}_p$ takes different forms dependent on selected rotation presentation (e.g. axis-angle, Cardan-, Euler- etc. angles). The measurable wire length s_i, on a sliding- or a winch drive, consists of the active wire-length l_i and the pulley arc of the contact length (Fig. 6). The basic kinematic relationships for the calibration thus include

$$\begin{aligned} s_i &= r_i\delta_i + l_i \\ \mathbf{p}_i &= \mathbf{a}_i + \mathbf{L}_i = \mathbf{p} + \mathbf{b}_i = \mathbf{p} + \mathbf{R}\tilde{\mathbf{b}}_i \\ \mathbf{L}_i &= \overrightarrow{A_iC_i} + \overrightarrow{C_iT_i} = r_i\left[\vec{e}_i \times \vec{n}_i - \frac{1}{l_i}l_i \times \vec{n}_i\right] \end{aligned} \tag{16}$$

where $\delta_i = \prec A_iT_i$ is the circular segment angle between points A_i and T_i (see Fig. 6). The kinematic *loop closure* between two wire chains i and j over the platform requires additionally

$$\mathbf{a}_i - \mathbf{a}_j + \mathbf{L}_i - \mathbf{L}_j = \mathbf{R}(\tilde{\mathbf{b}}_i - \tilde{\mathbf{b}}_j) \tag{17}$$

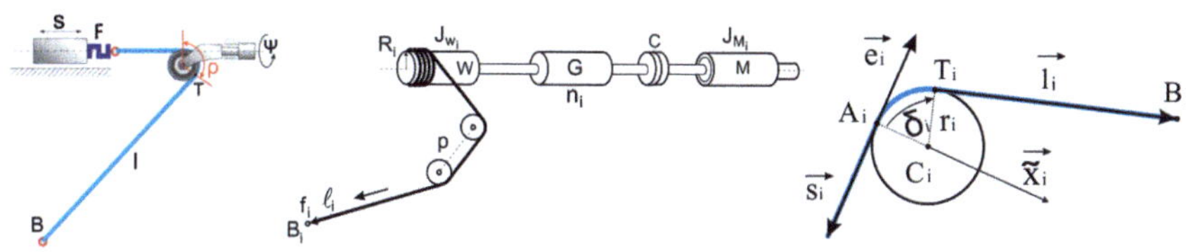

Fig. 6 Wire transmission chains with linear sliders (*left*) and winches (*right*)

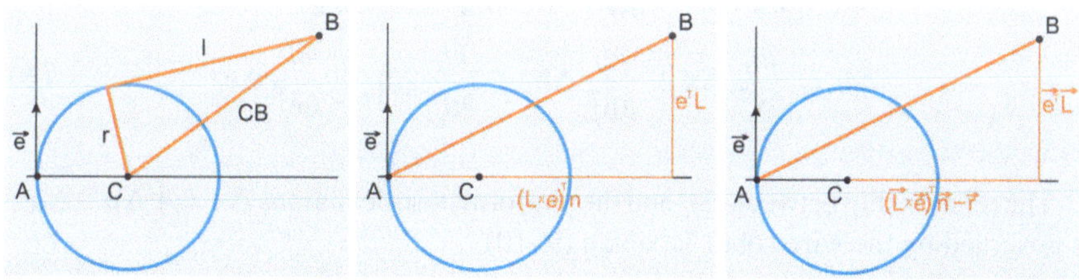

Fig. 7 Right-triangles relationships

The kinematic model parameters (16–17) uniquely define the position and orientation of platform, i.e. wire attachment points B_i, in the space. By means of these parameters two basic kinematic tasks *inverse* (*IK*) and *direct-forward kinematics* (*DK*) can be solved. The *IK* is trivial in CDPRs, mainly based on simple geometric considerations: starting from given $\mathbf{x}_p$, i.e. known B_i, determining the tangent on pulley (Thales' circle), following by computation of δ_i and finally s_i (16). Further transformation to corresponding motor angle or linear drive position is also easy and depends on specific transmission train (Fig. 7). As well known, *DK* that determines platform pose starting from measured s_i appears to be more complex in parallel manipulators, and has been usually solved by iterative numeric procedures [9] optimized for real-time applications (control).

In the reconfigurable systems, however, the fixed wire-robot parameters $(\vec{\mathbf{e}}_i, \vec{\mathbf{a}}_i, \tilde{\vec{\mathbf{b}}}_i)$ are unknown or uncertain and should be estimated before operation. The linearization of (16) around an initial nominal parameter-model configuration ($\vec{\mathbf{e}}_{i0}$, $\vec{\mathbf{a}}_{i0}$, $\tilde{\vec{\mathbf{b}}}_{i0}$) and an initial model pose ($\mathbf{p}_0$, $\mathbf{R}_0$) yields

$$\begin{aligned}\Delta\mathbf{a}_i + \Delta\mathbf{L}_i &= \Delta\mathbf{p} + \mathbf{R}_0\,\Delta\tilde{\mathbf{b}}_i + \underline{\underline{\Delta\mathbf{o}}}\,\mathbf{R}_0\tilde{\mathbf{b}}_{i0} \\ \Delta s_i &= r_i\Delta\delta_i + \Delta l_i \end{aligned} \qquad (18)$$

where $\underline{\underline{\Delta\mathbf{o}}}$ represents skew-symmetric matrix of infinitesimal platform rotations. The above equations provide a framework for a complete identification concerning both parameter and pose uncertainties.

The efficiency of the calibration can be improved by considering explicit dependency between wire-robot parameters ($\vec{\mathbf{e}}_i$, $\vec{\mathbf{a}}_i$, $\vec{\tilde{\mathbf{b}}}_i$) and cable length Δl_i (Δs_i) deviations

$$l_i = l_i\left(\mathbf{a}_i, \tilde{\mathbf{b}}_i, \mathbf{e}_i, \mathbf{p}, \mathbf{o}\right) = l_i\left(\mathbf{a}_i, \tilde{\mathbf{b}}_i, \mathbf{e}_i, \mathbf{x}\right) \tag{19}$$

and

$$\begin{aligned} dl_i &= \frac{\partial l_i}{\partial \mathbf{x}^T} d\mathbf{x} + \frac{\partial l_i}{\partial \tilde{\mathbf{b}}_i^T} d\tilde{\mathbf{b}}_i + \frac{\partial l_i}{\partial \mathbf{a}_i^T} d\mathbf{a}_i + \frac{\partial l_i}{\partial \mathbf{e}_i^T} d\mathbf{e}_i \\ \Delta l_i &= \frac{\partial l_i}{\partial \mathbf{x}^T} \Delta\mathbf{x} + \frac{\partial l_i}{\partial \tilde{\mathbf{b}}_i^T} \Delta\tilde{\mathbf{b}}_i + \frac{\partial l_i}{\partial \mathbf{a}_i^T} \Delta\mathbf{a}_i + \frac{\partial l_i}{\partial \mathbf{e}_i^T} \Delta\mathbf{e}_i \end{aligned} \tag{20}$$

The relationship between Δl_i and the platform pose deviations $\Delta\mathbf{x} = \left[\, \Delta\mathbf{p}^T \;\; \Delta\mathbf{o}^T \,\right]^T$ is governed by the wire-robot Jacobian (9–10).

$$\frac{\partial l_i}{\partial \mathbf{x}^T} = \left[\, \mathbf{l}_{i0}{}^T \;\; -\mathbf{l}_{i0}{}^T \underline{\underline{\mathbf{b}}}_i \,\right] \tag{21}$$

The Jacobian (10) and the equivalence between linear displacement $\Delta\mathbf{p}$, i.e. $\Delta\mathbf{p}_i$, and platform parameters deviations $\Delta\tilde{\mathbf{b}}_i$, given by $\Delta\mathbf{p}_i = \mathbf{R}\Delta\tilde{\mathbf{b}}_i$, yields

$$\frac{\partial l_i}{\partial \tilde{\mathbf{b}}_i^T} = \mathbf{l}_{i0}{}^T \tag{22}$$

The partial derivatives of l_i with respect to remaining parameter vectors may be computed based on the geometrical relations in right triangles (Fig. 7)

$$\begin{aligned} l_i^2 &= \overline{C_i B_i}^2 - r_i^2 \\ \overline{C_i B_i}^2 &= (\mathbf{L}_i \cdot \mathbf{e}_i)^2 + ((\mathbf{L}_i \times \mathbf{e}_i) \cdot \mathbf{n}_i - r_i)^2 \\ L_i^2 &= (\mathbf{L}_i \cdot \mathbf{e}_i)^2 + ((\mathbf{L}_i \times \mathbf{e}_i) \cdot \mathbf{n}_i)^2 \end{aligned} \tag{23}$$

yielding

$$l_i^2 = L_i^2 - 2\left((\mathbf{L}_i \times \mathbf{e}_i) \cdot \mathbf{n}_i\right)^2 \tag{24}$$

From (23–24), considering generalized partial derivative transformation

$$\frac{\partial l_i}{\partial \mathbf{u}_i^T} = \frac{\partial \left(l_i^2\right)^{1/2}}{\partial \mathbf{u}_i^T} = \frac{1}{2l_i} \frac{\partial \left(l_i^2\right)}{\partial \mathbf{u}_i^T} = \frac{1}{2l_i} \frac{\partial \left(L_i^2 - 2r_i\,(\mathbf{L}_i \times \mathbf{e}_i) \cdot \mathbf{n}_i\right)}{\partial \mathbf{u}_i^T} \tag{25}$$

and taking into account the relationships

$$\frac{\partial L_i^2}{\partial \mathbf{a}_i^T} = -2\mathbf{L}_i^T \text{ and } \frac{\partial\left((\mathbf{L}_i \times \mathbf{e}_i)\cdot \mathbf{n}_i\right)}{\partial \mathbf{a}_i^T} = -\left(\mathbf{e}_i \times \mathbf{n}_i\right)^T \tag{26}$$

yields the desired partial derivatives with respect to $\mathbf{a}_i$

$$\frac{\partial l_i}{\partial \mathbf{a}_i^T} = \frac{1}{l_i}\left(-\mathbf{L}_i^T + r_i(\mathbf{e}_i \times \mathbf{n}_i)^T\right) = -\frac{1}{l_i}\overrightarrow{C_i B_i}^T \tag{27}$$

In other words, an infinitesimal variation $\delta\, \mathbf{a}_i$ that is orthogonal to $\overrightarrow{C_i B_i}$ doesn't cause the variation of the cable length $\delta l_i = 0$ (Fig. 8).

Finally the variations δe_i which are compatible with other constraints (small rotations of e_i around A_i), cause

$$\frac{\partial l_i}{\partial \mathbf{e}_i^T} = \frac{r_i}{l_i}\frac{\mathbf{L}_i^T \mathbf{e}_i}{(\mathbf{L}_i \times \mathbf{e}_i)\cdot \mathbf{n}_i}\mathbf{L}_i^T \tag{28}$$

Due to constrained space the proof of (28), which is also based on relationship (23–25), as well as on the statement that $\mathbf{L}_i$ is not changed during e_i variations (points A_i and B_i remain fixed) (Fig. 5), is omitted. The deviations δe_i of pulley axes are related to small angular rotations (errors) around two orthogonal axes defined by the rotation vector $\delta\, o_i$ with respect to a local coordinate frame attached to the e_i. This relationship is defined by

$$\delta e_i = \delta o_i \times e_i = \underline{\underline{\delta o_i}}\, e_i = -\underline{\underline{e_i}}\, \delta o_i = -R_i \underline{\underline{\tilde{e}_i}}\, R_i^T \delta o_i = -R_i \underline{\underline{\tilde{e}_i}}\, \delta \tilde{o}_i \tag{29}$$

where "$\sim$" denotes vectors defined by projections into local frames, R_i is rotation matrix of this frame with respect to the base coordinate system. Substituting (29) into (28) defines parameter sensitivity of the l_i variations to small axis e_i deviations $\delta\tilde{\mathbf{o}}_i = [\,\delta\tilde{o}_{ix}\ \delta\tilde{o}_{iy}\ 0\,]$.

The parameter sensitivity model (17–29) provides a framework for the development of novel calibration procedures that is described in the following.

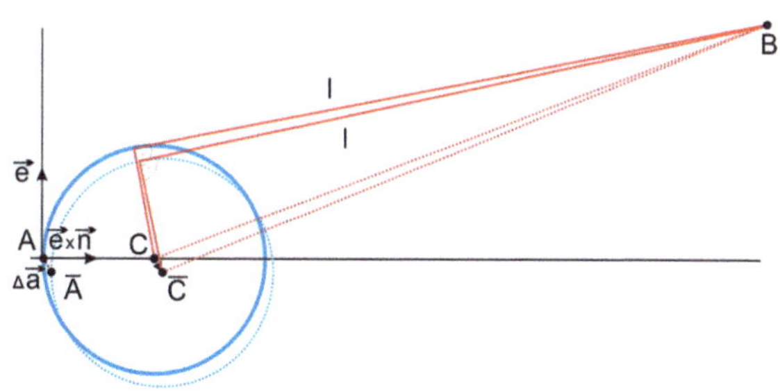

Fig. 8 Geometric proof of (26)—virtual displacement in direction orthogonal to $\vec{C_i B_i}$ doesn't produce wire-length changes (δl_i=0)

5 Calibration Procedure

Let us consider the extended-crane system in (Fig. 9) and the large over-constrained agricultural robots (Fig. 10). In principle, the calibration starts from a new wire-robot (on the example of an extended-crane) to be configured by attaching the wires to the common-workpiece (platform). At the beginning the winches have been fixed at known but uncertain locations defined by $\boldsymbol{a}_i$ w.r.t. a base frame of reference. The crane transports the workpiece to the assembly locations and the operator can manually attach the wires using local winch force/impedance (precisely damping) control. Thereby the wire-tension, i.e. force sensors, built in all winches (see [13] for more details), including also the main crane wire, have been utilized. After attachment, the tension wire control ensures the tension of all wires that are required to fix the workpiece in a desired, but also uncertain pose $\mathbf{x}_j$(j= 1...N$_x$), where N$_x$ denotes number of poses used in calibration. The local wires have been calibrated to provide an effective wire length measurements s_i(i= 1,...n) of the winches, for each pose. For the measurements of $\mathbf{x}_j$ after initial point calibration (details have been also omitted), various sensors such as IMU's with additional drift compensation sensors (including also internal implicit wire-robot sensors and constraints equations) and Kalman-filtering techniques may be used (it is not the focus of this paper).

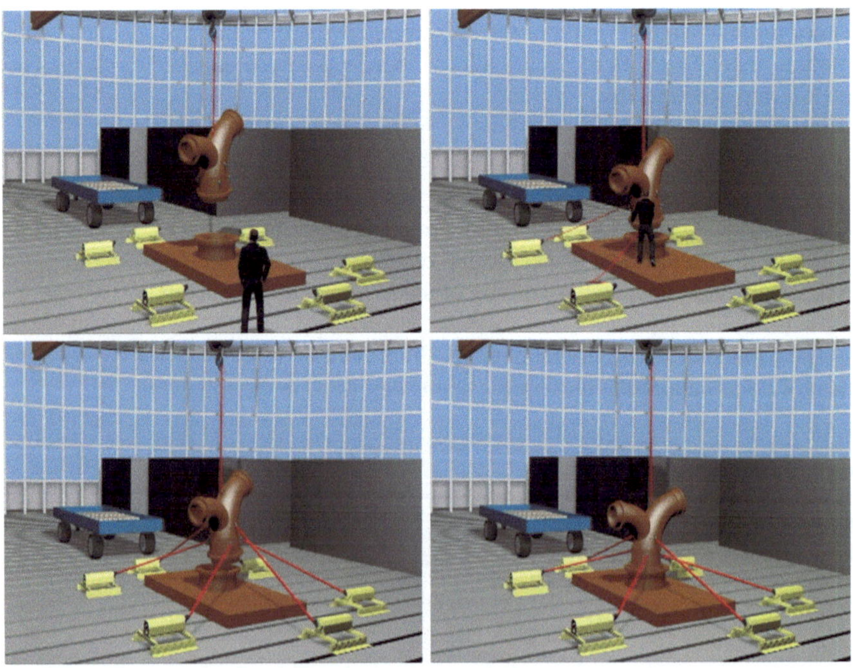

Fig. 9 Configuration of an extended crane system: transportation, wires attachment and tension

Fig. 10 Deviation model of a large span wire-robot

During calibration the platform moves to the poses $\mathbf{x}_j$, for example by combining the Cartesian space jogging and wire-tension control ensuring a steady-state stable pose. A previous selection of the points (calibration experiment design ensuring some optimality criterion, see [14]), may considerably improve the quality of identification, however, it is also out of the scope of this paper.

Calibration can be described briefly as the following problem. For the given set of measurements $M = \{\mathbf{x}_j, s_i(\mathbf{x}_j)\}$, and initial (model) parameters values $\hat{P} = \{\hat{\mathbf{a}}_i, \hat{\tilde{\mathbf{b}}}_i, \hat{\mathbf{e}}_i\}$, identify the system parameters $P = \{\mathbf{a}_i, \tilde{\mathbf{b}}_i, \mathbf{e}_i\}$ that in an optimal way fit the set of measurements and kinematic model constraints (17–21).

The calibration procedure involves the following linear approximation steps (***Algorithm 1***):

- Estimation of initial wire lengths based on the IK model

$$\hat{l}_{ij} = l_i^{-1}\left(\hat{\mathbf{a}}_i, \hat{\tilde{\mathbf{b}}}_i, \hat{\mathbf{e}}_i, \mathbf{x}_j\right)$$

- Estimation of wire active lengths based on total cable length measurements taking nominal values for the non-measurable wire contact arc $\delta_{ij} \approx \hat{\delta}_{ij}$ based on

$$\bar{l}_{ij} = s_{ij} - r_i \delta_{ij}$$

- Optimal fitting of parameters by minimizing cable length errors (squares) $\Delta l_{ij} = \bar{l}_{ij} - \hat{l}_{ij}$. Using derived explicit sensitivity model (26) this leads to solving a linear regression problem, in the considered case

$$\Delta \mathbf{a} = inv\left(\mathbf{R}^T \mathbf{R}\right) \mathbf{R}^T \Delta \mathbf{l}_{N_j}$$

where the regression matrix $\mathbf{R} \in \Re^{N_j \cdot n \times n \cdot 3}$ includes parameter Jacobians (10).

In the illustrative example, 20 platform measuring points have been selected nearly to the middle of the working space (which may be realized by practical tensions)

(Fig. 10). Instead of an extended crane, an over-constrained robot with n=8 wires has been considered. This robot represents an example of large agricultural wire-robot (Fig. 10) with the nominal span of 100 m and relatively large parameter perturbations. For the sake of simplicity only estimation of **a** will be considered. For this purpose a set of the following attachment points for simulating measurements has been selected

$$\mathbf{a}_0 = [50\ -50\ -50\ \ 50\ \ 50\ -50\ -50\ \ 50$$
$$50\ \ 50\ -50\ -50\ \ 50\ \ 50\ -50\ -50$$
$$25\ \ 25\ \ 25\ \ 25\ \ 0\ \ 0\ \ 0\ \ 0]$$

As initial model parameters the following vector has been selected with a relatively large initial deviation 2-norm of $\Delta a(0) \approx 4$ (m)

$$\hat{\mathbf{a}} = [46.9980\ -52.3952\ -51.0638\ \ 51.9303\ \ 48.0770\ -51.5475\ -50.4222$$
$$48.1406\ \ 52.0706\ \ 47.5845\ -53.0748\ -50.4700\ \ 52.5282\ \ 53.4440$$
$$-50.2576\ -48.4173\ \ 22.3897\ \ 28.2407\ \ 26.6444\ \ 22.3588\ \ -2.8097$$
$$-3.4949\ \ 1.3985\ \ 2.3109]$$

If we neglect the position measurements errors (ideal sensors case), the Algorithm 1 converges very fast, reducing the initial errors after only few steps to 2-norm $\Delta a \approx 10^{-10}$(m). The cable length (and contact arc angles δ_{ij}) residuum become also almost nullified.

However, this is a very simplified case that doesn't meet the practice, since the platform position measurements are also erroneous. Therefore the position errors have been randomly added to any measurements with again relative large errors, with uniform distribution errors and 2-norm $\Delta x(0) \approx 0.04$ (m). When the errors have been included, the linear regression fitting converges to an optimum with relatively large deviations and 2-norm $\Delta a(4) \approx 0.1987$ (m). Thereby relatively large residuum's of cable lengths 2-norm $\Delta l(4) \approx 0.0415$(m) (often of the same order as initial $\Delta \mathbf{x}$) has been achieved, where the index in parenthesis denotes number of iterations.

These residua cannot be reduced by further iterations since a local optimum has been reached. In order to advance the calibration, a logical improvement may be the estimation of both $\Delta \mathbf{a}$ and position errors measurements $\Delta \mathbf{x}$. However, this leads to a large linear optimization problem with commonly very bad conditioning and scaling of the regressor. The linear constrained fitting algorithms may limit parameters exploding, however, bad convergence and numerical errors remains typical problems. In order to tackle these limitations, a next *sequential optimization* step has been proposed.

The aim of the next sequential sub-estimation step (***Algorithm 2***) is to

- identify position errors by means of the wire robot Jacobian and thus to reduce (commonly after next few steps 2-norm $\Delta l < 0.0001$(m);
- reduce further the residuum Δa by performing sequential optimization with more stable fitting of smaller problems. Thereby both $\Delta \mathbf{x}$ and $\Delta \mathbf{a}$ gradients have to be computed iteratively.

As a result the cable length error can be considerably improved, while $\Delta\mathbf{a}$ reduction remains usually not significant (0.10 m) (dependent on selected start points).

To remove further $\Delta\mathbf{a}$ errors, finally the following sub-problem has to be solved (***Algorithm 3***)

- *self-calibration* by checking all possible closed loops with the wires based on (17–19) (there are totally 28 loop closure equations for each point in the considered case).

After this step finally the typical results have been achieved: 2-norm $\Delta\mathbf{a} < 0.01$ (m), while 2-norm $\Delta\mathbf{l} < 10^{-4}$(m) has been kept small by iterative position errors estimations.

Commonly all the above *Algorithms* steps require a few iterations (4–6) representing smaller sub-problems and are quite suitable for real-time applications.

$$\mathbf{a} = \begin{bmatrix} 50.0032 & -49.9968 & -49.9968 & 50.0032 & 50.0032 & -49.9968 & -49.9968 \\ 50.0032 & 49.9905 & 49.9905 & -50.0095 & -50.0095 & 49.9905 & 49.9905 \\ -50.0095 & -50.0095 & 25.0025 & 25.0025 & 25.0025 & 25.0025 & 0.0025 \\ & & & & 0.0025 & 0.0025 & 0.0025 \end{bmatrix}$$

It is worth mentioning that the calibration of pulley axes orientation errors $\delta\tilde{\mathbf{o}}_i$ appears to be sensitive to the wire contact arc $\delta_{ij} \approx \hat{\delta}_{ij}$ errors which are not measurable. Therefore the calibration of these angles based on both parameter sensitivity model and internal loop closure must be performed synchronously.

6 Conclusion

This paper has presented the detailed modeling of wire-robot kinematic models parameter sensitivity and application for calibration in general wire-robot systems with pulley elements. The novel algorithms include exact mathematical models to avoid complex numerical procedures and apply sequential gradient methodology to cope with large parameters fitting and bad conditioning problems.

The developed calibration procedures appear to be especially effective for variable structure CDPR systems, which cover classes of robots in which the common robot platform represents a working object to be manipulated, itself. Such systems are typical in novel CDPRs referred as *extended-cranes wire robots* under development at IPK. An illustrative example presents the performance of the developed calibration algorithms for large-scale cable robots with relatively higher parameter uncertainities.

The future work will focus on implementation in the mock-up extended crane system of which a presentation has been planned in September 2014.

Acknowledgments The work presented in this paper has been partially funded by Fraunhofer Community within WISA-ATLAS Project focusing on automated assembly of large-scale structures by means of wire robotic systems.

References

1. Surdilovic D, Bernhardt R, Schmidt T, Zhang J (2004) STRING-MAN: a novel wire robot for gait rehabilitation. In: Advances in rehabilitation robotics, Lecture Notes in Control and Information Science, Springer-Verlag, vol 306, 2004, pp 413–424
2. Wampler CW, Hollerbach JM, Arai T (1995) An implicit loop method for kinematic calibration and its application to closed-chain mechanisms. IEEE Trans Robot Autom 11(5):710–724
3. Traslosheros A, Sebastian JM, Torrijos J, Carelli R, Castillo E (2013) An inexpensive method for kinematic calibration of a parallel robot by using one hand-held camera as main sensor. Sensors 13:9941–9965
4. Miermeister P, Pott A, Verl A (2012) Auto-calibration method for overconstrained cable-driven parallel robots. In: 7th German conference on robotics (ROBOTIK 2012), Proceedings CD-ROM, Berlin VDE-Verlag, 2012, pp 301–306
5. Borgstrom PH, Jordan BL, Borgstrom BJ, Sukhatme GS, Batalin MA, Kaiser WJ (2009) NIMS-PL: a cable-driven robot with self-calibration capabilities. IEEE Trans Robot 25(5):1005–1015
6. Dit Sandretto JA, Daney D, Gouttefarde M, Baradat C (2012) Calibration of a fully-constrained parallel cable-driven robot. In: 19th CISM-IFToMM symposium on robot design, dynamics and Control (ROMANSY 2012), Paris
7. Miermeister P, Pott A (2013) Auto-calibration method for cable-driven parallel robots using force sensors. In: Latest Advances in Robot Kinematics, 1st international conference on cable-driven parallel robots, Stuttgart, 2012, Springer (2013)—Mechanisms and Machine Science, pp 269–276
8. Duan X, Qiu Y, Duan Q, Du J (2014) Calibration and motion control of a cable-driven parallel manipulator based triple-level spatial positioner, Hindawi Publishing Corporation, Advances in Mechanical Engineering, vol 2014, Article ID 368018, p 10
9. Merlet JP (2008) Analysis of wire elasticity for wire-driven parallel robots. In: Proceedings of EuCoMeS, 2nd European conference on mechanism science, Cassino (Italy), 2008
10. Surdilovic D, Radojicic J, Krüger J (2013) Geometric stiffness analysis of wire robots: a mechanical approach. In: 1st international conference on cable-driven parallel robots, Stuttgart, 2012, Springer (2013)—Mechanisms and Machine Science, pp 389–404
11. Radojicic J, Surdilovic D, Krüger J (2013) Application challenges of large-scale wire robots in agricultural plants. In: Proceedings of 2013 IFAC bio-robotics conference, vol 1, Part 1, pp 77–82
12. Lamberti L, Pappalettere C (2002) Design optimization of large-scale structures with sequential linear programming. Proc Inst Mech Eng C: J Mech Eng Sci 216:799
13. Vukobratovic M, Surdilovic D, Ekalo Y, Katic D (2009) Dynamics and robust control of robot-environment interaction, vol 2. New Frontiers in Robotics, Mar 2009, World Scientific Publishing, Singapore
14. Lizama E, Surdilovic D (1996) Designing G-optimal experiments for robot dynamics identification. In: Proceedings of IEEE international conference on robotics and automation, 1996, vol 1, pp 311–316

Part III
Control

Robust Internal Force-Based Impedance Control for Cable-Driven Parallel Robots

Christopher Reichert, Katharina Müller and Tobias Bruckmann

Abstract In this paper a robust internal force-based impedance controller for redundantly actuated cable-driven parallel robots (CDPRs) is proposed. The controller is governed by a computed-torque-control structure with a shaping of the internal forces resulting from the chosen mass matrix. These gains are intended to equip each manipulator with the feature of an impedance to enforce a dynamical relationship between the end-effector (EE) velocity and the internal forces. Non-linear effects like model uncertainties take a negative influence on the controller. A disturbance observer based on the generalized momentum approach is incooperted into the control scheme to impart the impedance controller the necessary robustness. To validate the described robust impedance control scheme, experiments with a 6-DOF CDPR with industrial BLDC-Motors are presented.

1 Introduction

It is known that cable-driven parallel robots (CDPR) have some advantages in comparison to serial kinematic manipulators in terms of high end-effector (EE) accelerations over a wide workspace. Since the platform is guided by lightweight cables which are conventionally wound up by winch drives attached to the base. Due to the unilateral properties of the cables, usually actuation redundancy is required to completely restrain the platform. Actuation redundancy can be achieved by additional winch drives without changing the degree-of-freedom (DOF) of theunderlying

C. Reichert (✉) · K. Müller · T. Bruckmann
Chair of Mechatronics, University of Duisburg-Essen, Duisburg, Germany
e-mail: reichert@imech.de

K. Müller
e-mail: mueller@imech.de

T. Bruckmann
e-mail: bruckmann@imech.de

A. Pott and T. Bruckmann (eds.), *Cable-Driven Parallel Robots*,
Mechanisms and Machine Science 32, DOI 10.1007/978-3-319-09489-2_10

mechanism. That means the degree-of-actuation (DOA) exceeds the DOF of the platform. Hence, this redundancy is used to enforce a desired tension level in the cable system and has a direct influence onto the stiffness of the mechanism. Moreover, it affects the force capabilities of a redundantly actuated CDPR in terms of acceleration and payload. These properties must be supported by the controller to take full advantage from the capabilities of a CDPR. In order to achieve accurate trajectory tracking, model-based controllers are required like the computed torque control (CTC) or augmented PD (APD). However, these approaches are only well suited for non-redundant CDPRs. There is a need to develop an efficient control scheme considering the regulation of the desired tension level in the cable system for redundantly actuated CDPRs. For this reason, force control strategies have to be integrated, taking a significant influence on the dynamical behavior of the CDPR. The authors in [1] provide an analytical overview of the involved dynamics and clarify possible influences which can lead to bandwidth limitations in robot force control. To overcome these limitations, the well-known impedance control approach introduced in [2] represents an elegant solution. In the past decade, the impedance control (IC) of cooperating manipulators became increasingly more popular for dynamic positioning. A common approach for fast trajectory tracking is the internal-force based impedance control approach [3] for instance.

The paper is organized as follows: First the dynamics of a CDPR are formulated in EE coordinates shown in Sect. 2. In Sect. 3 a disturbance observer for redundantly actuated CDPRs is described and its passive properties are shown. In the following Sect. 4 the derivation of the impedance control law is introduced. In Sect. 5 experiments on a 6-DOF CDPR with industrial BLDC-Motors are presented and summarized in Sect. 6.

2 Manipulator Dynamics

2.1 Formulation in EE Coordinates

A CDPR can be modeled as a multi-body system (MBS) consisting of a working platform (end-effector) with δ-DOFs and is constrained by m flexible cables in a parallel configuration. The EE is considered as a rigid body driven by constraint forces and process forces from the environment. As shown in [4], the equations of motion for CDPRs can be formulated by Lagrange's equation of the first kind, with n generalized (passive and active) joint coordinates $\mathbf{q} \in \mathbb{V}^n$ describing the motion of the mechanism and $\delta \in \mathbb{N}$ EE coordinates $\mathbf{x} = \left[\mathbf{r}\ \psi\right]^{\mathrm{T}}$ (task-space) according to the EE position $\mathbf{r}$ and orientation ψ. These equations can be obtained by cutting each kinematic chain of the mechanism at the EE and introducing redundant geometric and kinematic closure conditions

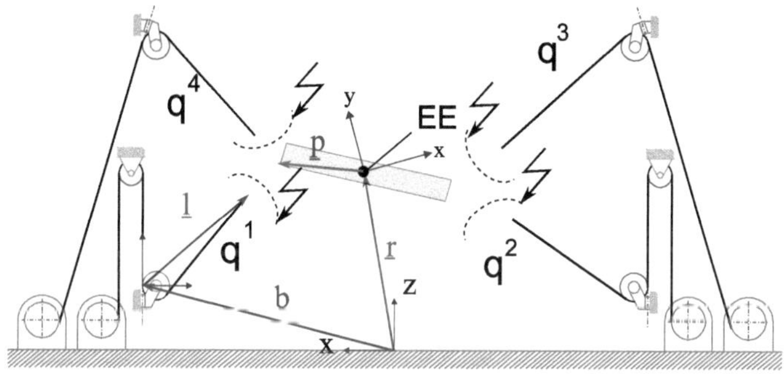

Fig. 1 Tree structure of a cable-driven parallel robot after opening the kinematic chains at the EE

$$\mathbf{0} = \mathbf{h}(\mathbf{q}, \mathbf{r}, \psi), \qquad \mathbf{h}(\mathbf{q}, \mathbf{r}, \psi) \in \mathbb{R}^n, \tag{1}$$

$$\mathbf{0} = \mathbf{J}(\mathbf{q})\dot{\mathbf{q}} + \mathbf{J}_X(\mathbf{r}, \psi)\dot{\mathbf{x}}, \quad \mathbf{J}(\mathbf{q}) = \frac{\partial \mathbf{h}}{\partial \mathbf{q}} \in \mathbb{R}^{n \times n}, \ \mathbf{J}_X(\mathbf{r}, \psi) = \frac{\partial \mathbf{h}}{\partial \mathbf{x}} \in \mathbb{R}^{n \times \delta}, \tag{2}$$

where $\mathbf{h}(\mathbf{q}, \mathbf{r}, \psi)$ defines the forward kinematics of each kinematic chain between the cable exit point and the EE's center of mass. Figure 1 demonstrates the cuts to be made for the visualization of the tree structure for the given mechanism and m fundamental loops can be constituted in the topological graph.

The m external control forces $\mathbf{u}$ generated by the winch drives directly controlling the mechanism along a predefined trajectory and the equations of motion become

$$\begin{bmatrix} \mathbf{D}\ddot{\mathbf{q}} + \mathbf{Q} \\ \mathbf{M}_X\ddot{\mathbf{x}} + \mathbf{K}_X\dot{\mathbf{x}} + \mathbf{Q}_X \end{bmatrix} = \begin{bmatrix} \mathbf{u} \\ \mathbf{0}_\delta \end{bmatrix} + \begin{bmatrix} \mathbf{J}^T \\ \mathbf{J}_X^T \end{bmatrix} \lambda, \tag{3}$$

where $\mathbf{D}$ is the inertia matrix related to each winch drive and $\mathbf{Q}$ includes all remaining potential forces, especially friction forces. The subscript X denotes elements for the particular linear EE equations of motion. Accordingly, the generalized mass matrix is $\mathbf{M}_X$ and the generalized Coriolis and centrifugal matrix are summarized in $\mathbf{K}_X$. In $\mathbf{Q}_x$ all generalized EE disturbances and the gravity forces are included.

Here, the Lagrange multipliers λ characterize the constraint forces. The set of all valid configurations of a CDPR, according to the constraints, defines the so-called configuration space (c-space) $V := \{\mathbf{q} \in \mathbb{V}^n | \mathbf{h}(\mathbf{q}, \mathbf{r}, \psi) = \mathbf{0}\}$ of the CDPR [4]. Based on the constraint Jacobian $\mathbf{J}$ which has locally full rank and its inverse is denoted as $\mathbf{J}^{-1}$, the generalized velocities can be expressed as

$$\begin{bmatrix} \dot{\mathbf{q}} \\ \dot{\mathbf{x}} \end{bmatrix} = \mathbf{F}\dot{\mathbf{x}}, \quad \mathbf{F} = \begin{bmatrix} -\mathbf{J}^{-1}\mathbf{J}_X \\ \mathbf{I}_\delta \end{bmatrix} \tag{4}$$

to parametrize all admissible configurations $\mathbf{q} \in \mathbb{V}^n$. In the next step, the so-called orthogonal complement $\mathbf{F}$ of $[\mathbf{J}\ \mathbf{J}_X]$ is defined, fulfilling $[\mathbf{J}\ \mathbf{J}_X]\,\mathbf{F} \equiv \mathbf{0}$. The defined orthogonal complement is used to elimante the unknown constraint forces λ by the projection of the equations of motion (3) onto the c-space V. Since the vector of external control forces $\mathbf{u}$ only comprises non-zero entries for the m actuated joints with generalized coordinates, a submatrix $\mathbf{A}$ of the orthogonal complement $\mathbf{F}$ can be identified, so that

$$\mathbf{F}^{\mathrm{T}}\begin{bmatrix}\mathbf{u}\\ \mathbf{0}_{\delta}\end{bmatrix} = \mathbf{A}^{\mathrm{T}}\mathbf{c}\,,$$

where $\mathbf{c} \equiv (c_1, \dots, c_m)$ is a vector of generalized control forces corresponding to the actuator coordinates. In the same way the joint accelerations can be yielded by the time derivative of Eq. (4): $\ddot{\mathbf{q}} = \mathbf{A}\ddot{\mathbf{x}} + \dot{\mathbf{A}}\dot{\mathbf{x}}$.

With the help of the parametrization in Eq. (4) and using its time derivative, the equations of motion in EE coordinates can be described by:

$$\overline{\mathbf{M}}(\mathbf{q})\,\ddot{\mathbf{x}} + \overline{\mathbf{K}}\,(\mathbf{q},\dot{\mathbf{q}})\,\dot{\mathbf{x}} + \overline{\mathbf{Q}}\,(\mathbf{q},\dot{\mathbf{q}},t) = \mathbf{A}^{\mathrm{T}}\,(\mathbf{q})\,\mathbf{c}\,, \tag{5}$$

where

$$\begin{aligned}
\overline{\mathbf{M}} &:= \mathbf{F}^{\mathrm{T}}\left[\operatorname{diag}\{\mathbf{D},\mathbf{M}_{\mathrm{X}}\}\right]\mathbf{F}\,,\\
\overline{\mathbf{K}} &:= \mathbf{F}^{\mathrm{T}}\left[\operatorname{diag}\{\mathbf{0}_m,\mathbf{K}_{\mathrm{X}}\}\,\mathbf{F} + \operatorname{diag}\{\mathbf{D},\mathbf{M}_{\mathrm{X}}\}\,\dot{\mathbf{F}}\right],\\
\overline{\mathbf{Q}} &:= \mathbf{F}^{\mathrm{T}}[\mathbf{Q}^{\mathrm{T}}\ \mathbf{Q}_{\mathrm{X}}^{\mathrm{T}}]^{\mathrm{T}}\,.
\end{aligned}$$

In Eq. (5), the matrix $\mathbf{A}^{\mathrm{T}}$ is the so-called structure matrix which allows to classify the DOA $\alpha = \operatorname{rank}(\mathbf{A})$. The degree of redundancy is defined as $\rho := m - \alpha$. Hence, the DOA determines how many generalized accelerations are affected by the controller [4]. To fulfill the requirements of a fully constrained CDPR, guaranteeing that the cables can be tensed within the wrench-feasible workspace, the DOA needs to lead to $\rho \geq 1$.

To investigate the passivity of the underlying mechanism, the equations of motion (5) are transformed into the joint space

$$\mathbf{M}(\mathbf{q})\,\ddot{\mathbf{q}}_{\mathrm{a}} + \mathbf{K}\,(\mathbf{q},\dot{\mathbf{q}})\,\dot{\mathbf{q}}_{\mathrm{a}} + \mathbf{Q}(\mathbf{q},\dot{\mathbf{q}},t) = \mathbf{c}\,, \tag{6}$$

with $\dot{\mathbf{x}} = \left(\mathbf{F}^{\mathrm{T}}\right)^{+}\dot{\mathbf{q}}_{\mathrm{a}}$. $\mathbf{F}$ is partitioned regarding to the actuated joints $\mathbf{q}_{\mathrm{a}}$. For that purpose, it has to be proven that

Proposition 1 $\dot{\mathbf{q}}_a^T\left(\dot{\mathbf{M}} - 2\mathbf{K}\right)\dot{\mathbf{q}}_a = 0$ *holds.*

Proof Given the skew symmetry of $\dot{\overline{\mathbf{M}}} - 2\overline{\mathbf{K}}$ [5] the following result holds:

$$\begin{aligned}
\dot{\mathbf{q}}_{\mathrm{a}}^{\mathrm{T}}\left(\dot{\mathbf{M}} - 2\mathbf{K}\right)\dot{\mathbf{q}}_{\mathrm{a}} &= \dot{\mathbf{q}}_{\mathrm{a}}^{\mathrm{T}}\left(\mathbf{F}^{\mathrm{T}}\right)^{+}\left(\dot{\overline{\mathbf{M}}} - 2\overline{\mathbf{K}}\right)\dot{\mathbf{q}}_{\mathrm{a}}\mathbf{F}^{+}\\
&\quad + \dot{\mathbf{q}}_{\mathrm{a}}^{\mathrm{T}}\dot{\mathbf{F}}^{+}\overline{\mathbf{M}}\mathbf{F}^{+}\dot{\mathbf{q}}_{\mathrm{a}} - \dot{\mathbf{q}}_{\mathrm{a}}^{\mathrm{T}}\mathbf{F}^{+}\overline{\mathbf{M}}\dot{\mathbf{F}}^{+}\dot{\mathbf{q}}_{\mathrm{a}} = 0\,.
\end{aligned} \tag{7}$$

Since the last two terms are scalars, it results $\dot{\mathbf{q}}_{\mathrm{a}}^{\mathrm{T}}\dot{\mathbf{F}}^{+}\overline{\mathbf{M}}\mathbf{F}^{+}\dot{\mathbf{q}}_{\mathrm{a}} = \dot{\mathbf{q}}_{\mathrm{a}}^{\mathrm{T}}\mathbf{F}^{+}\overline{\mathbf{M}}\dot{\mathbf{F}}^{+}\dot{\mathbf{q}}_{\mathrm{a}}$ which completes the proof. This property is shown in [6].

2.2 Inverse Dynamics in EE Coordinates

The inverse dynamics problem consists in determining the required cable forces for a given trajectory. The number of cables of a redundantly actuated CDPR exceeds its DOF by ρ. The structure matrix is supposed to have full rank δ. In case of redundantly actuated CDPRs ($\rho > 0$), $\mathbf{A}^{\mathrm{T}}$ is not square and its kernel (null-space) is of dimension ρ. Therefore Eq. (5) cannot be solved unambiguously for the control forces $\mathbf{c}$. Hence, it is possible to generate control forces lying in the kernel of $\mathbf{A}^{\mathrm{T}}$, which have no effect on the motion but can be used to generate a certain tension level. Let $\mathbf{c}^0 \in \mathbb{R}^m$ be a vector representing a desired tension level, a solution for $\mathbf{c}$ such that $\left[\mathbf{c}-\mathbf{c}^0\right]^{\mathrm{T}}\mathbf{W}\left[\mathbf{c}-\mathbf{c}^0\right] \rightarrow \min$ is

$$\left(\mathbf{A}^{\mathrm{T}}\right)_{\mathbf{W}}^{+}\left(\overline{\mathbf{M}}(\mathbf{q})\,\ddot{\mathbf{x}} + \overline{\mathbf{K}}\,(\mathbf{q},\dot{\mathbf{q}})\,\dot{\mathbf{x}} + \overline{\mathbf{Q}}(\mathbf{q},\dot{\mathbf{q}},t)\right) + \mathbf{N}_{\mathbf{A}^{\mathrm{T}},\mathbf{W}}\mathbf{c}^0\,. \tag{8}$$

Therewith $\left(\mathbf{A}^{\mathrm{T}}\right)_{\mathbf{W}}^{+} := \mathbf{W}^{-1}\mathbf{A}\left(\mathbf{A}^{\mathrm{T}}\mathbf{W}^{-1}\mathbf{A}\right)^{-1}$ is a weighted pseudo inverse, where $\mathbf{W}$ is a positive definite weighting matrix, with respect to the drive capabilities. The null-space projector $\mathbf{N}_{\mathbf{A}^{\mathrm{T}},\mathbf{W}}$ generates the tension level in the cable system close to the desired value of $\mathbf{c}^0$ [7]. In the field of redundantly actuated CDPRs, the desired tension level must be chosen in a way that the unilateral properties of cables are met. Based on this requirement, the authors in [8] delivers real-time capable force distribution algorithms based on geometrical approaches.

3 Disturbance Observer

3.1 Motivation

Within the design of model-based control schemes for CDPRs, it is desired to effectively compensate all generalized EE disturbances. Especially disturbances take a negative effect on the trajectory tracking behavior of the control scheme. Therefore, this step plays a central role in enhancing the performance of the control scheme regarding stability, robustness and control quality. For these reasons, the compensation of disturbances in the application of CDPRs represents an important subject.

3.2 Derivation of the Disturbance Observer

The proposed disturbance observer is inspired by the momentum-based collision detection algorithm described in [9]. The idea behind the disturbance observer is to consider the generalized EE disturbances as an error within the actuated winch drive system.

Including all disturbances, the equations of motion (5) can be described by

$$\overline{\mathbf{M}}(\mathbf{q})\,\ddot{\mathbf{x}} + \overline{\mathbf{K}}\,(\mathbf{q},\dot{\mathbf{q}})\,\dot{\mathbf{x}} + \overline{\mathbf{Q}}(\mathbf{q},\dot{\mathbf{q}},t) = \mathbf{A}^{\mathrm{T}}(\mathbf{q})\,\mathbf{f} - \overline{\eta}(t)\,. \tag{9}$$

Besides generalized EE disturbances, the disturbance observer is also able to reconstruct any uncertainties in the equations of motion of the underlying mechanism.

As proposed in [9] for the design of a disturbance observer, the generalized momentum $\mathbf{p}(t)$ is used:

$$\overline{\mathbf{p}}(t) = \overline{\mathbf{M}}(\mathbf{q})\,\dot{\mathbf{x}}\,. \tag{10}$$

The time derivative of the momentum equation is

$$\dot{\overline{\mathbf{p}}}(t) = \overline{\mathbf{M}}(\mathbf{q})\,\ddot{\mathbf{x}} + \dot{\overline{\mathbf{M}}}(\mathbf{q})\,\dot{\mathbf{x}}\,. \tag{11}$$

Furthermore, the accelerations in Eq. (11) can be replaced by the equations of motion (5) to get an expression for $\dot{\overline{\mathbf{p}}}(t)$. In combination with the measured momentum $\overline{\mathbf{p}}(t)$, the following description of the disturbance observer can be yielded:

$$\dot{\hat{\overline{\mathbf{p}}}}(t) = \mathbf{A}^{\mathrm{T}}\,(\mathbf{q})\,\mathbf{f} - \overline{\zeta}\,(\mathbf{q},\dot{\mathbf{q}},t) + \mathbf{K}_{\mathrm{R}}\left(\overline{\mathbf{p}}(t) - \hat{\overline{\mathbf{p}}}(t)\right)\,, \tag{12}$$

$$\hat{\overline{\eta}}(t) = \mathbf{K}_{\mathrm{R}}\left(\hat{\overline{\mathbf{p}}}(t) - \overline{\mathbf{p}}(t)\right)\,, \tag{13}$$

with

$$\overline{\zeta}\,(\mathbf{q},\dot{\mathbf{q}},t) = -\dot{\overline{\mathbf{M}}}(\mathbf{q})\,\dot{\mathbf{x}} + \overline{\mathbf{K}}\,(\mathbf{q},\dot{\mathbf{q}})\,\dot{\mathbf{x}} + \overline{\mathbf{Q}}(\mathbf{q},\dot{\mathbf{q}},t)\,. \tag{14}$$

By the positive definite gain matrix $\mathbf{K}_{\mathrm{R}}$ the observer error $\overline{\mathbf{r}}(t) = \overline{\mathbf{p}}(t) - \hat{\overline{\mathbf{p}}}(t)$ is feed-back and an asymptotic convergence of the observer error can be ensured.

By using Eq. (12), the following residual vector can be given to estimate generalized EE disturbances $\overline{\eta}(t)$:

$$\hat{\overline{\eta}}(t) = \mathbf{K}_{\mathrm{R}}\left[\int_0^t \left(\mathbf{A}^{\mathrm{T}}(\mathbf{q})\,\mathbf{f} - \overline{\zeta}\,(\mathbf{q},\dot{\mathbf{q}},t) - \hat{\overline{\eta}}(t)\right)\,\mathrm{d}t - \mathbf{p}(t)\right]\,. \tag{15}$$

For the implementation of Eq. (15), the measured EE velocities $\dot{\mathbf{x}}$ and forces $\mathbf{f}$ are required but neither EE accelerations $\ddot{\mathbf{x}}$ nor inversion of the mass matrix $\overline{\mathbf{M}}(\mathbf{q})$ are required.

The residual dynamics satisfy the following form

$$\dot{\hat{\overline{\eta}}}(t) = -\mathbf{K}_{\mathrm{R}}\hat{\overline{\eta}} + \mathbf{K}_{\mathrm{R}}\overline{\eta}\,, \tag{16}$$

which can be identified as a linear exponentially stable system driven by the fault $\overline{\eta}(t)$ [9]. Furthermore, the disturbance observer delivers decoupled generalized EE disturbances for each direction of movement. This fact can be shown by following transfer function

$$\frac{\hat{\overline{\eta}}_{\mathrm{i}}}{\overline{\eta}_{\mathrm{i}}} = \frac{\mathrm{K}_{\mathrm{R_i}}}{s + \mathrm{K}_{\mathrm{R_i}}}\,, \qquad i = 1, \ldots, n \tag{17}$$

having unitary gain. From this it is obvious that for

$$\mathbf{K}_{\mathrm{R}} \rightarrow \infty \quad \Rightarrow \quad \hat{\overline{\eta}} \approx \overline{\eta} \tag{18}$$

the disturbance observer yields the generalized EE disturbances $\overline{\eta}$ on its output.

3.3 Passivity

One advantage of the proposed disturbance observer is that it can be seamlessly integrated into the control scheme. This statement can be emphasized by following passivity analysis. A CDPR is passive with the following storage function

$$S_{\mathrm{EE}} = \frac{1}{2}\dot{\mathbf{x}}^{\mathrm{T}}\overline{\mathbf{M}}\dot{\mathbf{x}}\,. \tag{19}$$

Here the storage function S_{EE} represents the kinetic energy of the CDPR expressed in the Cartesian space. The time derivative of the storage function is given by:

$$\begin{aligned}
\dot{S}_{\mathrm{EE}} &= \dot{\mathbf{x}}^{\mathrm{T}}\overline{\mathbf{M}}\ddot{\mathbf{x}} + \frac{1}{2}\dot{\mathbf{x}}^{\mathrm{T}}\dot{\overline{\mathbf{M}}}\dot{\mathbf{x}}\,, \\
&= \dot{\mathbf{x}}^{\mathrm{T}}\left(\mathbf{A}^{\mathrm{T}}\mathbf{c} - \overline{\mathbf{K}}\dot{\mathbf{x}} - \eta\right) + \frac{1}{2}\dot{\mathbf{x}}^{\mathrm{T}}\dot{\overline{\mathbf{M}}}\dot{\mathbf{x}}\,, \\
&= \dot{\mathbf{x}}^{\mathrm{T}}\mathbf{A}^{\mathrm{T}}\mathbf{c} + \frac{1}{2}\dot{\mathbf{x}}^{\mathrm{T}}\left(\dot{\overline{\mathbf{M}}} - 2\overline{\mathbf{K}}\right)\dot{\mathbf{x}} - \dot{\mathbf{x}}^{\mathrm{T}}\eta\,.
\end{aligned} \tag{20}$$

In the following Eq. (20) is mapped into the joint space. By the help of Proposition 1 and using the matrix operation $\dot{\mathbf{x}}^{\mathrm{T}}\mathbf{A}^{\mathrm{T}} = \mathbf{A}\dot{\mathbf{x}}$ the following simplification can be written

$$\dot{S}_{\mathrm{q}} = \mathbf{c}^{\mathrm{T}}\dot{\mathbf{q}}_{\mathrm{a}} - \eta^{\mathrm{T}}\mathbf{q}_{\mathrm{a}}\,. \tag{21}$$

The first term in Eq. (21) represents the generated energy of the m winch drives. With the second term the dissipated energy due to the joint disturbances is described which is always negative semi-definite.

The next step is to demonstrate its ability to guarantee the stability of the CDPR under the compensation of disturbances. After this, the derivative of the storage function $\dot{S}_q$ from Eq. (21) takes up following form:

$$S_q = \dot{\mathbf{q}}_a^T \mathbf{c} + \dot{\mathbf{q}}_a^T \left(\hat{\eta} - \eta\right) \tag{22}$$

In case the term $\dot{\mathbf{q}}_a^T \left(\hat{\eta} - \eta\right)$ is negative semi-definite, no additional energy is introduced into the system and, moreover, a steady-state error caused by disturbances can be prevented.

The differences in the absolute values of $\hat{\eta} - \eta$ are always less than or equal to zero, due to the low-pass behavior of the disturbance observer [10]. Thus, the requirement from Eq. (22) is met and the disturbance observer behaves dissipative at any time.

4 Internal Force-Based Impedance Control

4.1 Motivation

Consider the CDPR in Fig. 1 as a multi-manipulator system handling an object. Therefore, a control algorithm for CDPRs must feature the characteristic of giving each manipulator knowledge of the others to form a cooperating interaction of the manipulators. Otherwise, without this knowledge contradicting (antagonistic) control forces appear and cause an undesired tension level in the cable system. Summarized following desirable properties should be provided within the design of the controller:

- Each manipulator features a compliance to eliminate contraditing control forces.
- Platform dynamics take no influence on tracking or steady state position errors.

All these properties can be considered in the design of a robust internal force-based impedance controller. This control approach represents a class of control algorithms where the controller directly gives each manipulator the necessarly compliance and as proposed in [3]. This remarkable feature imparts each manipulator the necessary robustness to cope with harsh interacting forces. Thus, impedance control becomes an attractive strategy for controlling CDPRs.

4.2 Derivation of the Control Law

The equation of motion for each winch drive is described by

$$\mathbf{u}_i = \mathbf{M}_i\left(\mathbf{q}\right)\ddot{\mathbf{q}}_i + \mathbf{Q}_i\left(\mathbf{q}_i, \dot{\mathbf{q}}_i\right) + \mathbf{f}_i\,, \tag{23}$$

where $\mathbf{M} := \mathbf{D} + \left(\mathbf{A}^{\mathrm{T}}\right)^{+} \mathbf{M}_{\mathrm{X}} (\mathbf{A})^{+}$ represents the joint space mass matrix. The compensation of disturbances $\mathbf{Q}$ ensures that the external control forces $\mathbf{u}$ are exactly generated by the winch drives. To maintain the desired tension level in the cable system the forces $\mathbf{f}$ have to be replaced by the inverse dynamics solution $\mathbf{c}$ (8).

When multiple manipulators are connected to a platform, the forces $\tilde{\mathbf{f}}$ generated by each manipulator expressed in the Cartesian space can be decomposed into motion-inducing forces $\tilde{\mathbf{f}}_{\mathrm{M}}$ and internal forces $\tilde{\mathbf{f}}_{\mathrm{I}}$, with [3]

$$\tilde{\mathbf{f}} = \tilde{\mathbf{f}}_{\mathrm{M}} + \tilde{\mathbf{f}}_{\mathrm{I}} . \tag{24}$$

Each manipulator only makes a point contact with the EE and, therefore, can only exert a force onto the EE. The net forces at the EE frame are related to the forces applied by the manipulators by

$$\tilde{\mathbf{f}}_{\mathrm{EE}} = \left[\mathbf{J}_{\mathrm{X1}} \cdots \mathbf{J}_{\mathrm{Xm}}\right] \tilde{\mathbf{f}} = \mathbf{J}_{\mathrm{X}} \tilde{\mathbf{f}} , \tag{25}$$

where $\mathbf{J}_{\mathrm{X}}$ defines the EE-to-manipulator Jacobian.

Internal forces $\tilde{\mathbf{f}}_{\mathrm{I}}$ produce no net forces on the EE, however, a tension in the cable system and, thus, they must be chosen to lie in the range of the null-space of $\mathbf{J}_{\mathrm{X}}$. That means internal forces $\tilde{\mathbf{f}}_{\mathrm{I}}$ are not affected by motion-inducing forces $\tilde{\mathbf{f}}_{\mathrm{M}}$ and can be regulated simultaneously. Using this fact, following decomposition method to compute the internal forces for each manipulator can be used [11]:

$$\tilde{\mathbf{f}}_{\mathrm{Ii}} = \left(\mathbf{I}_{\delta} - \mathbf{J}_{\mathrm{Xi}}^{\#} \mathbf{J}_{\mathrm{Xi}}\right) \tilde{\mathbf{f}}_{\mathrm{i}} , \tag{26}$$

where $\mathbf{J}_{\mathrm{X}}^{\#}$ is a generalized inverse of $\mathbf{J}_{\mathrm{X}}$. A valid solution is $\mathbf{J}_{\mathrm{X}}^{\#} = \mathbf{J}_{\mathrm{X}}^{\mathrm{T}} \left(\mathbf{J}_{\mathrm{X}} \mathbf{J}_{\mathrm{X}}^{\mathrm{T}}\right)^{-1}$ which is the Moore-Penrose pseudoinverse. The decompostion method defined in Eq. (26) results in zero interaction forces when $\tilde{\mathbf{f}}_{\mathrm{Ii}} = \mathbf{0}_{\delta}$ holds.

As described in [3] impedance control in the application of cooperating manipulators must enforce a relationship between the EE velocity and the internal forces on the manipulated EE comparable to the interaction forces between the manipulators. Otherwise, if the total forces imposed by the environment on the manipulator will be incorporated in the impedance relationship, platform dynamics will contribute to tracking and steady-state position errors.

Each manipulator is equipped with the following impedance describing a linear second-order function expressed in the Cartesian space

$$\overline{\mathbf{G}} \delta \ddot{\mathbf{x}} + \overline{\mathbf{B}} \delta \dot{\mathbf{x}} + \overline{\mathbf{C}} \delta \mathbf{x} = \delta \tilde{\mathbf{f}}_{\mathrm{Ii}} , \tag{27}$$

where $\delta \mathbf{x} = \mathbf{x}_{\mathrm{d}} - \mathbf{x}$ = pose errors of the platform,
$\delta \tilde{\mathbf{f}}_{\mathrm{Ii}} = \tilde{\mathbf{f}}_{\mathrm{Ii}} - \tilde{\mathbf{f}}_{\mathrm{Iid}}$ = internal force errors,
$\overline{\mathbf{G}}, \overline{\mathbf{B}}, \overline{\mathbf{C}}$ = desired mass-, damping- und stiffness matrices.

Here, index d represents a desired quantity.

One property of the choosen impedance is that separate force and position control loops are not required. Therefore, the impedance parameters can be chosen almost freely in contrast to the mentioned bandwidth limitations shown in [1]. Additionally, internal forces produce no net forces on the EE and therefore, the platform dynamics have no influence on tracking or steady state position errors.

To met the stability constraints shown in [3] the impedance mass $\overline{\mathbf{G}}$ should be chosen equal to the projected mass matrix $\overline{\mathbf{M}} := \mathbf{M}_{\mathrm{X}}+\mathbf{A}^{\mathrm{T}}\mathbf{D}\mathbf{A}$. It should be mentioned that the selected mass matrix affects all EE coordinates simultaneously.

Each joint acceleration is related to the EE acceleration by its structure matrix

$$\ddot{\mathbf{q}}_{\mathrm{i}} = \mathbf{A}_{\mathrm{i}}\ddot{\mathbf{x}}_{\mathrm{i}} + \dot{\mathbf{A}}_{\mathrm{i}}\dot{\mathbf{x}}_{\mathrm{i}} \,. \tag{28}$$

Solving Eq. (27) for $\ddot{\mathbf{x}}$, substituting into Eq. (28), and incorporating into each winch drive dynamic equation (23), yields the following control law for each manipulator:

$$\mathbf{u}_{\mathrm{i}} = \mathbf{M}_{\mathrm{i}} \left\{ \mathbf{A}_{\mathrm{i}} \left(\ddot{\mathbf{x}}_{\mathrm{d}} + \overline{\mathbf{G}}^{-1} \left[\overline{\mathbf{B}}\delta\dot{\mathbf{x}} + \overline{\mathbf{C}}\delta\mathbf{x} - \delta\tilde{\mathbf{f}}_{\mathrm{Ii}} \right] \right) + \dot{\mathbf{A}}_{\mathrm{i}}\dot{\mathbf{x}}_{\mathrm{i}} \right\} + \mathbf{Q}_{\mathrm{i}} + \mathbf{c}_{\mathrm{i}} \tag{29}$$

In general, the internal force-based impedance controller is governed by a CTC structure with a shaping of the internal forces resulting from the chosen mass matrix. Additionally, the compensation of generalized EE disturbances including effects as a variable mass matrix $\overline{\mathbf{M}}$ are handled by the proposed disturbance observer. The compensation is executed considering the inverse dynamics solution described in Eq. (8). Therefore, an energy optimal distribution of the generalized EE disturbance to the m winch drives can be performed.

The computational effort of the internal force-based impedance controller is quite low. Based on a desired trajectory $\mathbf{x}_{\mathrm{d}}$ and $\tilde{\mathbf{f}}_{\mathrm{Id}}$ as well as measuring the pose $\mathbf{x}$ and solving the inverse dynamics (8) to get an expression for $\mathbf{c}$, the required external control forces $\mathbf{u}$ can be computed. The internal forces $\tilde{\mathbf{f}}_{\mathrm{I}}$ are computed by using Eq. (26) from the forces $\mathbf{f}$ sensed at all winch drives.

5 Experimental Results

The presented robust internal force-based impedance controller was implemented on the 6-DOF SEGESTA prototype shown in Fig. 1. The CDPR is controlled by means of eight BLDC-Motors (Maxon RE 60) driving the winches with a nominal torque generation of $c_{nom} = 0.85$ Nm. Thus $m = 8$, i.e. the CDPR is redundantly actuated. The winches are placed at the vertices of a nearly symmetric cuboid with an overall dimension of 1.2 m x 1.4 m x 1 m (L x W x H). The mass of the EE is approximately 0.3 kg. Strain-gauge beam arrangements (Megatron KM302) are integrated into the winch drives and used to measure the cable forces. The applied angular encoders have a resolution of 2,000 increments per turn.

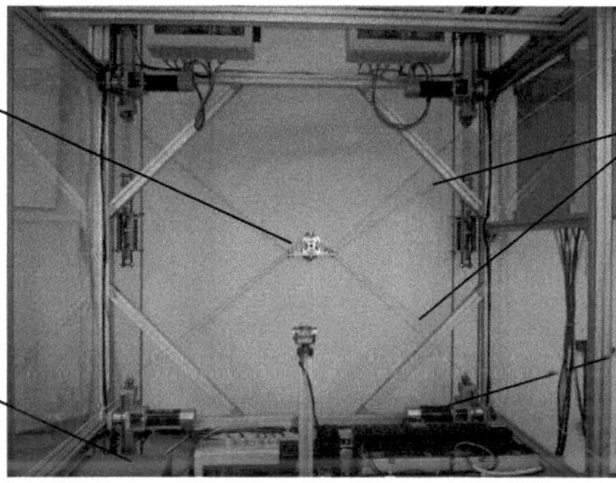

Fig. 2 SEGESTA prototype

Table 1 Coordinates of the choosen trajectory

Path	1	2	3	4	5
xz-pose in m	(0/0.5)	(−0.25/0.4)	(−0.25/0.65)	(0.25/0.4)	(0.25/0.65)

To measure the EE pose, the forward kinematics as proposed in [12] is used. Additionally, the cable stiffness to alter the cable length is in-cooperated in the forward kinematics. A high-gain observer as proposed in [13] is used to get all state variables to fulfill the requirements of the impedance controller. The BLDC-Motors have high stiction of about 0.05 Nm which makes it necessary to use an additional friction observer as proposed in [14] to support the internal force-based impedance controller. In addition, further unmodeled effects are included as a variable mass matrix $\overline{\mathbf{M}}$ and handled by the proposed disturbance observer. The disturbance observer is parametrized with $\mathbf{K}_\mathrm{R} = 125$. To compute a desired tension level in the cable system, the so-called Puncture Method presented in [8] is used due to its property to deliver nearly minimum cable forces. Here, the lower force limit $\mathbf{f}_{\min}$ is set to 5 N.

For the experiment, each manipulator impedance was chosen using the guidelines introduced in [3] examining the projected mass matrix $\overline{\mathbf{M}}$. The stiffness matrix $\overline{\mathbf{C}}$ was chosen such that an impedance bandwidth of approximately 3 Hz is achieved and $\overline{\mathbf{B}}$ was chosen to achieve critical damping: $\overline{\mathbf{C}} = 1000\overline{\mathbf{M}}$ and $\overline{\mathbf{B}} = 63\overline{\mathbf{M}}$. By lowering the desired mass matrix $\overline{\mathbf{G}}$, the internal forces errors can be minimized. However, as described in [3] there is a lower limit where the controller loses stability.

The EE was controlled along the path shown in Table 1 with a maximum acceleration of 3 g and the maximum velocity was 3.25 m/s.

The evolution of the EE errors, the generalized EE disturbances, the internal force errors expressed in the joint space and the forces of cable 1 and 8, respectively, during the run are shown in Figs. 3 and 4.

The results show that the controller has a good accuracy and the EE errors in Fig. 3 (left) are below 100 µm at the rest positions. During the run, error peaks of about 0.75 mm due to the high stiction can be observed. The results shown in Fig. 3 (right)

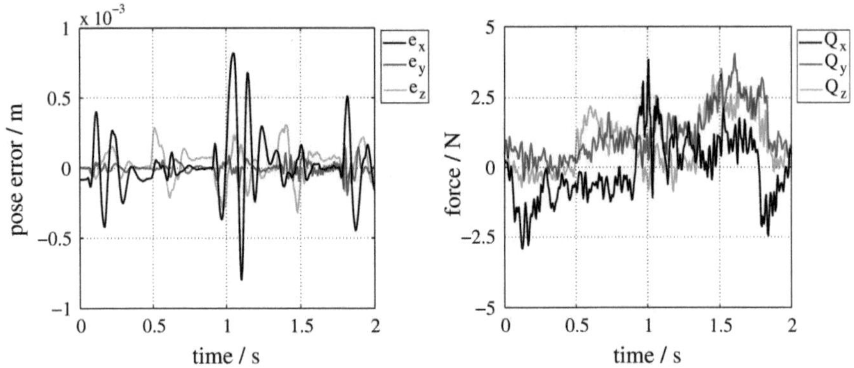

Fig. 3 EE errors (*left*) and generalized EE forces (*right*) when the CDPR is controlled along the EE path

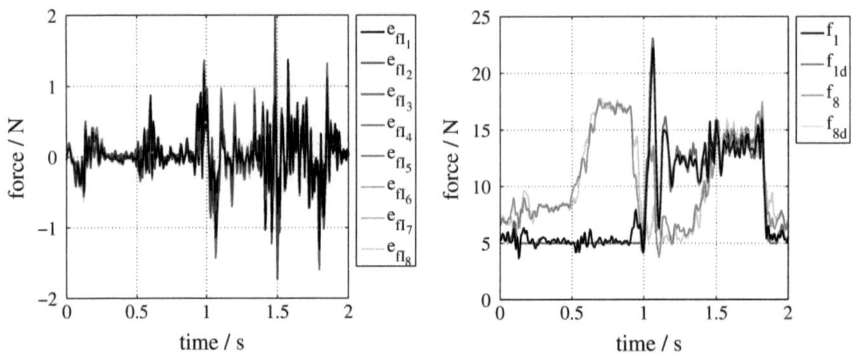

Fig. 4 Internal force errors (*left*), measured and desired forces (*right*) of cable 1 and 8 when the CDPR is controlled along the EE path

further demonstrate the importance of the disturbance observer, due to the elimination of steady state EE errors at the rest positions. Beyond that, the disturbance observer delivers the decoupled generalized EE disturbances for each direction of movement and, thereby, sets an improved tracking countering within the path segments in the y-direction. In addition, the internal force errors shown in Fig. 4 (left) are going to zero at steady state, due to the compensation of joint disturbances like friction forces. This is shown in the forces in Fig. 4 (right) which can be tracked very well.

6 Summary

In this paper the equations of motion for CDPRs in terms of EE coordinates are presented using a projection method. On this base, a disturbance observer estimating all generalized EE forces was developed and its passivity was proven. Considering CDPRs as a multi-manipulator system, a robust internal force-based impedance controller considering generalized EE disturbances has been derived. One advantage is that no explicit position and force control loops are necessary and the impedance parameters can be freely chosen. Its implementation is discussed and the feasibility is shown. Experimental results are reported for a 6-DOF CDPR, showing that an accurate trajectory tracking behavior and a desired tension level in the cable system can be guaranteed during the EE movement.

Acknowledgments The research leading to these results has received funding from the European Community's Seventh Framework Programme under grant agreement No. NMP2-SL-2011-285404 (CableBOT).

References

1. Eppinger SD, Seering WP (1987) Understanding bandwidth limitations in robot force control. In: Proceedings of IEEE international conference on robotics and automation, pp 904–909
2. Hogan N (1985) Impedance control: an approach to manipulation, part i—theory. ASME J Dyn Syst 107:1–7
3. Bonitz RG, Hsia TC (1996) Internal force-based impedance control for cooperating manipulators. IEEE Trans Rob Autom 12(1):78–89
4. Reichert C, Müller K, Bruckmann T (2014) Internal force-based impedance control for cable-driven parallel robots. In: New trends in mechanism and machine science, mechanisms and machine science (submitted)
5. Ortega R, Loria A, Nicklasson PJ, Sira-Ramirez H (1998) Passivity-based control of Euler-Lagrange systems, 1st edn. Springer, New York
6. Abdellatif H, Heimann B, Kotlarski J (2008) Passivity-based observer/controller design with desired dynamics compensation for 6 dofs parallel manipulators. In: IEEE/RSJ International Conference on Intelligent Robots and Systems, pp 2392–2397
7. Müller A (2011) Problems in the control of redundantly actuated parallel manipulators caused by geometric imperfections. Meccanica 46(1):41–49
8. Müller K, Reichert C, Bruckmann T (2014) Analysis of geometrical force calculation algorithms for cable-driven parallel robots with a threefold redundancy. In: Latest advances in robot kinematics, (submitted)
9. De Luca A, Mattone R (2003) Actuator failure detection and isolation using generalized momenta. In: IEEE International Conference on robotics and automation, pp 634–639
10. Reichert C, Hufnagel T, Schramm D (2012) Regelung redundant angetriebener parallelmanipulatoren mit reibungsbehafteten motoren. In: 16. VDI Getriebetagung, pp 118–130
11. Bonitz RG, Hsia TC (1994) Force decomposition in cooperating manipulators using the theory of metric spaces and generalized inverses. In: IEEE international conference on robotics and automation, pp 1521–1527
12. von Zitzewitz J, Rauter G, Vallery H, Morger A, Riener R (2010) Forward kinematics of redundantly actuated, tendon-based robots. In: IEEE/RSJ international conference on intelligent robots and systems, pp 2289–2294
13. Khalil HK (2009) High-gain observers in nonlinear feedback control. In: IEEE international conference on control and automation, pp 1527–1528
14. Le Tien L, Albu-Schäffer A, De Luca A, Hirzinger G (2008) Friction observer and compensation for control of robots with joint torque measurement. In: IEEE/RSJ international conference on intelligent robots and systems, pp 3789–3795

Adaptive Control of KNTU Planar Cable-Driven Parallel Robot with Uncertainties in Dynamic and Kinematic Parameters

Reza Babaghasabha, Mohammad A. Khosravi and Hamid D. Taghirad

Abstract This paper addresses the design and implementation of adaptive control on a planar cable-driven parallel robot with uncertainties in dynamic and kinematic parameters. To develop the idea, firstly, adaptation is performed on dynamic parameters and it is shown that the controller is stable despite the kinematic uncertainties. Then, internal force term is linearly separated into a regressor matrix in addition to a kinematic parameter vector that contains estimation error. In the next step to improve the controller performance, adaptation is performed on both the dynamic and kinematic parameters. It is shown that the performance of the proposed controller is improved by correction in the internal forces. The proposed controller not only keeps all cables in tension for the whole workspace of the robot, it is computationally simple and it does not require measurement of the end-effector acceleration as well. Finally, the effectiveness of the proposed control algorithm is examined through some experiments on KNTU planar cable-driven parallel robot and it is shown that the proposed control algorithm is able to provide suitable performance in practice.

1 Introduction

Cable-driven parallel robots have some positive features such as large workspace [1, 2], high speed manipulation [3], high payload to robot weight ratio [4], transportability and ease of assembly/disassembly. Despite all advantages listed for this class of robots, using cables in the robot structure introduces new challenges in the study of cable-driven parallel robots. Cables are able to apply only tensile forces, and in order to avoid structural failures, design of control algorithms should be performed

R. Babaghasabha (✉) · M.A. Khosravi (✉) · H.D. Taghirad (✉)
Advanced Robotics and Automated Systems (ARAS), Faculty of Electrical Engineering, Industrial Control Center of Excellence (ICCE), K. N. Toosi University of Technology, P.O. Box 6315-1355, Tehran, Iran
e-mail: Reza.bgha@mail.kntu.ac.ir

M.A. Khosravi
e-mail: Makh@ee.kntu.ac.ir

H.D. Taghirad
e-mail: Taghirad@kntu.ac.ir

A. Pott and T. Bruckmann (eds.), *Cable-Driven Parallel Robots*,
Mechanisms and Machine Science 32, DOI 10.1007/978-3-319-09489-2_11

such that the cables remain in tension for the whole workspace of the robot. These features make feedback control of the cable-driven parallel robots more challenging than that of the conventional parallel robots.

Motion control topologies of the cable-driven parallel robots may be classified into two categories, the ones that are formed in the cable length coordinates and the others that are designed in the task space coordinates. In [3, 5] decentralized controllers have been proposed in the cable length coordinates and their performance has been evaluated through some experiments. In this framework the length of the cables can be simply measured by the encoders. Therefore, the controllers can be implemented economically in practice. However, in applications with high accuracy and high bandwidth, using the cable lengths measurement in the control algorithms is not reliable due to the inherent flexibility of the cables. Moreover, the main control aim is to realize the trajectory tracking in the task space. Therefore, suitable control accuracy can be achieved only when the task space tracking error is used as the feedback to the controller without any kinematic transformation.

In task space control topologies, the pose of the end-effector shall be measured directly. This measurement requires high-tech and expensive instrumentation system such as laser ranging equipment or inertial measurement unit. For this reason, only few researches focus on implementation of the task space controllers in practice. In [6] laser ranging equipment has been used to measure the end-effector pose of the 6-DOF cable-driven parallel robot in the task space coordinates. In [7] the PID controller is designed and implemented in the task space coordinates. However, in order to avoid direct pose measurement, length of the cables is measured by the encoders and the pose of the end-effector is estimated by forward kinematic analysis. As mentioned earlier, using cable length information is not reliable due to the inherent flexibility of the cables. In addition, solving the forward kinematics equations of the robot in the feedback loop reduces the measurement accuracy and limits the controller bandwidth due to the complexity and multiplicity of the solutions. Computer vision is another way for direct measurement of the end-effector pose and it can be used as a suitable and economical alternative. In [8] Vision-based pose measurement is performed on a planar cable-driven parallel robot. But only the translational motion is measured and the end-effector rotation is not considered.

Classic controllers such as PID [7] have simple structures and do not require complete dynamic model of the cable-driven parallel robot. However, due to the lack of consideration of dynamic effects in the structure of the controller, it may only operate suitably in regulation problems, while have limited performance in tracking objectives. On the other hand, internal forces which are used in the controller to ensure that all cables remain in tension, shall span the null space of the Jacobian matrix appropriately. However, uncertainties in kinematic parameters affect on the internal forces and degrade performance of the controller even in regulation problems.

Inclusion of the dynamic behavior of the robot in design of nonlinear controllers [9–12] can improve the performance of the controller in tracking purposes. However, detailed information of the dynamic and kinematic models of the robot is required to implement this controllers. Furthermore, it shall be noted that precise knowledge on kinematics and dynamics of the robot is unavailable in practice, and this may significantly limit the tracking performance. To partially remedy this shortcoming, calibration methods may be applied to identify near true parameters. However, robot calibration process is time consuming and requires expensive and precise measurement units. These all shortcomings may be remedied by using an adaptive controller in task space coordinates in which the kinematic and dynamic parameters are simultaneously adapted.

As a representative of such method, in [8] robust PD controller with an adaptive compensation term has been used. It has been shown that measuring both the cables length and the pose of the end effector in real-time enables the designer to linearly separate the internal force term into a regressor matrix and a kinematic parameter vector. In this paper the performance of the controller has been evaluated only in regulation problem through some experiments on a translational system using the minimum number of cables under zero-gravity condition. In addition, the dynamic model of the robot has not been used in the controller structure and consequently, the adaptation has not been performed on dynamic parameters. Furthermore, a large number of kinematic parameters have been adapted and therefore, updating the kinematic parameters becomes difficult.

In this paper, an adaptive controller is designed in task space coordinates for a planar cable-driven parallel robot with uncertainties in dynamic and kinematic parameters. In the first section of controller design, adaptation is performed only on dynamic parameters and it is shown that the controller is stable despite kinematic uncertainties. Next, real-time measurement of the cable length through encoders and that of the pose of the end-effector through a CCD camera is used to linearly separate the internal force term into a regressor matrix and a kinematic parameter vector that contains estimation error. In this paper, vision-based pose measurement is chosen as a suitable and economical solution. Then, adaptation is performed on both the dynamic and kinematic parameters and it is shown that the performance of the proposed controller is significantly improved by the correction of the internal forces. In addition, the proposed controller is able to keep all the cables in tension for the whole workspace of the robot. Finally, to show the effectiveness of proposed control algorithm some experiments are performed on KNTU planar cable-driven parallel robot and it is shown that the proposed control algorithm is able to provide suitable performance in practice.

2 Robot Dynamics and Kinematics

The dynamic model of a planar cable-driven parallel robot without considering the flexibility of the cables can be written as [13]

$$\mathbf{M}(\mathbf{x})\ddot{\mathbf{x}} + \mathbf{C}(\mathbf{x},\dot{\mathbf{x}})\dot{\mathbf{x}} + \mathbf{G}(\mathbf{x}) = \mathbf{f} = -\mathbf{J}^T\tau \qquad (1)$$

in which, $\mathbf{x}$ denotes the generalized coordinates vector for pose of the end-effector, τ denotes the vector of cable forces and $\mathbf{f}$ is the vector of Cartesian wrench. $\mathbf{M}(\mathbf{x})$ denotes mass matrix of the robot, $\mathbf{C}(\mathbf{x},\dot{\mathbf{x}})$ denotes Coriolis and centrifugal terms,

$\mathbf{G}(\mathbf{x})$ denotes the vector of gravity terms and $\mathbf{J}$ denotes the Jacobian matrix of the robot. Some important properties of the dynamic equation described by (1) are listed as follows [13].

Property 1: The mass matrix $\mathbf{M}(\mathbf{x})$ is symmetric and uniformly positive definite for all $\mathbf{x}$.

Property 2: The matrix $\dot{\mathbf{M}}(\mathbf{x}) - 2\mathbf{C}(\mathbf{x},\dot{\mathbf{x}})$ is skew-symmetric, and therefore the expression $\mathbf{Z}^T(\dot{\mathbf{M}}(\mathbf{x}) - 2\mathbf{C}(\mathbf{x},\dot{\mathbf{x}}))Z$ is equal to zero, for all $\mathbf{Z}$.

Property 3: The dynamic model described by (1) may be represented by a linear regressor with respect to a set of dynamic parameters as

$$\mathbf{M}(\mathbf{x})\ddot{\mathbf{x}} + \mathbf{C}(\mathbf{x},\dot{\mathbf{x}})\dot{\mathbf{x}} + \mathbf{G}(\mathbf{x}) = \mathbf{Y}_D(\mathbf{x},\dot{\mathbf{x}},\ddot{\mathbf{x}})\mathbf{a} \tag{2}$$

where $\mathbf{Y}_D(\mathbf{x},\dot{\mathbf{x}},\ddot{\mathbf{x}})$ denotes the regressor matrix and $\mathbf{a}$ denotes dynamic parameters vector. In [13] kinematic and dynamic analysis for a planar cable-driven parallel robot have been reported in detail and it is shown that the dynamic model as described by (1) can be written into the linear regression form. KNTU planar cable-driven parallel robot consists of an end-effector that is connected by four cables to the base platform as shown in Fig. 1. Using massless rigid string model for the cable, the equations of motion for this robot may be written as follows:

$$\mathbf{M}(\mathbf{x})\ddot{\mathbf{x}} + \mathbf{G}(\mathbf{x}) = \mathbf{f} = -\mathbf{J}^T\tau \tag{3}$$

in which,

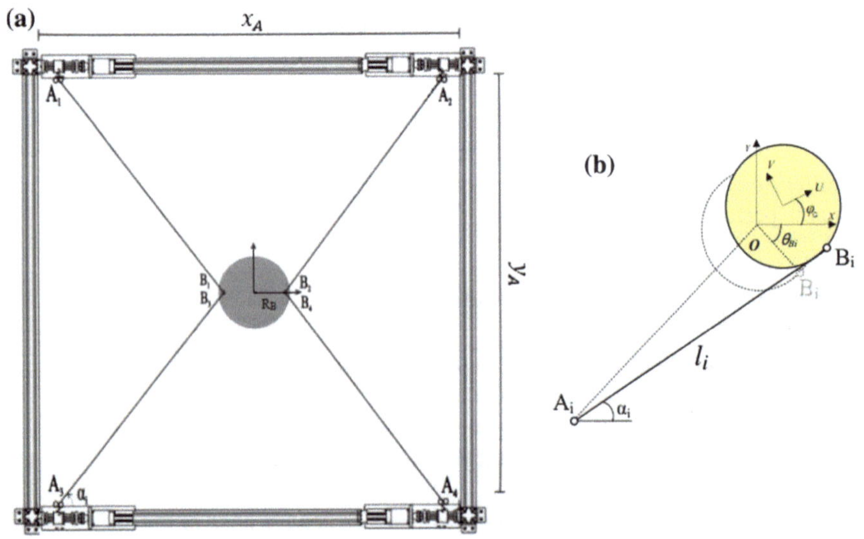

Fig. 1 **a** The schematics of KNTU cable-driven robot. **b** Kinematic configuration of mechanism

$$\mathbf{M} = \begin{bmatrix} m & 0 & 0 \\ 0 & m & 0 \\ 0 & 0 & I_z \end{bmatrix} \qquad \mathbf{G} = \begin{bmatrix} 0 \\ mg \\ 0 \end{bmatrix}$$

where m is the mass and I_z is the moment of inertia of the end-effector about its center of mass and g is the gravity acceleration. The Jacobian matrix of the robot can be expressed by

$$\mathbf{J}^T = \begin{bmatrix} S_{1x} & S_{2x} & S_{3x} & S_{4x} \\ S_{1y} & S_{2y} & S_{3y} & S_{4y} \\ S_{1z} & S_{2z} & S_{3z} & S_{4z} \end{bmatrix} \tag{4}$$

in which,

$$\begin{aligned} S_{ix} &= \frac{x_G - 0.5x_A + R_B \cos(\phi_i)}{l_i} \\ S_{iy} &= \frac{y_G - 0.5y_A + R_B \sin(\phi_i)}{l_i} \\ S_{iz} &= R_B \cos(\phi_i) S_{iy} - R_B \sin(\phi_i) S_{ix} \end{aligned}$$

and $\mathbf{x} = [x_G, y_G, \phi_G]$ denotes the generalized coordinate vector for the pose of the end-effector, x_A and y_A are the length and width of the rectangle that actuators have been attached on its vertices, l_i is the cables length, B_i is the attachment point of the cables on the end-effector which lies at the radial distance of R_B from center of the end-effector, $\phi_i = \phi_G + \theta_{B_i}$ is instantaneous orientation angle of the moving attachment point and θ_{B_i} is its absolute angular position, respectively.

3 Controller Design

3.1 Adaptation of Dynamic Parameters

In this section, an adaptive controller is proposed for the planar cable-driven parallel robots with uncertainties in dynamic and kinematic parameters. To derive the control and adaptation laws, consider the following Lyapunov function candidate as [14]

$$V(t) = \frac{1}{2}(\mathbf{S}^T \mathbf{M} \mathbf{S} + \tilde{\mathbf{a}}^T \Gamma_D \tilde{\mathbf{a}}) \tag{5}$$

in which,

$$\mathbf{S} = \dot{\tilde{\mathbf{x}}} + \Lambda \tilde{\mathbf{x}}$$

where $\mathbf{S}$ denotes a sliding surface, Γ_D and Λ are symmetric diagonal positive definite matrices, $\tilde{\mathbf{x}}=\mathbf{x}-\mathbf{x}_d$ is the tracking error vector and $\tilde{\mathbf{a}}=\hat{\mathbf{a}}-\mathbf{a}$ denotes the dynamic parameter estimation error vector. Differentiate $V(t)$ with respect to time:

$$\dot{V}(t)=\mathbf{S}^T\mathbf{M}\dot{\mathbf{S}}+\frac{1}{2}\mathbf{S}^T\dot{\mathbf{M}}\mathbf{S}+\tilde{\mathbf{a}}^T\Gamma_D\dot{\tilde{\mathbf{a}}} \tag{6}$$

$$\dot{V}(t)=\mathbf{S}^T\mathbf{M}(\ddot{\tilde{\mathbf{x}}}+\Lambda\dot{\tilde{\mathbf{x}}})+\mathbf{S}^T[\frac{1}{2}(\dot{\mathbf{M}}-2\mathbf{C})+\mathbf{C}]\mathbf{S}+\tilde{\mathbf{a}}^T\Gamma_D\dot{\tilde{\mathbf{a}}} \tag{7}$$

Using skew-symmetricity of $\dot{\mathbf{M}}-2\mathbf{C}$ and substitution from (1) yields

$$\dot{V}(t)=\mathbf{S}^T(-\mathbf{J}^T\tau-\mathbf{C}\dot{\mathbf{x}}-\mathbf{G}-\mathbf{M}\ddot{\mathbf{x}}_d+\mathbf{M}\Lambda\dot{\tilde{\mathbf{x}}}+\mathbf{C}\mathbf{S})+\tilde{\mathbf{a}}^T\Gamma_D\dot{\tilde{\mathbf{a}}} \tag{8}$$

Define the virtual reference trajectory as

$$\mathbf{x}_r=\mathbf{x}_d-\Lambda\int_0^t\tilde{\mathbf{x}}dt \tag{9}$$

Differentiate $\mathbf{x}_r$ twice with respect to time, and substitute into Eq. (8):

$$\dot{V}(t)=\mathbf{S}^T(-\mathbf{J}^T\tau-\mathbf{M}\ddot{\mathbf{x}}_r-\mathbf{C}\dot{\mathbf{x}}_r-\mathbf{G})+\tilde{\mathbf{a}}^T\Gamma_D\dot{\tilde{\mathbf{a}}} \tag{10}$$

For the purpose of stability of the closed loop system, let us define the control law as

$$\mathbf{f}=-\hat{\mathbf{J}}^T\tau=\hat{\mathbf{M}}\ddot{\mathbf{x}}_r+\hat{\mathbf{C}}\dot{\mathbf{x}}_r+\hat{\mathbf{G}}-\mathbf{K}_D\mathbf{S} \tag{11}$$

where $\mathbf{K}_D$ denotes symmetric diagonal positive definite matrix and $\hat{\mathbf{J}}^T$ denotes uncertain Jacobian matrix of the robot. In practice the attachment points are not precisely implemented in practice. Therefore, it is assumed to have an uncertain Jacobian matrix obtained from uncertain kinematic parameters. The general solution of (11) for τ is

$$\tau=\overline{\tau}+\mathbf{Q} \tag{12}$$

where $\overline{\tau}$ is the minimum solution of (11) and derived by using the pseudo-inverse of $\hat{\mathbf{J}}^T$ and is given by

$$\overline{\tau}=-\hat{\mathbf{J}}(\hat{\mathbf{J}}^T\hat{\mathbf{J}})^{-1}=-\hat{\mathbf{J}}^{\dagger}\mathbf{f} \tag{13}$$

and $\mathbf{Q}$ may be physically interpreted as the internal forces that spans the null space of $\hat{\mathbf{J}}^T$. Therefore,

$$\hat{\mathbf{J}}^T\mathbf{Q}=0 \tag{14}$$

Internal forces are used to ensure that all cables remain in tension within the whole workspace. Moreover, term can be used to increase the robot stiffness [15]. Now, substitute the control law (11) into Eq. (10). This yields to:

$$\begin{aligned}\dot{V}(t) &= \mathbf{S}^T[(\mathbf{J}^T\hat{\mathbf{J}}^{\dagger})\mathbf{f} - \mathbf{J}^T\mathbf{Q} - \mathbf{M}\ddot{\mathbf{x}}_r - \mathbf{C}\dot{\mathbf{x}}_r - \mathbf{G})] + \tilde{\mathbf{a}}^T\Gamma_D\dot{\hat{\mathbf{a}}}\\ &= \mathbf{S}^T[(\mathbf{J}^T\hat{\mathbf{J}}^{\dagger})\mathbf{f} + \mathbf{f} - \mathbf{f} - \mathbf{J}^T\mathbf{Q} + \hat{\mathbf{J}}^T\mathbf{Q} - \mathbf{M}\ddot{\mathbf{x}}_r - \mathbf{C}\dot{\mathbf{x}}_r - \mathbf{G})] + \tilde{\mathbf{a}}^T\Gamma_D\dot{\hat{\mathbf{a}}}\\ &= \mathbf{S}^T[\tilde{\mathbf{M}}\ddot{\mathbf{x}}_r + \tilde{\mathbf{C}}\dot{\mathbf{x}}_r + \tilde{\mathbf{G}} - \mathbf{K}_D\mathbf{S} + (\hat{\mathbf{J}}^T - \mathbf{J}^T)\mathbf{Q} - (\mathbf{I} - \mathbf{J}^T\hat{\mathbf{J}}^{\dagger})\mathbf{f}] + \tilde{\mathbf{a}}^T\Gamma_D\dot{\hat{\mathbf{a}}}\end{aligned} \tag{15}$$

in which

$$\tilde{\mathbf{M}} = \hat{\mathbf{M}} - \mathbf{M}, \quad \tilde{\mathbf{C}} = \hat{\mathbf{C}} - \mathbf{C}, \quad \tilde{\mathbf{G}} = \hat{\mathbf{G}} - \mathbf{G}$$

Since the matrices **M**,**C** and **G** are linear in terms of the manipulator parameters, we may separate the dynamic model of a planar cable-driven parallel robot into a linear regressor matrix and a vector of parameters as

$$\tilde{\mathbf{M}}\ddot{\mathbf{x}}_r + \tilde{\mathbf{C}}\dot{\mathbf{x}}_r + \tilde{\mathbf{G}} = \mathbf{Y}_D(\mathbf{x}, \dot{\mathbf{x}}, \dot{\mathbf{x}}_r, \ddot{\mathbf{x}}_r)\tilde{\mathbf{a}} \tag{16}$$

Substitution above equation into Eq. (15) yields to

$$\dot{V}(t) = \mathbf{S}^T[\mathbf{Y}_D(\mathbf{x}, \dot{\mathbf{x}}, \dot{\mathbf{x}}_r, \ddot{\mathbf{x}}_r)\tilde{\mathbf{a}} - \mathbf{K}_D\mathbf{S} + (\hat{\mathbf{J}}^T - \mathbf{J}^T)\mathbf{Q} - (\mathbf{I} - \mathbf{J}^T\hat{\mathbf{J}}^{\dagger})\mathbf{f}] + \tilde{\mathbf{a}}^T\Gamma_D\dot{\hat{\mathbf{a}}}. \tag{17}$$

Let us define the dynamic adaptation law as

$$\dot{\hat{\mathbf{a}}} = -\Gamma_D^{-1}\mathbf{Y}_D^T(\mathbf{x}, \dot{\mathbf{x}}, \dot{\mathbf{x}}_r, \ddot{\mathbf{x}}_r)\mathbf{S} \tag{18}$$

By using the above adaptation law the resulting expression of $\dot{V}(t)$ is reduced to

$$\dot{V}(t) = \mathbf{S}^T[-\mathbf{K}_D\mathbf{S} + (\hat{\mathbf{J}}^T - \mathbf{J}^T)\mathbf{Q} - (\mathbf{I} - \mathbf{J}^T\hat{\mathbf{J}}^{\dagger})\mathbf{f}] \tag{19}$$

The uncertain jacobian matrix $\hat{\mathbf{J}}^T$, and the internal forces may be assumed to be bounded by the following relations:

$$\|\hat{\mathbf{J}}^T - \mathbf{J}^T\| \leq \delta_1, \quad \|\mathbf{I} - \mathbf{J}^T\hat{\mathbf{J}}^{\dagger}\| \leq \delta_2, \quad \|\mathbf{Q}\| \leq \delta_3 \tag{20}$$

Therefore, the proposed adaptive controller can certainly stabilize the closed loop system by appropriate choice of large enough $\mathbf{K}_D$. However, uncertainties in the kinematic parameters affect on the internal forces and degrade the performance of the proposed controller.

3.2 Adaptation of Kinematic Parameters

In the previous section, the stability of the proposed adaptive controller was proved despite uncertainties in kinematic parameters. However, uncertainties in kinematic parameters may change the direction of the resultant internal force and may consequently degrade the performance of the controller. Furthermore, increasing the control gains may not effective to improve the performance of the controller and it causes greater impact of the internal forces which do not appropriately span the null space of the Jacobian matrix. In order to overcome this shortcoming, the important internal force term may be also separated into a linear regressor matrix and a kinematic parameter vector expressed by

$$(\hat{\mathbf{J}}^T - \mathbf{J}^T)\mathbf{Q} = \tilde{\mathbf{J}}^T\mathbf{Q} = \mathbf{Y}_K(\mathbf{x}, \mathbf{L}, \mathbf{Q})\tilde{\mathbf{b}} \tag{21}$$

where $\mathbf{Y}_K(\mathbf{x}, \mathbf{L}, \mathbf{Q})$ denotes the regressor matrix, and for a planar cable-driven parallel robot with four actuated cable driven limbs, $\mathbf{L} = [l_1, l_2, l_3, l_4]$ denotes the cables length, $\mathbf{Q} = [q_1, q_2, q_3, q_4]$ denotes the internal force vector and $\tilde{\mathbf{b}} = \hat{\mathbf{b}} - \mathbf{b}$ denotes the kinematic parameters estimation error vector. According to kinematic equation of KNTU planar cable-driven parallel robot and by some manipulations, Eq. (21) may be derived in an explicit form as

$$\mathbf{Y}_K(\mathbf{x}, \mathbf{L}, \mathbf{Q})\tilde{\mathbf{b}} = \begin{bmatrix} Y_{11} & 0 & Y_{13} & 0 & 0 \\ 0 & Y_{22} & Y_{23} & 0 & 0 \\ 0 & 0 & Y_{33} & Y_{34} & Y_{35} \end{bmatrix} \begin{bmatrix} x_A \\ y_A \\ R_B \\ x_A R_B \\ y_A R_B \end{bmatrix} \tag{22}$$

in which,

$$Y_{11} = \frac{1}{2}(\frac{q_1}{l_1} - \frac{q_2}{l_2} + \frac{q_3}{l_3} - \frac{q_4}{l_4}), \quad Y_{13} = \sum_{i=1}^{4} \frac{q_i \cos(\phi_i)}{l_i},$$

$$Y_{22} = \frac{1}{2}(-\frac{q_1}{l_1} - \frac{q_2}{l_2} + \frac{q_3}{l_3} + \frac{q_4}{l_4}), \quad Y_{23} = \sum_{i=1}^{4} \frac{q_i \sin(\phi_i)}{l_i},$$

$$Y_{33} = \sum_{i=1}^{4} \frac{q_i}{l_i}(y_G \cos(\phi_i) - x_G \sin(\phi_i)), \quad Y_{34} = \frac{1}{2}\sum_{i=1}^{4}(-1)^i[\frac{q_i \sin(\phi_i)}{l_i}],$$

$$Y_{35} = \frac{1}{2}(-\frac{q_1 \cos(\phi_1)}{l_1} - \frac{q_2 \cos(\phi_2)}{l_2} + \frac{q_3 \cos(\phi_3)}{l_3} + \frac{q_4 \cos(\phi_4)}{l_4}).$$

Now, one can propose a new Lyapunov function candidate as following:

$$V(t) = \frac{1}{2}(\mathbf{S}^T\mathbf{M}\mathbf{S} + \tilde{\mathbf{a}}^T\Gamma_D\tilde{\mathbf{a}} + \tilde{\mathbf{b}}^T\Gamma_K\tilde{\mathbf{b}}) \tag{23}$$

in which, Γ_K is a symmetric positive definite matrix. Differentiate $V(t)$ with respect to time, and according to the formulations given in previous section, substitute the control law and the dynamic adaptation law into Eq. (23). By some manipulation and using Eq. (21), one may obtain the following relation.

$$\dot{V}(t) = \mathbf{S}^T[-\mathbf{K}_D\mathbf{S} + \mathbf{Y}_K(\mathbf{x}, \mathbf{L}, \mathbf{Q})\tilde{\mathbf{b}} - (\mathbf{I} - \mathbf{J}^T\hat{\mathbf{J}}^\dagger)\mathbf{f}] + \tilde{\mathbf{b}}^T\Gamma_K\dot{\tilde{\mathbf{b}}} \quad (24)$$

By considering the kinematic adaptation law as

$$\dot{\hat{\mathbf{b}}} = -\Gamma_K^{-1}\mathbf{Y}_K^T(\mathbf{x}, \mathbf{L}, \mathbf{Q})\mathbf{S} \quad (25)$$

the resulting expression of $\dot{V}(t)$ yields to

$$\dot{V}(t) = \mathbf{S}^T[-\mathbf{K}_D\mathbf{S} - (\mathbf{I} - \mathbf{J}^T\hat{\mathbf{J}}^\dagger)\mathbf{f}] \quad (26)$$

Expression (26) is negative semi-definite by choosing $\mathbf{K}_D$ large enough and it shows that the proposed adaptive controller can stabilize the system. Moreover, adaptation of kinematic parameters with adaptation law in Eq. (25) will significantly improve the performance of the proposed controller by adjusting the direction of the resultant internal forces. If the estimation error of the kinematic parameters is zero, the resulting expression of $\dot{V}(t)$ yields to

$$\dot{V}(t) = -\mathbf{S}^T\mathbf{K}_D\mathbf{S} \leq 0 \quad (27)$$

Therefore, the trajectories of the closed loop system will eventually converges to the sliding surface,

$$\mathbf{S} = \dot{\tilde{\mathbf{x}}} + \Lambda\tilde{\mathbf{x}} = 0 \quad (28)$$

and therefore, the proposed adaptive controller guarantees zero steady-state tracking error for the pose of the end-effector. But it is notable that the dynamic and kinematic parameters can not necessarily converge to their exact values. This means that the tracking error will remain uniformly ultimately bounded (UUB) by choosing $\mathbf{K}_D$ large enough.

4 Experimental Results

In order to verify the effectiveness of the proposed adaptive controller, it is applied to KNTU planar cable-driven parallel robot. This manipulator is under investigation for high speed and wide workspace applications in Advanced Robotics and Automated Systems (ARAS) group of K. N. Toosi University of Technology (KNTU).

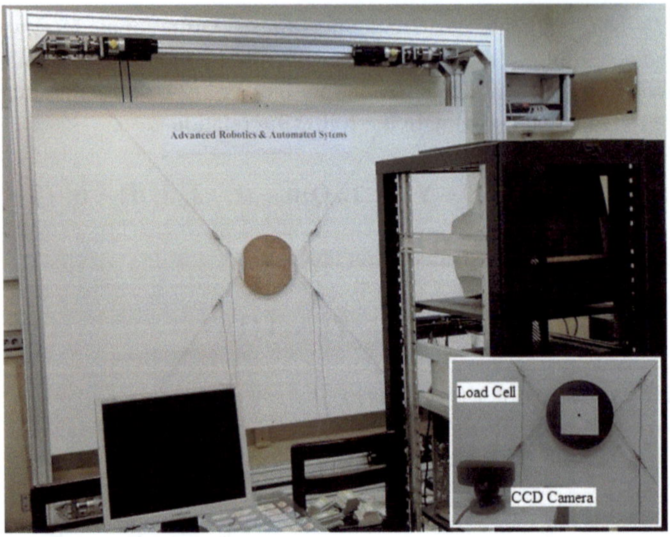

Fig. 2 KNTU planar cable-driven parallel robot

4.1 KNTU Planar Cable-Driven Parallel Robot

KNTU planar cable-driven parallel robot consists of four actuated cable driven limbs with three degrees of freedom planar motion which is shown in Fig. 2. Actuators are located on the vertices of a rectangle with dimension of 2.24×2.1 m and the radial distance of attachment points of the cables from center of the end-effector is considered as $R_B = 15$ cm. Moreover, mass and moment of inertia of the end-effector is considered as $m = 5$ kg and $I_z = 0.1\,\mathrm{kg\,m^2}$, respectively. All of the dynamic and kinematic parameters mentioned above, considered to be uncertain. In addition, a CCD camera with resolution of 320×240 pixels and frame rate of 100 fps at the distance of 1.12 m from the plane of motion of the end effector is used to directly measure the pose of the end-effector. In order to determine the pose of the end effector, at least four coplanar non-aligned features on the object are required [16]. For this reason, a square marker is used for fast and accurate tracking and the pose of the end-effector is measured by extracting corners using Harris corner detector [17] in real-time. The resolution of the measurement of the position and orientation of the end-effector are 0.1 mm and 0.2°, respectively. Furthermore, The sampling time of the control loop is one millisecond which provides real time execution of the proposed controller.

4.2 Control Scheme

To have a desirable performance in position and orientation tracking, it is necessary to have ideal torque sources as the actuators. In practice, however, the actuator drivers

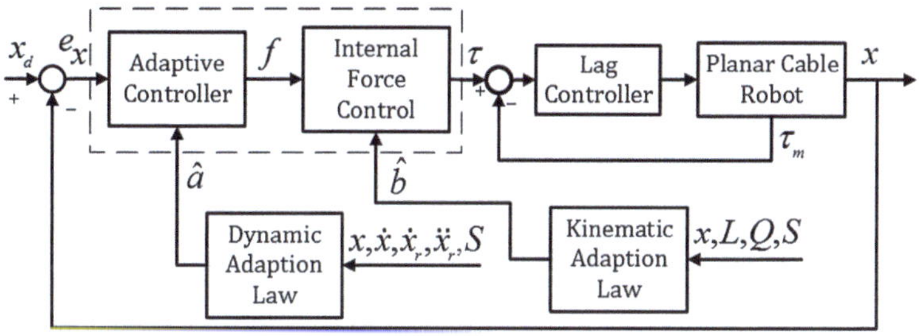

Fig. 3 Block diagram of proposed adaptive controller

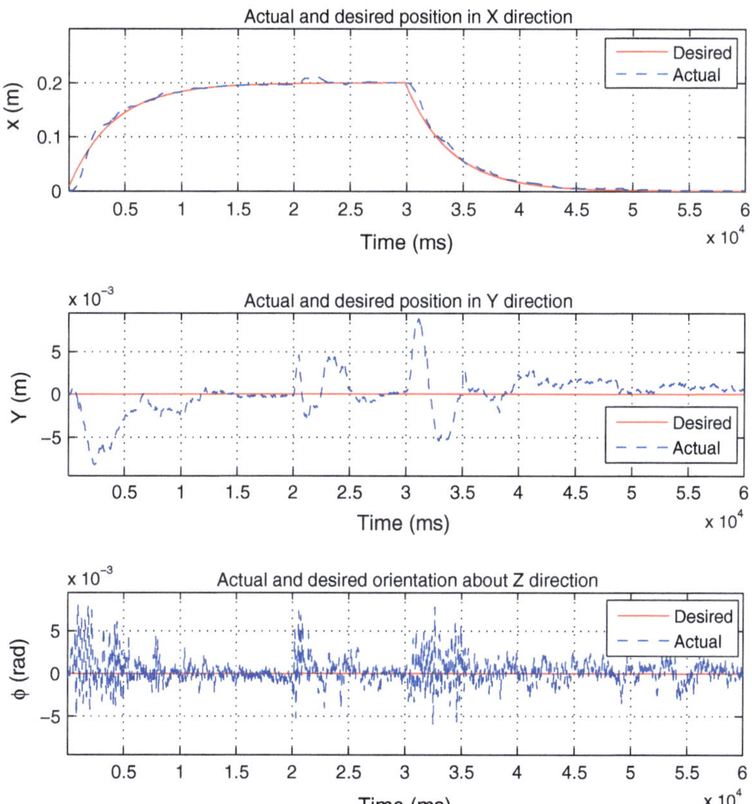

Fig. 4 Actual and desired position and orientation of the end-effector in the first experiment

suffer from a number of limitations, and cannot perform as ideal torque sources. In order to overcome this shortcoming, cascade control scheme is implemented in the experiments. The cascade control strategy uses two control loops, called outer and inner loops as shown in Fig. 3. In the proposed control algorithm, while adaptive

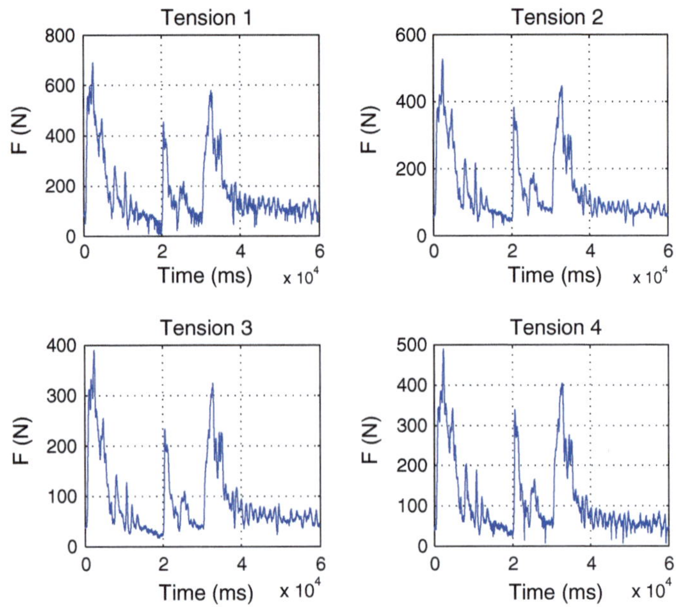

Fig. 5 Cables tension in the first experiment

controller controls the pose of the end-effector in the outer loop, a lag controller is designed in the inner loop to regulate the cables tension, and prevents them from tearing. Inputs of adaptive controller are position and orientation errors and dynamic parameter estimation vector, while its outputs are the required Cartesian wrench. The calculated output wrench is transformed into positive actuator forces through the internal force control block. This block uses kinematic parameter estimation vector to find the internal forces that span appropriately the null space of the Jacobian matrix. Next, the resulting desired positive tensions are compared to the actual tensions measured by the load cells located near the end-effector attachment points. TLL500 from Transducer Techniques is used to measure the cable tensions in the setup. The proposed controller is implemented using RT-LAB software [18] and the equations of internal forces are solved by CFSQP implemented as an s-function in Simulink.

4.3 Results

In order to verify the effectiveness of the proposed adaptive controller, two disjointed motions in translation and rotation are considered. In the first experiment, the following exponential trajectory is considered in x direction, while the end-effector attempts to maintain $y = 0$ and $\phi = 0$ during the motion.

$$x_d = \begin{cases} 0.2(1 - e^{-0.25t}) & t < 30 \\ 0.2e^{-0.25(t-30)} & t \geq 30 \end{cases}$$

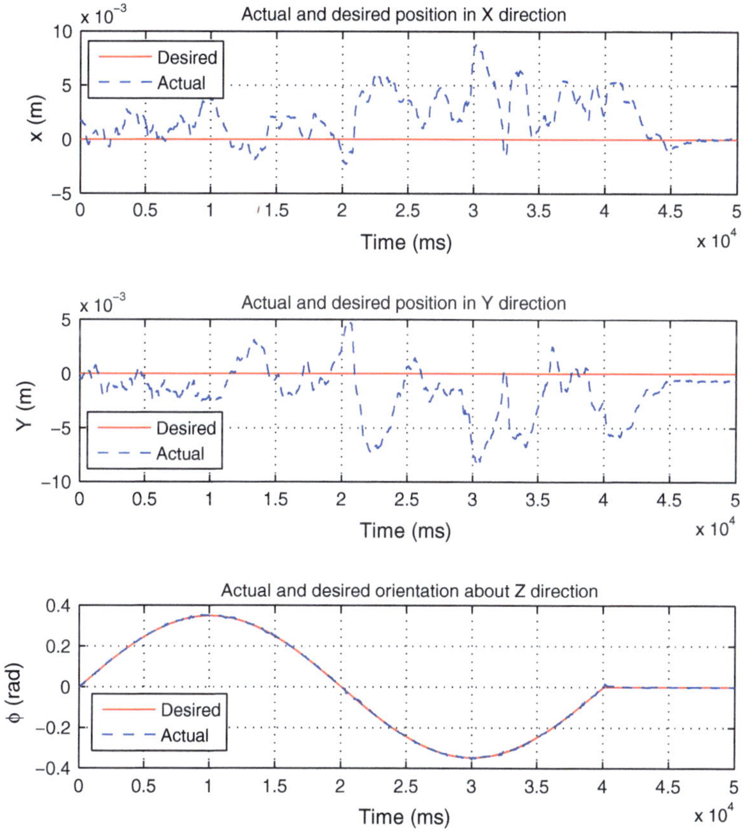

Fig. 6 Actual and desired position and orientation of the end-effector in the second experiment

The results of implementation using the proposed controller in tracking desired trajectories are given in Figs. 4, 5. Figure 4 illustrates the tracking errors in three directions, while Fig. 5 provides the cable forces measured by the load cells located near the end-effector attachment points. The controller gain is selected such that the stability of the controller is guaranteed, as $K_D = 1200$. As it is seen in Fig. 4 the pose of the end-effector can suitably track the desired trajectories and the errors are very small and in order of 10^{-3}. Moreover, as it is shown in Fig. 5 all cables remain in tension during the robot maneuvers. In the second experiment, the following sinusoidal trajectory is considered in ϕ direction, and it is considered that the end-effector has no motion in the other directions.

$$\phi_d = \begin{cases} \frac{\pi}{9}\sin\left(\frac{\pi}{20}t\right) & t < 40 \\ 0 & t \geq 40 \end{cases}$$

The tracking errors in three directions is shown in Fig. 6, while Fig. 7 provides the cables tension. As it is seen in Fig. 6 the proposed controller provide the suitable per-

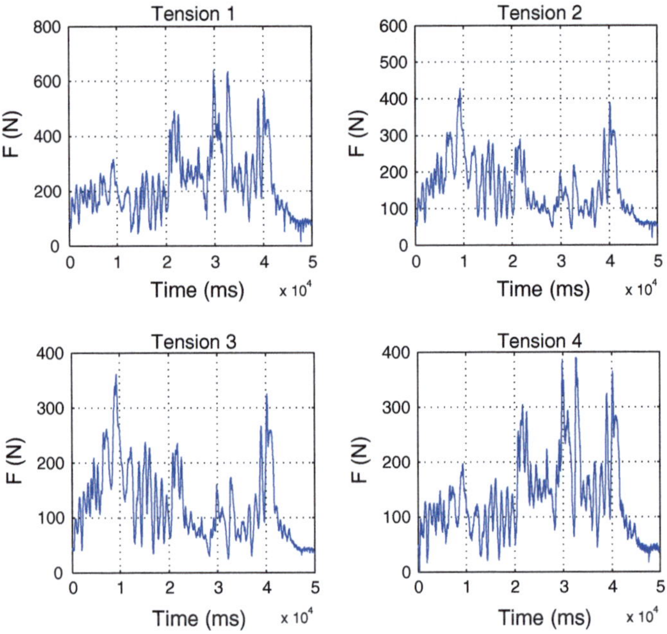

Fig. 7 Cables tension in the second experiment

formance in tracking and the errors are very small and in order of 10^{-3}. Furthermore, Fig. 7 shows that all cables remain in tension during the robot movements.

Small errors in tracking the desired trajectories are observed in this experiment. These errors may arise from the elasticity of the cables that are neglected in this analysis. The cable elasticity may lead to positioning errors, especially at high speed maneuvers. Future research is currently proceed to reduce these errors by considering elasticity of the cables in modeling and control of the robot.

5 Conclusion

In this paper, an adaptive controller is designed and implemented on a planar cable-driven parallel robot. Since some of the kinematic and dynamic parameters of the robot are uncertain, adaptation is performed on both the kinematic and dynamic parameters. It is shown that adaptation of the kinematic parameters improves the performance of the proposed controller by adjusting the direction of the resultant internal force. In addition, the controller keeps all cables in tension for the whole workspace of the robot. In order to show the effectiveness of the proposed controller, several experiments on a three degrees of freedom planar cable-driven parallel robot are performed with different desired trajectories and suitable tracking performances of some of the experiments are reported.

References

1. Taghirad HD, Nahon M (2008) Kinematic analysis of a macro-micro redundantly actuated parallel manipulator. Adv Robot 22(6–7):657–687
2. Cone LL (1985) Skycam: an aerial robotic camera system. Byte 10(10):122–132
3. Kawamura S, Kino H, Won C (2000) High-speed manipulation by using parallel wire-driven robots. Robotica 18(01):13–21
4. Albus J et al (1993) The nist robocrane. J Rob Syst 10(5):709–724
5. Fang S, Franitza D, Torlo M, Bekes F, Hiller M (2004) Motion control of a tendon-based parallel manipulator using optimal tension distribution. IEEE/ASME Trans Mechatron 9(3):561–568
6. Zi B, Duan BY, Du JL, Bao H (2008) Dynamic modeling and active control of a cable-suspended parallel robot. Mechatronics 18(1):1–12
7. Khosravi MA, Taghirad HD (2014) Robust PID control of fully-constrained cable driven parallel robots. Mechatronics 24(2):87–97
8. Kino H et al (2007) Robust pd control using adaptive compensation for completely restrained parallel-wire driven robots: translational systems using the minimum number of wires under zero-gravity condition. IEEE Trans Rob 23(4):803–812
9. Oh S-R, Agrawal SK (2005) Cable suspended planar robots with redundant cables: controllers with positive tensions. IEEE Trans Rob 21(3):457–465
10. Oh S-R, Agrawal SK (2006) Generation of feasible set points and control of a cable robot. IEEE Trans Rob 22(3):551–558
11. Oh S-R et al (2005) A dual-stage planar cable robot: dynamic modeling and design of a robust controller with positive inputs. J Mech Des 127:612
12. Williams RL, Gallina P, Vadia J (2003) Planar translational cable-direct-driven robots. J Robotic Syst 20(3):107–120
13. Taghirad HD (2013) Parallel robots: mechanics and control. CRC Press, Boca Raton
14. Slotine JJE, Li W (1987) On the adaptive control of robot manipulators. Int J Rob Res 6(3): 49–59
15. Behzadipour S, Khajepour A (2006) Stiffness of cable-based parallel manipulators with application to stability analysis. J Mech Des 128:303
16. Yuan JS-C (1989) A general photogrammetric method for determining object position and orientation. IEEE Trans Rob Autom 5(2):129–142
17. Harris C, Stephens M (1988) A combined corner and edge detector. In: Alvey vision conference, vol 15, p 50. Manchester, UK
18. Opal-RT Company (2005) Rt-lab version 8 user guide

Dynamic Analysis and Control of Fully-Constrained Cable Robots with Elastic Cables: Variable Stiffness Formulation

Mohammad A. Khosravi and Hamid D. Taghirad

Abstract In this paper dynamic analysis and control of fully-constrained parallel cable robots are studied in detail. In dynamic analysis, it is assumed that the dominant dynamics of cable can be approximated by linear axial spring. Furthermore, variable stiffness formulation for the cables is employed in modeling process. To overcome vibrations caused by inevitable elasticity of cables, a composite control law is proposed based on singular perturbation theory. Using the proposed control algorithm the dynamics of the cable robot is divided into two subsystems namely slow and fast. Then, based on the results of singular perturbation theory, stability analysis of the total system is performed. Finally, the effectiveness of the proposed composite control law is investigated through several simulations on a planar parallel cable robot.

1 Introduction

Although the robots are extensively used in industries, their application in long-reach robotics such as inspection and repair in shipyards, is limited and still is in infancy. Cable robots are a special class of parallel robots in which the rigid links are replaced by cables. Cable robots possess some useful properties such as large workspace capability, transportability and ease of assembly/disassembly, reconfigurability and economical structure and maintenance [1]. Consequently, cable robots are exeptionally suitable for many applications such as, handling of heavy materials [2], high speed manipulation [3, 4], cleanup of disaster areas [5], very large workspace applications [6, 7], and interaction with hazardeous environment [8].

However, replacing the rigid links by cables, introduces many new challenges in the study of cable robots compared to that of conventional robots, amongst them control is the most critical one. Unlike the rigid links, cables can only apply tensile

M.A. Khosravi (✉) · H.D. Taghirad
Advanced Robotics and Automated Systems (ARAS), Faculty of Electrical and Computer Engineering, K.N. Toosi University of Technology, Tehran, Iran
e-mail: makh@ee.kntu.ac.ir

H.D. Taghirad
e-mail: Taghirad@kntu.ac.ir

A. Pott and T. Bruckmann (eds.), *Cable-Driven Parallel Robots*,
Mechanisms and Machine Science 32, DOI 10.1007/978-3-319-09489-2_12

forces and not compressive forces. Due to this physical limitation, well-known control theories can not be used directly for the cable robots and they must be modified so that thay can provide positive tension for the cables. Dynamic behavior of the cables is another major challenge in mechanical design and control of this class of robots. Cables are usually elastic elements and may encounter some unavoidable situations such as elongation and vibration. In applications which require high bandwidth or high stiffness of the system, vibration may be a serious concern [9]. In terms of control, proposed control algorithms for this class of robot shall be designed to damp such vibrations.

Control of cable robots has received limited attention compared to that of conventional robots. With assumption of massless rigid string model for the cable, several efforts have been devoted to finding an efficient control strategy with positive cable tensions. Lyapunov based control [4], PID control [10], computed torque method [11], sliding mode [12] and adaptive PD control [13] are some control schemes being used in the control of cable robots. Kawamura et al. have proposed a PD controller accompanied with gravity compensation and internal forces in the cable-length coordinates [4]. They have analyzed the stability of motion based on Lyapunov theorem and vector closure conditions. Alp and Agrawal used PD control with gravity compensation in task space coordinates and analyzed asymptotic stability of the system [14]. Inverse Dynamics Control (IDC) or computed torque technique is another control scheme which is used in [11]. Fang et al. used nonlinear feed forward control laws in the cable length coordinates [15]. They proposed optimal tension distribution algorithm to compensate dynamic errors.

In these studies cables are treated as ideal massless rigid strings and cables elasticity is not considered. However, in practice especially in high-speed applications, this assumption may affect overal performance of the robot. In these cases, it is important to model both the static and dynamic effects of flexible cables. But, modeling the dynamic effects of elastic cables is an extremely comprehensive task. It is also important to note that the obtained models must not only be sufficiently accurate, they have to be suitable for controller synthesis, as well. Therefore, in practice it is recommended to include only dominant dynamic effects in the dynamic analysis. Ottaviano and Castelli have analyzed the effects of cables' mass and elasticity and their effects on pose capability of the cable robots [16]. They have shown that cables masses can be neglected if the ratio of the end-effector to cables masses is large or generally, the ratio of the end-effector wrenches to the cables tensions is small. Using natural frequencies of the system, Diao and Ma in [9] have shown that in fully–constrained cable robots transversal vibrations of cables have very limited effects on the total vibration of the end-effector and can be ignored compared to that of axial flexibility. Therefore, dominant dynamic characteristics of the cable can be modeled by an axial spring in dynamic modeling of fully-constrained cable robots.

According to these results, in this paper linear axial spring is used to model dominant dynamics of cable and by this means a more precise model of fully-constrained cable robots is derived for the controller design and stability analysis of such robots.

Cables elasticity may affect the precision of the cable robot, requiring appropriate control strategies. Efficient control design for the cable robots is very complicated when cable dynamic characteristics is considered and research on this topic is in infancy. Meunier et al. used multi loop control scheme for a large cable mechanism, in which the inner loop deals with cable model. This loop uses H_∞ controller and gain scheduling technique for adaptation of H_∞ with cable lengths, and in the outer loop an inverse dynamics control in addition to a PID controller is used [17]. However, in this research stability analysis of the closed–loop system has not been performed. Using elastic massless model for cable in [1], a new model for the cable robots is derived and a new control algorithm is proposed. This control algorithm is formed in cable length space and uses internal force concept. Stability of the closed-loop system is analyzed through Lyapunov theory and vector closure conditions. Using singular perturbation theory a new control algorithm has been developed in [18]. In this research cables are modeled by linear axial spring but with constant stiffness and stability analysis is performed based on the singular perturbation theory [19].

The structure of this paper is as follows. First, dynamics of cable robot with ideal rigid cables is elaborated and a new control algorithm is proposed for it. In the following sections, dynamics of cable robots with elastic cables is derived using variable stiffness formaulation for the cables. A composite control structure is proposed for this model, which consists of a rigid control term according to corresponding slow or rigid model of the system and a corrective term for vibrational damping. Then, using singular perturbation theory and Tikhonov's theorem total stability of the system is analyzed and sufficient conditions for the asymptotic stability of the closed-loop system are derived. Finally, to demonstrate the effectiveness of the proposed controller, simulation results on a planar cable robot are examined in details.

2 Control of Rigid Cable Driven Parallel Robots

In this section we assume that the elasticity of cables can be ignored and cables behave as massless rigid strings. Based on this assumption the standard model for the dynamics of n-cable parallel robot with actuators is [1, 10]

$$\mathbf{M}_{eq}(\mathbf{x})\ddot{\mathbf{x}} + \mathbf{C}_{eq}(\mathbf{x}, \dot{\mathbf{x}})\dot{\mathbf{x}} + \mathbf{G}_{eq}(\mathbf{x}) = \mathbf{J}^T\mathbf{u} \tag{1}$$

In which,

$$\begin{cases} \mathbf{M}_{eq}(\mathbf{x}) = r\mathbf{M}(\mathbf{x}) + r^{-1}\mathbf{J}^T\mathbf{I}_m\mathbf{J} \\ \mathbf{C}_{eq}(\mathbf{x}, \dot{\mathbf{x}}) = r\mathbf{C}(\mathbf{x}, \dot{\mathbf{x}}) + r^{-1}\mathbf{J}^T\mathbf{I}_m\dot{\mathbf{J}} \\ \mathbf{G}_{eq}(\mathbf{x}) = r\mathbf{G}(\mathbf{x}) \end{cases} \tag{2}$$

where $\mathbf{x} \in \mathbf{R}^6$ is the vector of generalized coordinates, $\mathbf{M}(\mathbf{x})$ is the inertia matrix, $\mathbf{I}_m$ is diagonal matrix of actuator inertias reflected to the cable side of the gears, $\mathbf{C}(\mathbf{x}, \dot{\mathbf{x}})$ represents the Coriolis and centrifugal terms, $\mathbf{G}(\mathbf{x})$ is the gravitational terms, r is radius of cable drums and $\mathbf{u}$ represents the input torque. $\mathbf{J}$ represents the Jacobian matrix of the system and relates $\dot{\mathbf{x}}$ to derivative of the cable length vector $\dot{\mathbf{L}} = \mathbf{J}\dot{\mathbf{x}}$. Although these equations are nonlinear and complex, they have some properties which are beneficial in the controller design [1].

Property 1 Inertia matrix $\mathbf{M}_{eq}(\mathbf{x})$ is symmetric and positive definite.
Property 2 Matrix $\dot{\mathbf{M}}_{eq}(\mathbf{x}) - 2\mathbf{C}_{eq}(\mathbf{x}, \dot{\mathbf{x}})$ is skew symmetric.

2.1 Control Algorithm

Given a two times continuously differentiable reference trajectory $\mathbf{x}_d$ for (1), consider the following control law for $\mathbf{J}^T\mathbf{u}$

$$\mathbf{J}^T\mathbf{u} = \mathbf{M}_{eq}(\mathbf{x})\ddot{\mathbf{x}}_d + \mathbf{C}_{eq}(\mathbf{x}, \dot{\mathbf{x}})\dot{\mathbf{x}}_d + \mathbf{G}_{eq}(\mathbf{x}) + \mathbf{K}_p(\mathbf{x}_d - \mathbf{x}) + \mathbf{K}_v(\dot{\mathbf{x}}_d - \dot{\mathbf{x}}) \quad (3)$$

where $\mathbf{M}_{eq}$, $\mathbf{C}_{eq}$ and $\mathbf{G}_{eq}$ are defined as (2) and $\mathbf{K}_p$, $\mathbf{K}_v$ are diagonal matrices of positive gains for the PD control part of the proposed control scheme. Since in the fully-constrained cable robots the Jacobian matrix of the manipulator is non-square and the system is redundantly actuated, (3) is an underdetermined system of equations and has many solutions if $\mathbf{J}^T\mathbf{J}$ is invertible. In this case, the general solution of (3) is,

$$\mathbf{u} = \overline{\mathbf{u}} + \mathbf{Q} \quad (4)$$

Here, $\overline{\mathbf{u}}$ is the minimum solution of (3) derived by using the pseudo-inverse of $\mathbf{J}^T$ and is given by

$$\overline{\mathbf{u}} = \mathbf{J}(\mathbf{J}^T\mathbf{J})^{-1}(\mathbf{M}_{eq}(\mathbf{x})\ddot{\mathbf{x}}_d + \mathbf{C}_{eq}(\mathbf{x}, \dot{\mathbf{x}})\dot{\mathbf{x}}_d + \mathbf{G}_{eq}(\mathbf{x}) + \mathbf{K}_p(\mathbf{x}_d - \mathbf{x}) + \mathbf{K}_v(\dot{\mathbf{x}}_d - \dot{\mathbf{x}})) \quad (5)$$

and, $\mathbf{Q}$ spans the null space of $\mathbf{J}^T$ and must satisfy

$$\mathbf{J}^T\mathbf{Q} = \mathbf{0} \quad (6)$$

$\mathbf{Q}$ can be physically interpreted as *internal forces*. It means that this term does not contribute into motion of the end-effector and only provides positive tension in the cables. With this notation, the proposed control scheme can be implemented according to Fig. (1). In this paper we assume that the system always satisfies the vector closure conditions [4] and at all times, positive internal forces can be produced such that the cables are in tension.

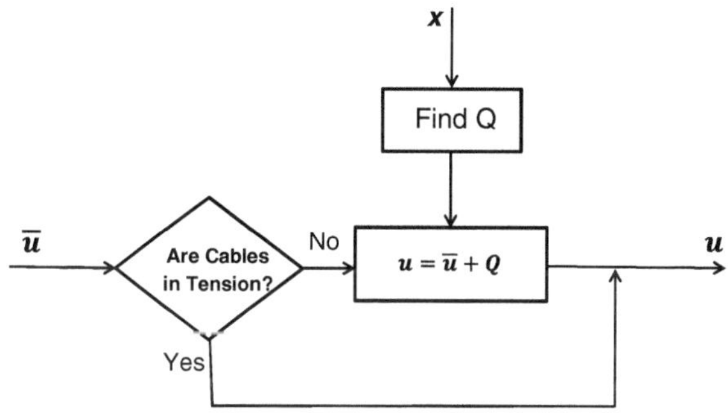

Fig. 1 Internal force control structure

2.2 *Stability Analysis*

Substituting (4) in (1) and using (6), we can write the closed loop system as

$$\mathbf{M}_{eq}(\mathbf{x})\ddot{\mathbf{e}} + \mathbf{C}_{eq}(\mathbf{x}, \dot{\mathbf{x}})\dot{\mathbf{e}} + \mathbf{K}_p\mathbf{e} + \mathbf{K}_v\dot{\mathbf{e}} = \mathbf{0} \tag{7}$$

where

$$\mathbf{e} = \mathbf{x}_d - \mathbf{x}$$

Consider the following Lyapunov function for the closed loop system (7)

$$V_R = \frac{1}{2}\dot{\mathbf{e}}^T\mathbf{M}_{eq}(\mathbf{x})\dot{\mathbf{e}} + \frac{1}{2}\mathbf{e}^T\mathbf{K}_p\mathbf{e} \tag{8}$$

V_R is positive for all $\mathbf{e} \neq \mathbf{0}$ because based on property 1, $\mathbf{M}_{eq}$ is positive definite. The time derivative of Lyapunov function V is given by [18]

$$\dot{V}_R = -\dot{\mathbf{e}}^T\mathbf{K}_v\dot{\mathbf{e}} \quad \leq 0 \tag{9}$$

As it has been fully elaborated in [18] this controller can suitably stabilize the rigid system asymptotically.

3 Robot with Elastic Cables

As mentioned earlier, in practice the overal performance of the cable robot may be affected by vibrations caused by inevitable elasticity of the cables [4, 9]. Thus, elasticity of cables must be considered in the modeling process and control schemes should be designed so that they stabilize the system and damp the vibrations efficiently. New research results show that in fully-constrained cable robots, dominant dynamics of cables are longitudinal vibration [9]. Therefore, axial spring model may suitably used to describe the effects of dominant dynamics of cable.

In order to model a general cable driven robot with n elastic cables assume that: $L_{1i} : i = 1, 2, ..., n$ denotes the length of i-th cable with tension which can be measured by pot-string. $L_{2i} : i = 1, 2, ..., n$ denotes the cable length of the i-th actuator which may be measured by shaft encoder. If the system is rigid, then $L_{1i} = L_{2i}, \forall i$. Let us denote:

$$\mathbf{L} = (L_{11}, L_{12}, \cdots, L_{1n}, L_{21}, L_{22}, \cdots, L_{2n}) = (\mathbf{L}_1^T, \mathbf{L}_2^T) \tag{10}$$

In a cable driven robot, the stiffness of cables is a function of cable lengths which are changing during the motion of the robot. Using linear axial spring model with Young's modulus E and cross-sectional area A for the cables, the instantaneous potential energy of i-th cable is

$$p_i = \frac{EA(L_{1i} - L_{2i})^2}{2L_{2i}}$$

With this notation the total potential energy of the system can be expressed by: $P = P_0 + P_1$, in which, P_0 denotes the potential energy of the rigid robot and P_1 denotes the potential energy of the cables and can be formulated by:

$$P_1 = \frac{1}{2}(\mathbf{L}_1 - \mathbf{L}_2)^T \mathbf{K}(\mathbf{L}_2)(\mathbf{L}_1 - \mathbf{L}_2) \tag{11}$$

where, $\mathbf{K}$ is the stiffness matrix of the cables during the motion which is a function of $\mathbf{L}_2$. Assume that all cables have the same Young's modulus E and cross-sectional area A,[1] then

$$\mathbf{K}(\mathbf{L}_2) = EA \cdot \text{diag}^{-1}(\mathbf{L}_2) \tag{12}$$

Furthermore, kinetic energy of the system is

$$T = \frac{1}{2}\dot{\mathbf{x}}^T \mathbf{M}(\mathbf{x})\dot{\mathbf{x}} + \frac{1}{2}\dot{\mathbf{q}}^T \mathbf{I}_m \dot{\mathbf{q}} \tag{13}$$

In which, $\mathbf{x}$ denotes the generalized coordinates in Cartesian space, $\mathbf{q}$ is the motor shaft position vector, $\mathbf{M}(\mathbf{x})$ is the mass matrix and $\mathbf{I}_m$ is the actuator moments of inertia. Using Euler-Lagrange formulation and some manipulations, final equations of motion can be written in the following form:

[1] This assumption does not reduce the generality of problem, since for the general case it can be easily reached by variable scaling.

$$\mathbf{M}(\mathbf{x})\ddot{\mathbf{x}} + \mathbf{N}(\mathbf{x}, \dot{\mathbf{x}}) = EA\mathbf{J}^T \text{diag}^{-1}(\mathbf{L}_2) \cdot (\mathbf{L}_2 - \mathbf{L}_1) \tag{14}$$

$$\mathbf{I}_m\ddot{\mathbf{q}} + r\,EA \cdot \ \text{diag}^{-1}(\mathbf{L}_2) \ \cdot \ (\mathbf{L}_2 - \mathbf{L}_1) + H(\mathbf{L}_2, \mathbf{L}_1) = \mathbf{u} \tag{15}$$

in which,

$$\mathbf{N}(\mathbf{x}, \dot{\mathbf{x}}) = \mathbf{C}(\mathbf{x}, \dot{\mathbf{x}})\dot{\mathbf{x}} + \mathbf{G}(\mathbf{x}) \ \ , \ \ \mathbf{L}_2 - \mathbf{L}_0 = r\mathbf{q}$$

$$\mathbf{H}(\mathbf{L}_2, \mathbf{L}_1) = -\frac{r}{2}EA \ \cdot \text{diag}^{-2}(\mathbf{L}_2) \ \cdot \text{diag}(\mathbf{L}_2 - \mathbf{L}_1) \cdot (\mathbf{L}_2 - \mathbf{L}_1)$$

In these equations $\mathbf{L}_0$ is the vector of cables length at $\mathbf{x} = 0$ and $\mathbf{J}$ is the Jacobian matrix of the system and relates $\dot{\mathbf{x}}$ to derivative of the cable length vector $\dot{\mathbf{L}}_1 = \mathbf{J}\dot{\mathbf{x}}$, and other parameters are defined as before. For notational simplicity we assumed that all Young's modulus E and cross-sectional A of cables are the same and EA is large with respect to other system parameters. To idealize the assumption of large cable stiffness and small cable damping, we assume that EA is of order $O(1/\varepsilon^2)$ (ε is a small scalar parameter), and furthermore, the damping terms in the cables are neglected.

Equations (14) and (15) represent cable robot as a nonlinear and coupled system. This representation includes both rigid and flexible subsystems and their interactions. The model of cable robot with elastic cables will be reduced to (1) as EA tends to infinity. Furthermore, the new formulation preserves the properties of rigid model (1), such as positive definiteness of inertia matrix and skew symmetry property.

3.1 Control

In this section we show that the control law (3) which is proposed under assumption of perfect rigidity, can be extended for cable robot with elastic cables. First, consider a composite control law by adding a damping term to the control law (3) in the form of

$$\mathbf{u} = \mathbf{u}_r + \mathbf{K}_d(\dot{\mathbf{L}}_1 - \dot{\mathbf{L}}_2) \tag{16}$$

in which, $\mathbf{u}_r$ is given by (3) in terms of $\mathbf{x}$. Furthermore, $\mathbf{K}_d$ is a constant positive diagonal matrix whose diagonal elements are designed to remain in order of $O(1/\varepsilon)$. Note that

$$\mathbf{L}_2 = r\mathbf{q} + \mathbf{L}_0 \Longrightarrow \dot{\mathbf{L}}_2 = r\dot{\mathbf{q}} \ , \ \ \ddot{\mathbf{L}}_2 = r\ddot{\mathbf{q}} \tag{17}$$

Apply control law (16) in (15) and define variable $\mathbf{z}$ as

$$\mathbf{z} = EA(\mathbf{L}_2 - \mathbf{L}_1) \tag{18}$$

The closed loop dynamic equation reduces to

$$r^{-1}\mathbf{I}_m\ddot{\mathbf{z}} + \mathbf{K}_d\dot{\mathbf{z}} + r\,\mathrm{diag}^{-1}(\mathbf{L}_2)\cdot\left[EA\mathbf{I} - \frac{1}{2}\mathrm{diag}^{-1}(\mathbf{L}_2)\,\mathrm{diag}(\mathbf{z})\right]\mathbf{z} = EA(\mathbf{u}_r - r^{-1}\mathbf{I}_m\ddot{\mathbf{L}}_1) \tag{19}$$

By the assumption on EA and the choice for $\mathbf{K}_d$ we may write

$$EA = \frac{K_1}{\varepsilon^2} \quad ; \quad \mathbf{K}_d = \frac{\mathbf{K}_2}{\varepsilon} \tag{20}$$

where, K_1 , $\mathbf{K}_2$ are $O(1)$. Therefore, (19) can be written as

$$\varepsilon^2 r^{-1}\mathbf{I}_m\ddot{\mathbf{z}} + \varepsilon\mathbf{K}_2\dot{\mathbf{z}} + r\,\mathrm{diag}^{-1}(\mathbf{L}_2)\cdot\left[K_1\mathbf{I} - \frac{\varepsilon^2}{2}\mathrm{diag}^{-1}(\mathbf{L}_2)\mathrm{diag}(\mathbf{z})\right]\mathbf{z} = K_1(\mathbf{u}_r - r^{-1}\mathbf{I}_m\ddot{\mathbf{L}}_1) \tag{21}$$

Now Eqs. (14) and (21) can be rewritten together:

$$\mathbf{M}(\mathbf{x})\ddot{\mathbf{x}} + \mathbf{C}(\mathbf{x}, \dot{\mathbf{x}})\dot{\mathbf{x}} + \mathbf{G}(\mathbf{x}) = \mathbf{J}^T\mathrm{diag}^{-1}(\mathbf{L}_2)\mathbf{z} \tag{22}$$

$$\varepsilon^2 r^{-1}\mathbf{I}_m\ddot{\mathbf{z}} + \varepsilon\mathbf{K}_2\dot{\mathbf{z}} + r\,\mathrm{diag}^{-1}(\mathbf{L}_2).\left[K_1\mathbf{I} - \frac{\varepsilon^2}{2}\mathrm{diag}^{-1}(\mathbf{L}_2)\mathrm{diag}(\mathbf{z})\right]\mathbf{z} = K_1(\mathbf{u}_r - r^{-1}\mathbf{I}_m\ddot{\mathbf{L}}_1) \tag{23}$$

System (22) and (23) is written in a singularly perturbed form. The variables $\mathbf{z}$ and $\dot{\mathbf{z}}$ have the interpretation of 'fast' variables while the end-effector position variables $\mathbf{x}$ and $\dot{\mathbf{x}}$ (or $\mathbf{L}_1$, $\dot{\mathbf{L}}_1$) are representing 'slow' variables. Using the results of singular perturbation theory, elastic system (22) and (23) can be approximated by two subsystems, namely the quasi-steady state or slow subsystem and boundary layer or fast subsystem. With $\varepsilon = 0$, Eq. (23) yields

$$\bar{\mathbf{z}} = r^{-1}\mathrm{diag}(\bar{\mathbf{L}}_1)(\bar{\mathbf{u}}_r - r^{-1}\mathbf{I}_m\ddot{\bar{\mathbf{L}}}_1) \quad , \quad \bar{\mathbf{L}}_1 = \bar{\mathbf{L}}_2 \tag{24}$$

in which, the overbar variables represent the variables when $\varepsilon = 0$. Substitute (24) into (22).

$$\mathbf{M}(\bar{\mathbf{x}})\ddot{\bar{\mathbf{x}}} + \mathbf{C}(\bar{\mathbf{x}}, \dot{\bar{\mathbf{x}}})\dot{\bar{\mathbf{x}}} + \mathbf{G}(\bar{\mathbf{x}}) = r^{-1}\mathbf{J}^T(\bar{\mathbf{u}}_r - r^{-1}\mathbf{I}_m\ddot{\bar{\mathbf{L}}}_1)$$

Substitute $\ddot{\bar{\mathbf{L}}}_1 = \mathbf{J}\ddot{\bar{\mathbf{x}}} + \dot{\mathbf{J}}\dot{\bar{\mathbf{x}}}$ in above equation:

$$(r\mathbf{M}(\bar{\mathbf{x}}) + r^{-1}\mathbf{J}^T\mathbf{I}_m\mathbf{J})\ddot{\bar{\mathbf{x}}} + (r\mathbf{C}(\bar{\mathbf{x}}, \dot{\bar{\mathbf{x}}})\dot{\bar{\mathbf{x}}} + r^{-1}\mathbf{J}^T\mathbf{I}_m\dot{\mathbf{J}}\dot{\bar{\mathbf{x}}}) + r\mathbf{G}(\bar{\mathbf{x}}) = \mathbf{J}^T\bar{\mathbf{u}}_r$$

or

$$\mathbf{M}_{eq}(\bar{\mathbf{x}})\ddot{\bar{\mathbf{x}}} + \mathbf{C}_{eq}(\bar{\mathbf{x}}, \dot{\bar{\mathbf{x}}})\dot{\bar{\mathbf{x}}} + \mathbf{G}_{eq}(\bar{\mathbf{x}}) = \mathbf{J}^T \bar{\mathbf{u}}_r \tag{25}$$

Equation (25) is called the quasi-steady state or slow subsystem. Note that (25) is the rigid model (1) in terms of $\bar{\mathbf{x}}$. Using Tikhonov's theorem [19], for $t > 0$ the elastic force $\mathbf{z}(t)$ and the end-effector position $\mathbf{x}(t)$ satisfy

$$\mathbf{z}(t) = \bar{\mathbf{z}}(t) + \boldsymbol{\eta}(\tau) + O(\varepsilon) \tag{26}$$

$$\mathbf{x}(t) = \bar{\mathbf{x}}(t) + O(\varepsilon) \tag{27}$$

where, $\tau = t/\varepsilon$ is the fast time scale and $\boldsymbol{\eta}$ is the fast state variable and satisfies boundary layer equation

$$r^{-1}\mathbf{I}_m \frac{d^2\boldsymbol{\eta}}{d\tau^2} + \mathbf{K}_2 \frac{d\boldsymbol{\eta}}{d\tau} + rK_1 \text{diag}^{-1}(\mathbf{L}_1)\boldsymbol{\eta} = \mathbf{0} \tag{28}$$

Considering these results, elastic system (22) and (23) can be approximated up to $O(\varepsilon)$ as

$$\mathbf{M}(\mathbf{x})\ddot{\mathbf{x}} + \mathbf{C}(\mathbf{x}, \dot{\mathbf{x}})\dot{\mathbf{x}} + \mathbf{G}(\mathbf{x}) = \mathbf{J}^T \text{diag}^{-1}(\mathbf{L}_1)(\bar{\mathbf{z}} + \boldsymbol{\eta}(\tau))$$

$$r^{-1}\mathbf{I}_m \frac{d^2\boldsymbol{\eta}}{d\tau^2} + \mathbf{K}_2 \frac{d\boldsymbol{\eta}}{d\tau} + rK_1 \text{diag}^{-1}(\mathbf{L}_1)\boldsymbol{\eta} = \mathbf{0}$$

According to (24)

$$\mathbf{M}_{eq}(\mathbf{x})\ddot{\mathbf{x}} + \mathbf{C}_{eq}(\mathbf{x}, \dot{\mathbf{x}})\dot{\mathbf{x}} + \mathbf{G}_{eq}(\mathbf{x}) = \mathbf{J}^T(\mathbf{u}_r + r\,\text{diag}^{-1}(\mathbf{L}_1)\boldsymbol{\eta}(\tau)) \tag{29}$$

$$r^{-1}\mathbf{I}_m \tfrac{d^2\boldsymbol{\eta}}{d\tau^2} + \mathbf{K}_2 \tfrac{d\boldsymbol{\eta}}{d\tau} + rK_1 \text{diag}^{-1}(\mathbf{L}_1)\boldsymbol{\eta} = \mathbf{0} \tag{30}$$

Since the gain $\mathbf{K}_2$ can be chosen suitably such that the boundary layer (28) becomes asymptotically stable, it follows that, with sufficiently small values of ε, the response of the elastic system (14) and (15) with the composite control (16), will be nearly the same as the response of rigid system (1) with the rigid control $\mathbf{u}_r$ alone, after some initially damped transient of fast variables $\boldsymbol{\eta}(t/\varepsilon)$.

3.2 Stability Analysis

Control of rigid model and its stability analysis were discussed in previous section. It was demonstrated that the boundary layer or fast subsystem (28) is asymptotically stable, due to damping term. Separate stability of boundary layer and quasi-steady state subsystems does not generally guarantee that the total system is stable [19]. In this section the stability of the total system is analyzed, based on stability analysis of the subsystems.

Consider dynamic equations of elastic system (29) and (30), and control law (4) from previous section. Then

$$\mathbf{M}_{eq}(\mathbf{x})\ddot{\mathbf{e}} + (\mathbf{C}_{eq}(\mathbf{x}, \dot{\mathbf{x}}) + \mathbf{K}_v)\dot{\mathbf{e}} + \mathbf{K}_p\mathbf{e} = -r\mathbf{J}^T \text{diag}^{-1}(\mathbf{L}_1)\boldsymbol{\eta}(t/\varepsilon) \tag{31}$$

$$r^{-1}\mathbf{I}_m \frac{d^2\boldsymbol{\eta}}{dt^2} + \mathbf{K}_d \frac{d\boldsymbol{\eta}}{dt} + rEA\text{diag}^{-1}(\mathbf{L}_1)\boldsymbol{\eta} = \mathbf{0} \tag{32}$$

Let us denote $\mathbf{y} = \left[\mathbf{e}^T, \dot{\mathbf{e}}^T\right]^T$ and $\mathbf{h} = \left[\boldsymbol{\eta}^T, \dot{\boldsymbol{\eta}}^T\right]^T$, in which $\mathbf{e} = \mathbf{x}_d - \mathbf{x}$, then one may write

$$\dot{\mathbf{y}} = \mathbf{A}\mathbf{y} + \mathbf{B}[\mathbf{I} \;\; \mathbf{0}]\,\mathbf{h} \tag{33}$$

$$\dot{\mathbf{h}} = \tilde{\mathbf{A}}\mathbf{h} \tag{34}$$

where

$$\mathbf{A} = \begin{bmatrix} \mathbf{0} & \mathbf{I} \\ -\mathbf{M}_{eq}^{-1}\mathbf{K}_p & -\mathbf{M}_{eq}^{-1}(\mathbf{K}_v + \mathbf{C}_{eq}) \end{bmatrix}$$

$$\mathbf{B} = \begin{bmatrix} \mathbf{0} \\ -r\mathbf{M}_{eq}^{-1}\mathbf{J}^T \text{diag}^{-1}(\mathbf{L}_1) \end{bmatrix}$$

$$\tilde{\mathbf{A}} = \begin{bmatrix} \mathbf{0} & \mathbf{I} \\ -r^2 EA\mathbf{I}_m^{-1}\text{diag}^{-1}(\mathbf{L}_1) & -r\mathbf{I}_m^{-1}\mathbf{K}_d \end{bmatrix}$$

Lemma 1 *There is a positive definite matrix* $\mathbf{K}_d$ *such that the closed-loop system described with* (34) *is asymptotically stable.*

Proof Consider the following Lyapunov function

$$V_F = \mathbf{h}^T\mathbf{W}\mathbf{h} \quad , \quad \mathbf{W} = \frac{1}{2}\begin{bmatrix} r^2\mathbf{K}_d & r\mathbf{I}_m \\ r\mathbf{I}_m & \mathbf{I}_m \end{bmatrix} \tag{35}$$

In order to have positive definite $\mathbf{W}$, according to the Shur complement, it is sufficient that $\mathbf{K}_d > \mathbf{I}_m$. Differentiate V_F along trajectories of (34)

$$\dot{V}_F = -\mathbf{h}^T\mathbf{S}\mathbf{h} \tag{36}$$

in which,

$$\mathbf{S} = \begin{bmatrix} r^3 EA\,\text{diag}^{-1}(\mathbf{L}_1) & 0.5r^2 EA\,\text{diag}^{-1}(\mathbf{L}_1) \\ 0.5r^2 EA\,\text{diag}^{-1}(\mathbf{L}_1) & r(\mathbf{K}_d - \mathbf{I}_m) \end{bmatrix}$$

According to Shur complement, $\mathbf{S}$ is positive definite, if and only if,

$$\begin{aligned} &\mathbf{K}_d > \mathbf{I}_m \\ &\mathbf{K}_d - (\mathbf{I}_m + .25EA\,\text{diag}^{-1}(\mathbf{L}_1)) > 0 \end{aligned} \tag{37}$$

in which, $\mathbf{K}_d$ and $\mathbf{I}_m$ are diagonal positive definite matrices. Regarding to workspace constraints

$$0 < \gamma \leq L_{1i} \leq \beta \qquad (i = 1, 2, \cdots, n) \tag{38}$$

where γ and β are positive constants. Thus, if

$$K_{di} > I_{mi} + .25EA\gamma^{-1} \qquad (i = 1, 2, \cdots, n) \tag{39}$$

then, $\dot{V}_F$ becomes negative definite and closed-loop system described by (34) is asymptotically stable.

Theorem 1 *There exist positive definite controller gains* $\mathbf{K}_v$*, and* $\mathbf{K}_d$*, to stabilize the closed-loop system* (33) *and* (34) *asymptotically.*

Proof Consider the following composite Lyapunov function

$$V(\mathbf{y}, \mathbf{h}) = V_R + V_F \tag{40}$$

where, V_R is the Lyapunov function for the rigid subsystem, and V_F is the Lyapunov function for the fast subsystem (28). Differentiate $V(\mathbf{y}, \mathbf{h})$ along trajectories of (33) and (34):

$$\dot{V}(\mathbf{y}, \mathbf{h}) = \dot{V}_R + \dot{V}_F = -\dot{\mathbf{e}}^T\mathbf{K}_v\dot{\mathbf{e}} - r\dot{\mathbf{e}}^T\mathbf{J}^T\mathrm{diag}^{-1}(\mathbf{L}_1)\boldsymbol{\eta} - \mathbf{h}^T\mathbf{S}\mathbf{h} \tag{41}$$

According to Rayleigh-Ritz inequality,

$$-\mathbf{h}^T\mathbf{S}\mathbf{h} \leq -\lambda_{min}(\mathbf{S})\|\mathbf{h}\|^2 \tag{42}$$

Furthermore,

$$\begin{gathered} -\dot{\mathbf{e}}^T\mathbf{K}_v\dot{\mathbf{e}} \leq -\lambda_{min}(\mathbf{K}_v)\|\dot{\mathbf{e}}\|^2 \\ -r\dot{\mathbf{e}}^T\mathbf{J}^T\mathrm{diag}^{-1}(\mathbf{L}_1)\boldsymbol{\eta} \leq r|\dot{\mathbf{e}}^T\mathbf{J}^T\mathrm{diag}^{-1}(\mathbf{L}_1)\boldsymbol{\eta}| \leq r\gamma^{-1}\sigma_{max}(\mathbf{J}^T)\|\dot{\mathbf{e}}\|\|\mathbf{h}\| \end{gathered} \tag{43}$$

in which, λ_{min} and σ_{max} denote the smallest eigenvalue and the largest singular value of the corresponding matrices, respectively. Using above inequalities, one may write

$$\dot{V}(\mathbf{y}, \mathbf{h}) \leq -\lambda_{min}(\mathbf{K}_v)\|\dot{\mathbf{e}}\|^2 + r\gamma^{-1}\sigma_{max}(\mathbf{J}^T)\|\dot{\mathbf{e}}\|\|\mathbf{h}\| - \lambda_{min}(\mathbf{S})\|\mathbf{h}\|^2$$

Or

$$\dot{V}(\mathbf{y}, \mathbf{h}) \leq [\, \|\dot{\mathbf{e}}\| \;\; \|\mathbf{h}\| \,] \cdot \begin{bmatrix} -\lambda_{min}(\mathbf{K}_v) & 0.5r\gamma^{-1}\sigma_{max}(\mathbf{J}^T) \\ 0.5r\gamma^{-1}\sigma_{max}(\mathbf{J}^T) & -\lambda_{min}(\mathbf{S}) \end{bmatrix} \begin{bmatrix} \|\dot{\mathbf{e}}\| \\ \|\mathbf{h}\| \end{bmatrix}$$

in order to guarantee $\dot{V}(\mathbf{y}, \mathbf{h}) \leq 0$, it is necessary to have

$$\lambda_{min}(\mathbf{K}_v)\lambda_{min}(\mathbf{S}) > 0.25r^2\gamma^{-2}\sigma^2_{max}(\mathbf{J}^T) \tag{44}$$

Condition (44) may be simply satisfied by choosing appropriate values for $\mathbf{K}_v$ in (3) and $\mathbf{K}_d$ for the fast subsystem. Negative semi-definiteness of $\dot{V}(\mathbf{y}, \mathbf{h})$ implies that $\mathbf{y}$ and $\mathbf{h}$ are bounded. This indicates that $\ddot{V}(\mathbf{y}, \mathbf{h})$ is bounded. Hence, $\dot{V}(\mathbf{y}, \mathbf{h})$ is uniformly continuous. Using Barbalat's lemma one may conclude that $\dot{\mathbf{e}} \to 0$ and $\mathbf{h} \to 0$ as $t \to \infty$. Now, according to uniform continuity of $\ddot{\mathbf{e}}$, it can be concluded that $\ddot{\mathbf{e}} \to 0$ as $t \to \infty$. As a result the total closed-loop system (33) and (34) becomes asymptotically stable.

4 Case Study

To show the effectiveness of the proposed control algorithm a simulation study has been performed. In the following simulation study, the results of the closed–loop performance of a planar cable driven manipulator are investigated. Our model of a planar cable robot consists of an end-effector that is connected by four cables to the base platform shown in Fig. (2a). As it is shown in Fig. (2), A_i denote the fixed base points of the cables, B_i denote points of connection of the cables on the end-effector, L_i denote the cable lengths, and α_i denote the cable angles. The position of the center of mass of the end-effector $\mathbf{P}$, is denoted by $\mathbf{P} = [x_p, y_p]$, and the orientation of the manipulator end-effector is denoted by ϕ with respect to the fixed coordinate frame. Hence, the manipulator possesses three degrees of freedom $\mathbf{x} = [x_p, y_p, \phi]$, with one degree of actuator redundancy. Furthermore, the Jacobian matrix of the manipulator,

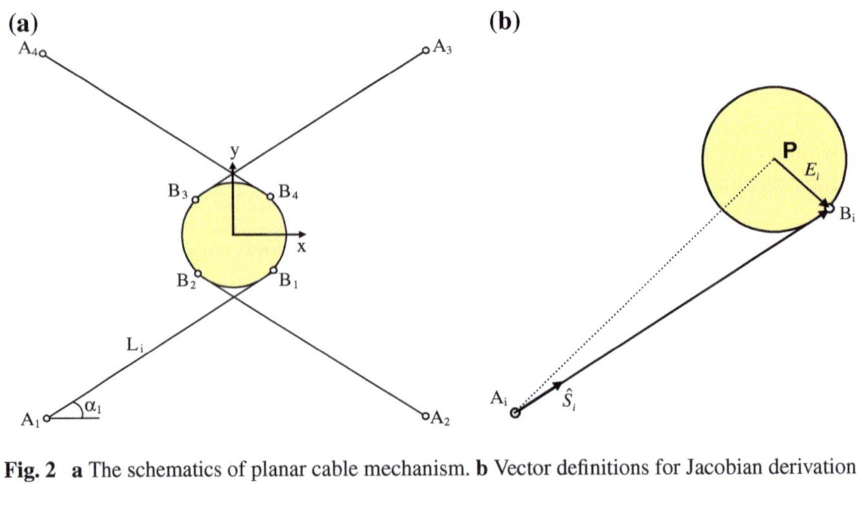

Fig. 2 **a** The schematics of planar cable mechanism. **b** Vector definitions for Jacobian derivation

Table 1 Inertial parameter of the planar cable robot

Parameter	Symbol	Nominal value
End-effector mass	m	2.5 Kg
End-effector inertia	I_z	$0.03\,\mathrm{Kg.m}^2$
Actuator inertia	I_{mi}	$0.6\,\mathrm{Kg.m}^2$
Gear ratio	N	50
Gravity acceleration	g	$9.8\,\mathrm{m/s}^2$
Drum radius	r	3.5 cm

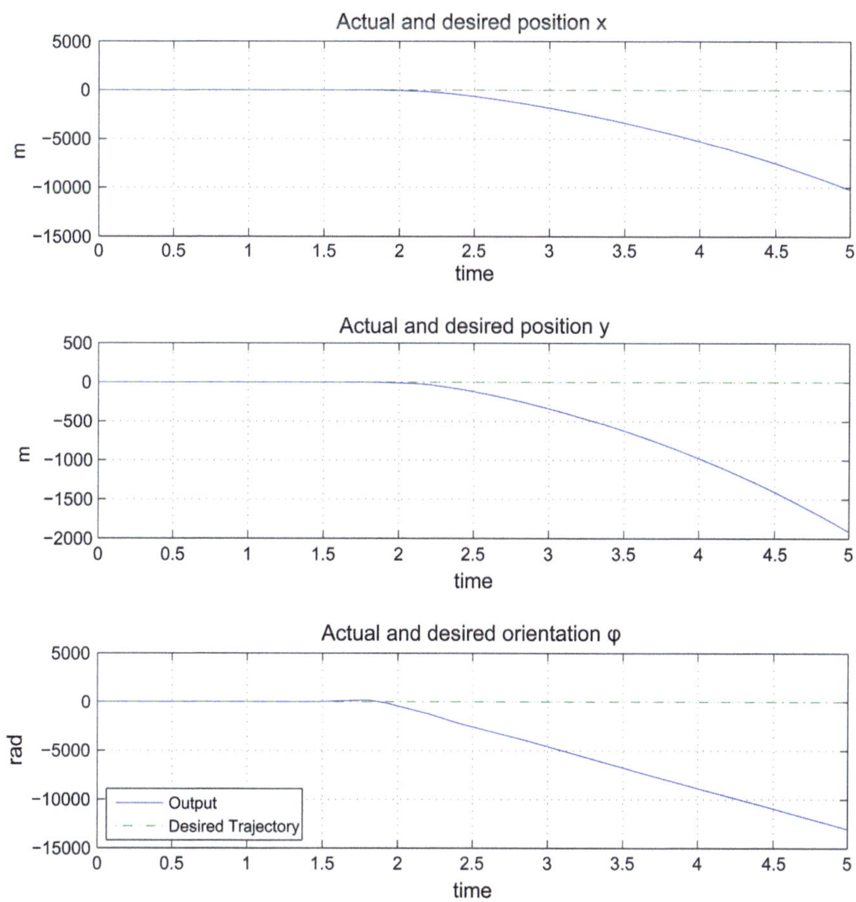

Fig. 3 The closed-loop system experiences instability, if only rigid controller $\mathbf{u}_r$ is applied

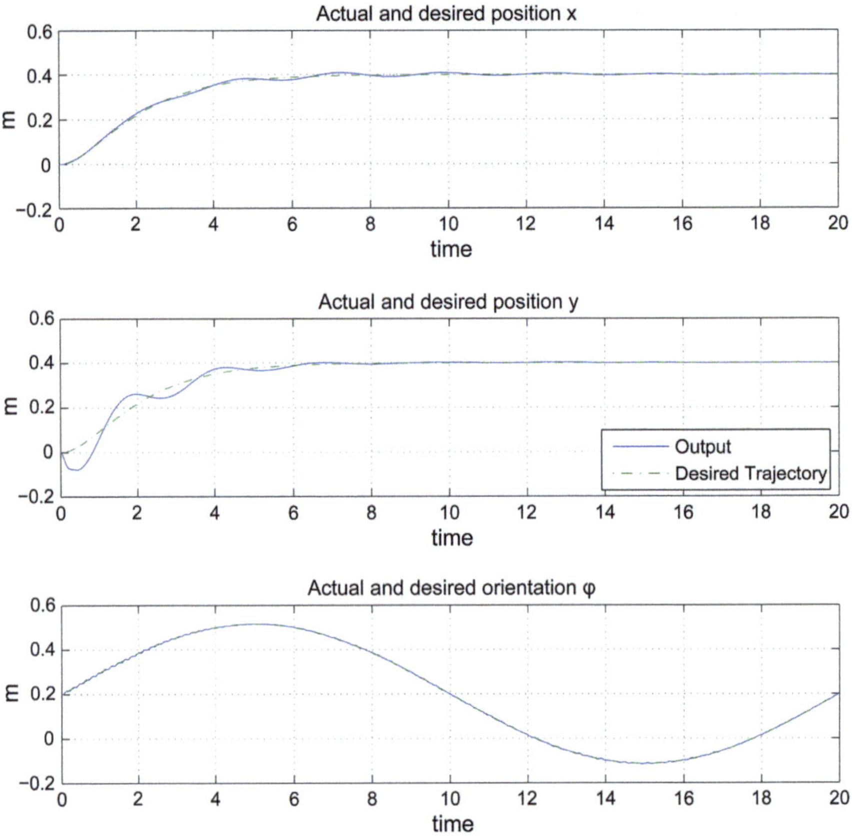

Fig. 4 Suitable tracking performance of the closed-loop system to smooth reference trajectories; Proposed control algorithm

which relates the length variable velocities $\dot{\mathbf{L}}_1$ to the end-effector velocities $\dot{\mathbf{x}}$, is given by

$$\dot{\mathbf{L}}_1 = \mathbf{J}\dot{\mathbf{x}} \tag{45}$$

Note that the Jacobian matrix $\mathbf{J}$ is a non-square 4×3 matrix, since the manipulator is redundantly actuated.

The equations of motion for the end-effector can be written in the following form

$$\mathbf{M}\ddot{\mathbf{x}} + \mathbf{G} = \mathbf{F}$$

in which, $\mathbf{x} = [x_p, y_p, \phi]$, and by considering flexibility in the cables, according to (14) and (15)

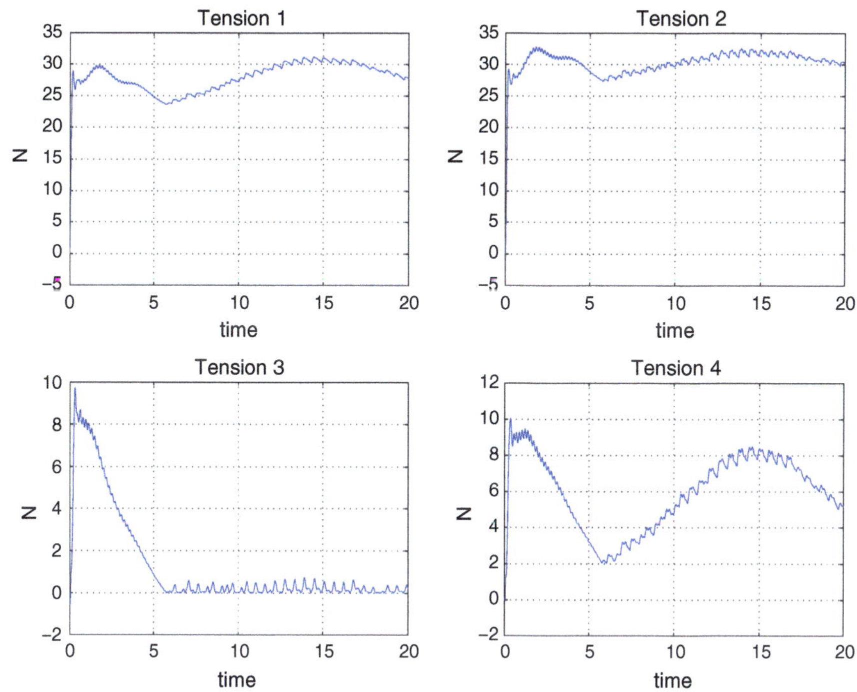

Fig. 5 Simulation results showing the cables tension for smooth reference trajectories

$$\mathbf{F} = \mathbf{J}^T EA \cdot \text{diag}^{-1}(\mathbf{L}_2)(\mathbf{L}_2 - \mathbf{L}_1)\ , \ \ \mathbf{L}_2 = r\mathbf{q} + \mathbf{L}_0$$

$$\mathbf{I}_m \ddot{\mathbf{q}} + rEA \cdot \text{diag}^{-1}(\mathbf{L}_2)(\mathbf{L}_2 - \mathbf{L}_1) - \frac{r}{2}EA \cdot \text{diag}^{-2}(\mathbf{L}_2)\text{diag}(\mathbf{L}_2 - \mathbf{L}_1)(\mathbf{L}_2 - \mathbf{L}_1) = \mathbf{u}$$

and,

$$\mathbf{M} = \begin{bmatrix} m & 0 & 0 \\ 0 & m & 0 \\ 0 & 0 & I_z \end{bmatrix} \quad \text{and} \quad \mathbf{G} = \begin{bmatrix} 0 \\ mg \\ 0 \end{bmatrix}$$

All mechanical parameters of the cable robot are given in Table (1). In order to demonstrate a high flexible system EA is intentionally chosen very low (EA = 1,000). To show the effectiveness of the proposed composite control algorithm suppose that the system is at the origin and has to track the following smooth reference trajectories in x, y, and ϕ coordinates,

$$\begin{aligned} x_d &= 0.4 + 2e^{-t} - 2.4e^{-t/1.2} \\ y_d &= 0.4 + 2e^{-t} - 2.4e^{-t/1.2} \\ \phi_d &= 0.1\pi \sin(0.1\pi t) \end{aligned}$$

in which, the task space variables x_p, y_p and ϕ reach a final value of 0.4, 0.4 and 0.1π from the zero states, respectively. The controller is based on (16) and consists of rigid control $\mathbf{u}_r$ given by (3) and the corrective term. Controller gain matrices are chosen as $\mathbf{K}_p = 200\ \mathbf{I}_{3\times3}$, $\mathbf{K}_v = 25\ \mathbf{I}_{3\times3}$, and $\mathbf{K}_d = 250\ \mathbf{I}_{4\times4}$ to satisfy the stability conditions. In the first step, rigid control $\mathbf{u}_r$ alone is applied to the manipulator. As is illustrated in Fig. (3), the manipulator experiences instability if only the rigid control $\mathbf{u}_r$ is applied to the system. The main reason for instability is the divergence of its fast variables. Figure (4) illustrates dynamic behavior of the closed-loop system with the proposed control algorithm. Internal force $\mathbf{Q}$ is used whenever at least one cable becomes slack (or $L_{1i} < L_{2i}$, $i = 1, \ldots, 4$) to ensure that the cables remain in tension. Although, the system is very flexible the proposed control algorithm can suitably stabilize the system. As it is seen in this figure, position and orientation outputs track the desired values very well and the steady state errors are very small, while as it is shown in Fig. (5), all cables are in tension for the whole maneuver. The simulation results clearly show the effectiveness of the proposed control algorithm to stabilize the system, while achieving positive tension in the cables.

5 Conclusions

In this paper modeling and control of parallel cable robots with elastic cables are examined in detail. In the modeling process of this class of manipulators cables are modeled by linear axial springs, and the model of fully constrained cable driven robot is derived using Euler-Lagrange approach. Since in this type of robots cables must remain in tension in the whole workspace, the notion of internal force is introduced and it is used in the proposed control algorithm. The proposed composite control algorithm consists of three components. Rigid control according to the rigid model of the system, the internal force to ensures that all cables are in tension, and a damping term in cable length space to stabilize fast subsystem. Then, the model of the system is formulated in the standard form of singular perturbation theory and fast and slow variables are separated and incorporated in the stability analysis. The stability of the closed-loop system is analyzed through Lyapunov second method, and it is shown that the proposed composite controller is capable to stabilize the system in presence of flexible cables. Finally, the performance of the proposed composite controller is examined by simulation results performed on a planar cable robot.

References

1. Khosravi MA, Taghirad HD (2011) Dynamic analysis and control of cable driven robots with elastic cables. Trans Can Soc Mech Eng 35(4):543–557
2. Bostelman R, Albus J, Dagalakis N, Jacoff A (1994) Applications of the NIST robocrane. In: Proceedings of the 5th international symposium on robotics and manufacturing, pp 403–410

3. Maeda K, Tadokoro S, Takamori T, Hiller M, Verhoeven R (1999) On design of a redundant wire-driven parallel robot WARP manipulator. In: Proceedings of IEEE international conference on robotics and automation, pp 895–900
4. Kawamura S, Kino H, Won C (2000) High-speed manipulation by using parallel wire-driven robots. Robotica 18(3):13–21
5. Roberts R, Graham T, Lippitt T (1998) On the inverse kinematics, statics, and fault tolerance of cable-suspended robots. J Rob Syst 15(10):649–661
6. Taghirad HD, Nahon MA (2008) Kinematic analysis of a macro-micro redundantly actuated parallel manipulator. Adv Rob 22(6–7):657–687
7. Yao R, Tang X, Wang J, Huang P (2010) Dimensional optimization design of the four-cable-driven parallel manipulator in FAST. IEEE/ASME Trans Mechatron 15(6):932–941
8. Riechel A, Bosscher P, Lipkin H, Ebert-Uphoff I (2004) Concept paper: cable-driven robots for use in hazardous environments. In: Proceedings of 10th international topical meeting on robot remote system in hazardous environment, Gainesville, pp. 310–317
9. Diao X, Ma O (2009) Vibration analysis of cable-driven parallel manipulators. Multibody Syst Dyn 21:347–360
10. Khosravi MA, Taghirad HD (2012) Experimental performance of robust PID controller on a planar cable robot. In: Cable-driven parallel robots, Springer, Stuttgart, pp 337–352
11. Williams RL, Gallina P, Vadia J (2003) Planar translational cable direct driven robots. J Rob Syst 20(3):107–120
12. Ryoek S, Agrawal S (2006) Generation of feasible set points and control of a cable robot. IEEE Trans Rob 22(3):551–558
13. Kino H, Yahiro T, Takemura F (2007) Robust PD control using adaptive compensation for completely restrained parallel-wire driven robots. IEEE Trans Rob 23(4):803–812
14. Alp A, Agrawal S (2002) Cable suspended robots: feedback controllers with positive inputs. In: Proceedings of the American control conference, pp 815–820
15. Fang S, Franitza D, Torlo M, Bekes F, Hiller M (2004) Motion control of a tendon-based parallel manipulator using optimal tension distribution. IEEE/ASME Trans Mechatron 9(3):561–568
16. Ottaviano E, Castelli G (2010) A study on the effects of cable mass and elasticity in cable-based parallel manipulators. In: ROMANSY 18 robot design, dynamics and control, vol 524. Chapter I, pp 149–156
17. Meunier G, Boulet B, Nahon M (2009) Control of an overactuated cable-driven parallel mechanism for a radio telescope application. IEEE Trans Control Syst Tech 17(5):1043–1054
18. Khosravi MA, Taghirad HD (2014) Dynamic modeling and control of parallel robots with elastic cables: singular perturbation approach. IEEE Trans Rob. doi:10.1109/TRO.2014.2298057
19. Kokotovic P, Khalil HK, O'Reilly J (1986) Singular perturbation methods in control: analysis and design. Academic Press, New York

Adaptive Terminal Sliding Mode Control of a Redundantly-Actuated Cable-Driven Parallel Manipulator: CoGiRo

Gamal El-Ghazaly, Marc Gouttefarde and Vincent Creuze

Abstract This paper presents an extended adaptive control scheme via terminal sliding mode (TSM) for cable-driven parallel manipulators (CDPM). Compared with linear hyperplane-based sliding mode control, TSM is able to guarantee high-precision and robust tracking performances which arise from its main feature of finite-time convergence. This motivates applying TSM to robotic manipulators in general and, as presented in this paper, to CDPM in particular. The scheme presented in this paper extends early developed TSM control schemes which are based on partial knowledge of system dynamics. Instead, making use of the property that the dynamic models of mechanical manipulators are linear in inertial parameters, an adaptive control law is synthesised based on an appropriate choice of Lyapunov function which guarantees finite-time convergence to neighborhood of sliding mode. A key challenge of the control of CDPM is that cable tensions must be admissible, i.e. lying in a non-negative range of admissible values. As long as cable tensions are admissible, the overall dynamics of CDPM can be easily written in either actuator space or operational space which in turn facilitates control system design. The extended adaptive control scheme has been applied to a large redundantly actuated CDPR prototype, CoGiRo. Simulation results show the effectiveness of the proposed control method.

1 Introduction

A cable-driven parallel manipulator (CDPM) is a particular type of robot in which the motion of the platform (end-effector) is provided by varying the lengths of the cables. Compared with classical parallel manipulators which are characterised by

G. El-Ghazaly (✉) · M. Gouttefarde · V. Creuze
Laboratoire d'Informatique, de Robotique et de Microélectronique de Montpellier (LIRMM, CNRS-UM2), 161 Rue Ada, 34392 Montpellier, France
e-mail: gamal.elghazaly@lirmm.fr

M. Gouttefarde
e-mail: marc.gouttefarde@lirmm.fr

V. Creuze
e-mail: vincent.creuze@lirmm.fr

A. Pott and T. Bruckmann (eds.), *Cable-Driven Parallel Robots*,
Mechanisms and Machine Science 32, DOI 10.1007/978-3-319-09489-2_13

their limited workspace, CDPMs can have a very large workspace and are capable of manipulating heavy loads at high operating speed [1] which are useful characteristics for many potential industrial applications. For crane-like applications [2] such as pick and place tasks [3], suspended CDPM are usually used as they reduce the risk of cable collisions in the workspace area. Cable-suspended CDPM are not fully-constrained and cannot have wrench closure over the entire workspace. However, a suspended CDPM uses the mobile platform and payload weights to keep the cables in tension. Furthermore, using more cables than DOFs in cable-suspended CDPM can allow the robot to have a significantly wider workspace [3–5]. In this paper, a 6-DOF suspended CDPM prototype with two degrees of actuation redundancy is used to evaluate the performance of the proposed TSM control scheme.

Many challenges arise in CDPMs among which control design is a difficult one. These challenges stem from the fact that cables can only apply tension forces imposing a strict constraint on control inputs. For redundantly actuated CDPM, an infinite number of cable tension distributions exist for a given external wrench applied to the platform. The corresponding redundancy resolution has attracted the attention of many researchers. The proposed algorithms to resolve redundancy can be categorised into two main classes namely, iterative algorithms and non-iterative algorithms. Iterative algorithms are usually based on optimization and are not always suitable for real time implementations [6–8]. On the other hand, non-iterative algorithms give a solution in a reasonable amount of time and can be more easily implemented in real-time [5, 9, 10].

Compared to the great deal of attention that has been paid to motion control of rigid-link manipulators, only relatively few results on the control of CDPM exist in the literature. Gholami et al. [11] proposed an operational-space PD controller for trajectory tracking. Kawamura et al. proposed a PD controller with gravity compensation in [1]. The PD controller is designed in the actuator-space with Lyapunov stability. A robust Lyapunov-based design of PID tracking controller has been introduced for a fully-constrained planar CDPM in [12]. A PD controller with adaptive compensation has been applied to fully-constrained CDPM in [13]. Computed-torque control methods appeared in [14]. To make use of the advantages of both the actuator-space and operational-space formulations, a dual space adaptive controller has been proposed in [15].

Sliding mode control (SMC) has been found effective to deal with dynamic systems with uncertainties, time-varying parameters, and bounded external disturbances [16]. The main idea of SMC is to force the states of the system to stay in a chosen switching manifold satisfying a desired dynamic behavior. The choice of linear hyperplanes as switching manifolds guarantee asymptotic stability, i.e., the closed-loop error converges to the neighborhood of the origin as the time approaches infinity. Although classical SMC gave reasonable performance for robotic manipulators,only few results for CDPMs exist in the literature [17]. The performance of classical SMC could be enhanced if the closed-loop errors are forced to reach the origin in finite time. SMC with finite-time stabilization are called terminal sliding mode (TSM). Several ideas of designing TSM based control schemes have been developed to achieve finite-time stabilization in [18–21]. The control schemes based on those ideas are able to guarantee high-precision and robust tracking performances due to finite-time convergence even with high uncertainty. This property justifies adopting TSM based control schemes for robotic manipulators and motivates us to use it for CDPMs. TSM control schemes developed for rigid-link manipulators may not exploit their properties very well. Even adaptive TSM control schemes developed for rigid-link manipulators do not rely on the linearity of inertial parameters [22]. In [18], a robust TSM control has been developed for robotics manipulators. However, the main problem of this scheme is that controller singularity may be reached in sliding mode. A non-singular TSM has been proposed for robotic manipulators in [19] where a new terminal sliding manifold has been proposed to avoid controller singularity. However, this control scheme did not eliminate chattering in control inputs. A continuous and non-singular TSM control scheme has been proposed also for robotic manipulators in [20] to avoid both chattering and singularity. However, the proposed schemes in [18–20] did not consider the aforementioned property of the dynamic model of robotic manipulators and therefore cannot be directly applied to control CDPM for which high uncertainty may appear specially when manipulating payloads of varying sizes and weights. Instead, adaptive TSM control scheme has been proposed also for robotic manipulators in [23] where the Coriolis, centrifugal and gravity terms are approximated by polynomials of first order. The role of the adaptive mechanism is to estimate the coefficients of these polynomials which approximate the manipulator dynamics. However, [23] did not estimate the true inertial parameters of the manipulator which may give relatively large control inputs specially when loads are introduced.

This paper extends an early developed non-singular continuous TSM control scheme for robotic manipulators proposed in [20] by incorporating an adaptive mechanism to estimate dynamic parameters and by applying it to CDPM. The scheme in [20] is based on the assumption that the nominal values of inertial matrix components as well as of Coriolis, centrifugal and gravity forces are known and their corresponding uncertainties have known bounds. It is worth noting that the CDPMs considered in this paper are designed to carry heavy payloads having different weights so that the mobile platform inertial parameters are subjected to large variations. Therefore, if the scheme in [20] is directly applied, very high control input torques may be required especially during loading situations. The most appropriate solution to this problem is to estimate online the dynamic parameters of the CDPM via an adaptive mechanism. In this paper, the overall dynamics of CDPMs (both winches and platform) is formulated to be linear in all inertial and friction parameters. Thereby, it facilitates the design of the adaptive control scheme. The whole control scheme is synthesised and analysed based on Lyapunov stability theory. Moreover, it guarantees the finite-time convergence of the closed-loop system. To show its effectiveness, the proposed control scheme is applied to a large redundantly actuated CDPM, CoGiRo [3].

The paper is organised as follows. In Sect. 2, kinematic and dynamic modeling of CDPMs are presented and some properties of the dynamic model are recalled. Section 3 focuses on the proposed adaptive terminal sliding mode control scheme where finite-time convergence stability of the overall closed-loop system is analysed via Lyapunov direct method. Finally, to show the effectiveness of the proposed control

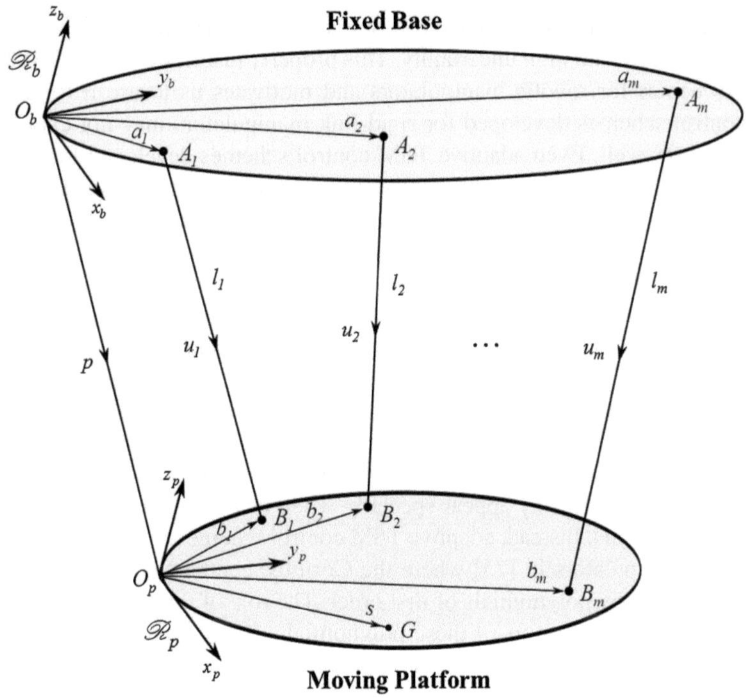

Fig. 1 Schematic diagram of a general CDPM

scheme, a simulation study of a pick-and-place task is performed on the CoGiRo CDPM in Sect. 4. Section 5 concludes this paper.

2 Kinematic and Dynamic Modeling of CDPM

In this section, kinematic and dynamic modeling of CDPM are presented. We assume that the elasticity of cables is negligible and cables almost behave as massless rigid strings. Modeling of CDPM has been presented in many research work before e.g. [3, 24–26]. However, in this section, our main objective is to present the overall dynamics of CDPM (both *winches* and *platform*) in compact form and linear in all dynamic parameters. This compact form facilitates the design of adaptive control schemes in general [22, 27] and is particularly suitable for the adaptive control scheme introduced in this paper. The notations of a general CDPM for kinematic modeling is shown in Fig. 1.

2.1 Kinematic Model

As shown in Fig. 1, a CDPM is composed of a moving platform linked to a fixed base through cables with varying lengths via winches which are generally attached to the base structure. Let $\mathscr{R}_b$ and $\mathscr{R}_p$ denote two frames assigned to the base and moving platform, respectively. In the case of CDPM with m cables, let cable i connects the point A_i at which it extends from the base frame to point B_i at which it is attached to the moving platform. The position and orientation $\mathscr{R}_p$ with respect to $\mathscr{R}_b$ is defined by the configuration vector $\mathbf{x} = [\, p^T \; \varphi^T \,]^T$ where $p = [\, x \; y \; z]^T$ is the position vector and $\varphi = [\, \phi \, \theta \, \psi]^T$ is a vector representing the orientation of $\mathscr{R}_p$ with respect to $\mathscr{R}_b$ using Euler angles. The rotation matrix from $\mathscr{R}_b$ to $\mathscr{R}_p$ denoted by $\mathbf{R}_b^p$ is represented by three successive rotations as

$$\mathbf{R}_b^p(\phi, \theta, \psi) = \mathbf{Rot}(x, \phi)\mathbf{Rot}(y, \theta)\mathbf{Rot}(z, \psi). \tag{1}$$

Using the definition of elementary rotations around coordinate axes, $\mathbf{R}_b^p$ can be written as

$$\mathbf{R}_b^p(\phi, \theta, \psi) = \begin{bmatrix} C\theta C\psi & -C\theta S\psi & S\theta \\ C\phi S\psi + S\phi S\theta C\psi & C\phi C\psi - S\phi S\theta S\psi & -S\phi C\theta \\ S\phi S\psi - C\phi S\theta C\psi & S\phi C\psi + C\phi S\theta S\psi & C\phi C\theta \end{bmatrix}, \tag{2}$$

where $C(\cdot) = \cos(\cdot)$ and $S(\cdot) = \sin(\cdot)$. If q_i denotes motor i angular position and r_i represents motor i angular position to cable length transmission ratio, then the cable length l_i can be expressed as

$$l_i = r_i q_i. \tag{3}$$

Let $\mathbf{q} = \left[q_1 \; q_2 \; \dots \; q_m\right]^T \in \mathbb{R}^m$, and $\mathbf{L} = \left[l_1 \; l_2 \; \dots \; l_m\right]^T \in \mathbb{R}^m$ be two vectors gathering angular position variables and cable lengths, respectively. Then, one can write (3) in a vector matrix form as

$$\mathbf{L} = \mathbf{R}\mathbf{q}, \tag{4}$$

where $\mathbf{R}$ is a diagonal matrix containing the transmission ratios r_i, i.e.,

$$\mathbf{R} = diag\{r_1, \; r_2, \; \dots, \; r_m\}. \tag{5}$$

Let a_i denotes the position vector corresponding to point A_i expressed in $\mathscr{R}_b$, b_i denotes the position vector corresponding to point B_i expressed in $\mathscr{R}_p$, and u_i is a vector along the direction of cable i and has the same magnitude of the cable length. Given the position and orientation of the moving platform $\mathbf{x}$, the loop closure of cable i expressed in $\mathscr{R}_b$ can be written as follows

$$u_i = p - a_i + \mathbf{R}_b^p b_i. \tag{6}$$

The length of cable i is then computed as

$$l_i^2 = u_i^T u_i = [p + \mathbf{R}_b^p b_i - a_i]^T [p + \mathbf{R}_b^p b_i - a_i]. \tag{7}$$

Differentiating (7) with respect to time together with (4) gives

$$\dot{\mathbf{L}} = \mathbf{R}\dot{\mathbf{q}} = \mathbf{J}(\mathbf{x})\dot{\mathbf{x}}, \tag{8}$$

where $\mathbf{J}(\mathbf{x})$ defines a $6 \times m$ Jacobian matrix of the CDPM. The second-order derivative of (7) with respect to time is then written as

$$\ddot{\mathbf{L}} = \mathbf{R}\ddot{\mathbf{q}} = \mathbf{J}(\mathbf{x})\ddot{\mathbf{x}} + \dot{\mathbf{J}}(\mathbf{x}, \dot{\mathbf{x}})\dot{\mathbf{x}}. \tag{9}$$

2.2 Platform Dynamics

Let $\mathbf{S} = \left[\mathrm{S}_x\ \mathrm{S}_y\ \mathrm{S}_z\right]^{\mathrm{T}}$ be a position vector of the platform COM G in $\mathscr{R}_p$. Let $\mathbf{I}_p = \mathbf{I}_o - \mathrm{M}\hat{\mathbf{S}}\hat{\mathbf{S}}$ be the inertia tensor of the platform with respect to the origin of $\mathscr{R}_p$ and expressed in the same frame, where $\mathbf{I}_o$ is the inertia tensor with respect to COM of the platform, M is the total platform mass, and $\hat{\mathbf{S}}$ denotes the 3×3 skew-symmetric matrix associated to $\mathbf{S}$. Let us denote the matrix $\mathbf{I}_p$ as

$$\mathbf{I}_p = \begin{bmatrix} \mathrm{XX}\ \mathrm{XY}\ \mathrm{XZ} \\ \mathrm{XY}\ \mathrm{YY}\ \mathrm{YZ} \\ \mathrm{XZ}\ \mathrm{YZ}\ \mathrm{ZZ} \end{bmatrix}. \tag{10}$$

Let **MS** be a vector representing the first moment of the moving platform which can be expressed as

$$\mathbf{MS} = \mathrm{M}\left[\mathrm{S}_x\ \mathrm{S}_y\ \mathrm{S}_z\right]^T = \left[\mathrm{MX}\ \mathrm{MY}\ \mathrm{MX}\right]^T. \tag{11}$$

The dynamics of the platform can be written in the following form

$$\mathbf{A}(\mathbf{x})\ddot{\mathbf{x}} + \mathbf{C}(\mathbf{x}, \dot{\mathbf{x}})\dot{\mathbf{x}} + \mathbf{Q}(\mathbf{x}) = -\mathbf{J}^{\mathrm{T}}\mathbf{t}, \tag{12}$$

where $\mathbf{A}(\mathbf{x})$ denotes the generalized inertia matrix, $\mathbf{C}(\mathbf{x}, \dot{\mathbf{x}})$ is the Coriolis and centrifugal matrix, $\mathbf{Q}(\mathbf{x})$ is a vector of gravity forces, and $\mathbf{t}$ is a vector of cable tensions. It is worth noting that the dynamics of the platform is linear in the inertial parameters i.e. in $\mathbf{I}_p$, **MS**, and M [22]. Let $\chi_p \in \mathbb{R}^{10}$ be a vector collecting the inertial parameters of the platform as

$$\chi_p = \left[\text{XX XY XZ YY YZ ZZ MX MY MZ M}\right]^T. \tag{13}$$

Consequently, one can easily express the dynamics of the platform in the following regression form

$$\boldsymbol{\Phi}_p(\mathbf{x}, \dot{\mathbf{x}}, \ddot{\mathbf{x}})\chi_p = -\mathbf{J}^T\mathbf{t}. \tag{14}$$

The details of the regressor $\boldsymbol{\Phi}_p(\mathbf{x}, \dot{\mathbf{x}}, \ddot{\mathbf{x}})$ is given in [15].

2.3 Winches Dynamics

The dynamics of the winches can be written as follows [15]

$$\boldsymbol{\Gamma}_a = \mathbf{I}_a\ddot{\mathbf{q}} + \mathbf{F}_v\dot{\mathbf{q}} + \mathbf{F}_c\text{sign}(\dot{\mathbf{q}}) + \mathbf{R}\mathbf{t} \tag{15}$$

where $\mathbf{I}_a \in \mathbb{R}^{m\times m}$, $\mathbf{F}_v \in \mathbb{R}^{m\times m}$, and $\mathbf{F}_c \in \mathbb{R}^{m\times m}$ are diagonal matrices denoting, respectively, the inertia, the viscous friction coefficients, and the dry friction coefficients of the motors, drums and other other rotating parts. Let I_a, F_v, and $\text{F}_c \in \mathbb{R}^m$ be vectors corresponding to the diagonal elements of $\mathbf{I}_a$, $\mathbf{F}_v$, and $\mathbf{F}_c$, respectively. Then, one can gather all dynamic parameters of the winches in one vector $\chi_a \in \mathbb{R}^{3m}$ defined as follows

$$\chi_a = \left[\text{I}_a^T\ \text{F}_v^T\ \text{F}_c^T\right]^T \tag{16}$$

Now, the winch dynamics can be written in a regression form as follows

$$\boldsymbol{\Gamma}_a = \boldsymbol{\Phi}_a(\mathbf{q}, \dot{\mathbf{q}}, \ddot{\mathbf{q}})\chi_a + \mathbf{R}\mathbf{t} \tag{17}$$

where $\boldsymbol{\Phi}_a(\mathbf{q}, \dot{\mathbf{q}}, \ddot{\mathbf{q}})$ is given by

$$\boldsymbol{\Phi}_a(\mathbf{q}, \dot{\mathbf{q}}, \ddot{\mathbf{q}}) = \left[\,\text{diag}(\ddot{\mathbf{q}}) \,|\, \text{diag}(\dot{\mathbf{q}}) \,|\, \text{diag}(\text{sign}(\dot{\mathbf{q}}))\right] \tag{18}$$

As cable tensions must be within admissible non-negative ranges i.e. $0 \leq \mathbf{t}_{min} \leq \mathbf{t} \leq \mathbf{t}_{max}$, the limits of the actuator torques should be within a lower and upper limits defined respectively as follows

$$\boldsymbol{\Gamma}_a^{min} = \boldsymbol{\Phi}_a(\mathbf{q}, \dot{\mathbf{q}}, \ddot{\mathbf{q}})\chi_a + \mathbf{R}\mathbf{t}_{min} \tag{19}$$

and

$$\boldsymbol{\Gamma}_a^{max} = \boldsymbol{\Phi}_a(\mathbf{q}, \dot{\mathbf{q}}, \ddot{\mathbf{q}})\chi_a + \mathbf{R}\mathbf{t}_{max} \tag{20}$$

In this paper, the dynamics of winches will be projected onto the operational space and then cable tensions will not be directly accessible. Instead, the tension distribution algorithm will be applied on the bounds defined by Eqs. (19) and (20).

2.4 CDPM Dynamics in Operational Space

Using the first-order and second-order differential models defined by Eqs. (8) and (9), one can compute the winch regression matrix in terms of either the platform or the winch acceleration, velocity, and position vectors e.g. $\mathbf{\Phi}_a(\mathbf{q}, \dot{\mathbf{q}}, \ddot{\mathbf{q}}) \Leftrightarrow \mathbf{\Phi}_a(\mathbf{x}, \dot{\mathbf{x}}, \ddot{\mathbf{x}})$. Using Eq. (17), the cable tension can be expressed as

$$\mathbf{t} = \mathbf{R}^{-1}(\mathbf{\Gamma}_a - \mathbf{\Phi}_a(\mathbf{x}, \dot{\mathbf{x}}, \ddot{\mathbf{x}})\chi_a) \tag{21}$$

Substituting cable tensions in Eq. (21) into platform dynamics given by Eq. (14) yields

$$\Phi_p(\mathbf{x}, \dot{\mathbf{x}}, \ddot{\mathbf{x}})\chi_p - \mathbf{J}^T\mathbf{R}^{-1}\Phi_a(\mathbf{x}, \dot{\mathbf{x}}, \ddot{\mathbf{x}})\chi_a = -\mathbf{J}^T\mathbf{R}^{-1}\mathbf{\Gamma}_a \tag{22}$$

Equation (22) can be written in a more compact form as

$$\Phi(\mathbf{x}, \dot{\mathbf{x}}, \ddot{\mathbf{x}})\chi = -\mathbf{J}^T\mathbf{R}^{-1}\mathbf{\Gamma}_a \tag{23}$$

where

$$\Phi(\mathbf{x}, \dot{\mathbf{x}}, \ddot{\mathbf{x}}) = \left[\Phi_p(\mathbf{x}, \dot{\mathbf{x}}, \ddot{\mathbf{x}}) \;\; - \mathbf{J}^T\,\mathbf{R}^{-1}\Phi_a(\mathbf{x}, \dot{\mathbf{x}}, \ddot{\mathbf{x}})\right] \text{ and } \chi = \left[\chi_p^T \;\; \chi_a^T\right]^T \tag{24}$$

Solving Eq. (23) for the actuator torques $\mathbf{\Gamma}_a$ taking into account the torque limits defined by Eqs. (19) and (20), gives the following general solution

$$\mathbf{\Gamma}_a = \mathbf{R}\mathbf{W}^{+}\mathbf{\Phi}(\mathbf{x}, \dot{\mathbf{x}}, \ddot{\mathbf{x}})\chi + \mathbf{R}\mathbf{N}\lambda, \tag{25}$$

where $\mathbf{W}^{+} \in \mathbb{R}^{m\times 6}$ and $\mathbf{N} \in \mathbb{R}^{m\times m-6}$ are the Moore-Penrose pseudo-inverse and the null-space basis matrix of $-\mathbf{J}^T$, respectively. $\lambda \in \mathbb{R}^{m-6}$ is an arbitrary vector to be obtained such that cable tensions (and actuator torques) are admissible. For the sake of controller synthesis and analysis, the dynamic model of the CDPM expressed in terms of an inertia matrix, Coriolis, centrifugal, gravity forces, and friction forces is needed. Straightforward manipulations after substituting Eq. (15) into Eq. (12), and considering an external disturbances $\mathbf{F}_d$, allow us to write the dynamics of the CDPM in the following form

$$\mathbf{A}_{eq}(\mathbf{x})\ddot{\mathbf{x}} + \mathbf{C}_{eq}(\mathbf{x}, \dot{\mathbf{x}})\dot{\mathbf{x}} + \mathbf{Q}_{eq}(\mathbf{x}) + \mathbf{F}_{eq}(\mathbf{x}, \dot{\mathbf{x}}, \ddot{\mathbf{x}}) = \mathbf{F}_{ext} + \mathbf{F}_d \tag{26}$$

where $\mathbf{A}_{eq}(\mathbf{x}) = \mathbf{A}(\mathbf{x}) - \mathbf{J}^T\mathbf{R}^{-1}\mathbf{I}_a\mathbf{R}^{-1}\mathbf{J}$, $\mathbf{C}_{eq}(\mathbf{x}, \dot{\mathbf{x}}) = \mathbf{C}(\mathbf{x}, \dot{\mathbf{x}}) - \mathbf{J}^T\mathbf{R}^{-1}\mathbf{I}_a\mathbf{R}^{-1}\dot{\mathbf{J}}$, $\mathbf{Q}_{eq}(\mathbf{x}) = \mathbf{Q}(\mathbf{x})$, $\mathbf{F}_{eq}(\mathbf{x}, \dot{\mathbf{x}}, \ddot{\mathbf{x}}) = -\mathbf{J}^T\mathbf{R}^{-1}\mathbf{F}_v\mathbf{R}^{-1}\mathbf{J}\dot{\mathbf{x}} - \mathbf{J}^T\mathbf{R}^{-1}\mathbf{F}_c\text{sign}(\mathbf{R}^{-1}\mathbf{J}\dot{\mathbf{x}})$, and $\mathbf{F}_{ext} = -\mathbf{J}^T\mathbf{R}^{-1}\boldsymbol{\Gamma}_a$.

2.5 Properties of the Dynamic Model

Some properties of the dynamic model (26) are recalled [12]. These properties are necessary for the controller synthesis and stability analysis.

Property 1 The inertia matrix $\mathbf{A}_{eq}(\mathbf{x})$ is a positive-definite symmetric matrix and bounded away from singularity whatever the uncertainties in inertial parameters. The boundedness of $\mathbf{A}_{eq}(\mathbf{x})$ implies

$$\lambda_m\mathbf{I}_6 \leq \mathbf{A}_{eq}(\mathbf{x}) \leq \lambda_M\mathbf{I}_6 \tag{27}$$

or

$$\lambda_m \leq \|\mathbf{A}_{eq}(\mathbf{x})\| \leq \lambda_M \tag{28}$$

where $\mathbf{I}_6$ is the 6×6 identity matrix and λ_m and λ_M are positive scalars.

Property 2 The external disturbance force vector $\mathbf{F}_d$ is assumed to be bounded i.e.

$$\|\mathbf{F}_d\| \leq \zeta \tag{29}$$

where ζ is a positive constant.

3 Adaptive TSM Control of CDPM

In this section, an adaptive TSM control is synthesized for the CDPM dynamic model given in (26). The main idea of TSM is first presented assuming that the dynamic parameters of the CDPM are known. Then, the idea is extended to the adaptive case where the dynamic parameters are assumed to be unknown and must be estimated.

3.1 TSM Control

For CDPM dynamics given by (26), let $\mathbf{x}_d$, $\dot{\mathbf{x}}_d$, and $\ddot{\mathbf{x}}_d$ be a desired trajectory and define $\mathbf{e} = \mathbf{x} - \mathbf{x}_d$ as the tracking error. The control objective is to design a feedback control law $\mathbf{F}_{ext}$ such that the platform posture $\mathbf{x}$ tracks the desired trajectory $\mathbf{x}_d$ in finite time.

In order to have a terminal convergence of tracking errors, a sliding surface is defined as follows [18]

$$\mathbf{s} = \dot{\mathbf{e}} + \beta \mathbf{e}^{p/q} \tag{30}$$

where $\beta > 0$ is a design constant parameter, p and q are positive odd integers satisfying $p > q$. In order to ensure that the terminal sliding mode exists on the switching surface and equilibrium is reached in finite time, the following η-reachability condition must be satisfied [28]

$$\frac{1}{2}\frac{\mathrm{d}}{\mathrm{dt}}\mathbf{s}^{\mathrm{T}}\mathbf{s} < -\eta \left|\mathbf{s}\right| \tag{31}$$

where $\eta > 0$ is a constant. If the dynamic parameters χ of the CDPM are known and the bounds on disturbance ζ are also known, then, one can choose the following control law to satisfy the η-reachability condition [19].

$$\mathbf{F}_{ext} = \mathbf{\Phi}(\mathbf{x}, \dot{\mathbf{x}}, \mathbf{v})\chi \tag{32}$$

where $\mathbf{v}$ is an auxiliary control input defined by

$$\mathbf{v} = \ddot{\mathbf{x}}_d + \beta \frac{q}{p} \mathbf{e}^{p/q-1}\dot{\mathbf{e}} + (\rho + \eta)\mathrm{sign}(\mathbf{s}) \tag{33}$$

where $\rho = \lambda_m \zeta$. If $\mathbf{s}(0) \neq \mathbf{0}$, then the sliding mode $\mathbf{s} = \mathbf{0}$ will be reached in finite time t_r which satisfies $t_r \leq \left|\mathbf{s}\right| / \eta$. When the sliding mode is reached, the system dynamics is described by the following nonlinear differential equation

$$\dot{\mathbf{e}} + \beta \mathbf{e}^{p/q} = 0 \tag{34}$$

Equation (34) has an equilibrium at $\mathrm{e} = 0$ which is a globally finite-time stable attractor. And the convergence time for any initial condition $\mathrm{x} = \mathrm{x}(t_r)$ is finite and given by

$$t_s \leq \frac{p}{\beta(p-q)} \left|\mathbf{x}(t_r)\right|^{1-q/p} \tag{35}$$

However, the choice of such terminal sliding surface may cause singularity at convergence due to fractional power p/q. In addition, the control laws (32) and (33) is discontinuous and may cause chattering [19].

3.2 Non-Singular TSM Control

In order to avoid the singularity problem of (30), another choice of terminal sliding mode is defined as follows [20]

$$\mathbf{s} = \mathbf{e} + \beta \left|\dot{\mathbf{e}}\right|^{\gamma} \operatorname{sign}(\dot{\mathbf{e}}) \tag{36}$$

where $\beta > 0$ and $1 < \gamma < 2$.

Remark 1 The terminal sliding mode defined by (36) has a global finite-time equilibrium, i.e., for any given initial condition $\mathbf{e}(0) \neq 0$, the variable $\mathbf{e}$ reaches zero in a finite time t_r given by [20]

$$t_r \leq \frac{\gamma \beta^{1/\gamma}}{(\gamma - 1)} \left|\mathbf{e}(0)\right|^{(\gamma-1)/\gamma} \tag{37}$$

Remark 2 The terminal sliding mode defined by (36) is continuous and time differentiable. Its first time derivative is written as [20]

$$\dot{\mathbf{s}} = \dot{\mathbf{e}} + \beta\gamma \left|\dot{\mathbf{e}}\right|^{\gamma-1} \ddot{\mathbf{e}} \tag{38}$$

Remark 3 Finite time stability is guaranteed for a given Lyapunov function candidate $\mathrm{V}(\mathbf{s})$ if it satisfies [20]

$$\dot{\mathrm{V}}(\mathbf{s}) + a\mathrm{V}(\mathbf{s}) + b\mathrm{V}^{c}(\mathbf{s}) \leq 0; \quad a > 0,\ b > 0, \ \text{ and } 0 < c < 1 \tag{39}$$

Moreover, the finite reaching time is given by

$$t_r \leq \frac{1}{a(1-c)} \ln \frac{a\mathrm{V}^{c}(\mathbf{s}(0)) + b}{b} \tag{40}$$

Now, if we assume that the dynamic parameters of CDPM, i.e., χ are known and the disturbance $\mathbf{F}_d = 0$, then to design a non-singular TSM controller, let us choose the following Lyapunov function candidate

$$\mathrm{V} = \frac{1}{2}\mathbf{s}^{\mathrm{T}}\mathbf{s} \tag{41}$$

The time derivative of the non-singular TSM (38) can be rewritten as follows

$$\dot{\mathbf{s}} = \beta\gamma \left|\dot{\mathbf{e}}\right|^{\gamma-1} (\beta^{-1}\gamma^{-1} \left|\dot{\mathbf{e}}\right|^{2-\gamma} \operatorname{sign}(\dot{\mathbf{e}}) + \ddot{\mathbf{e}}) \tag{42}$$

Let us choose a control law which is defined by

$$\mathbf{F}_{ext} = \mathbf{\Phi}(\mathbf{x}, \dot{\mathbf{x}}, \mathbf{v})\chi - \mathbf{K}_1\mathbf{s} - \mathbf{K}_2 \left|\mathbf{s}\right|^{\rho} \operatorname{sign}(\mathbf{s}) \tag{43}$$

where $\mathbf{K}_1 = \operatorname{diag}(k_{11},\ k_{12},\ \dots\ k_{16})$, $\mathbf{K}_2 = \operatorname{diag}(k_{21},\ k_{22},\ \dots\ k_{26})$, with $k_{ij} > 0$, $i \in \{1, 2\}$, $j \in \{1, 2, \dots, 6\}$, and $\rho < 1$, and

$$\mathbf{v} = \ddot{\mathbf{x}}_d - \frac{1}{\beta\gamma}\left|\dot{\mathbf{e}}\right|^{2-\gamma}\operatorname{sign}(\dot{\mathbf{e}}) \tag{44}$$

The control laws (43) and (44) reduces the time-derivative of V to

$$\dot{\mathrm{V}} = -\mathbf{s}^T\mathbf{H}_1\mathbf{s} - \mathbf{s}^T\mathbf{H}_2\left|\mathbf{s}\right|^{\rho}\operatorname{sign}(\mathbf{s}) \tag{45}$$

where

$$\mathbf{H}_1 = \operatorname{diag}(h_{11}, h_{12}, \ldots h_{16}) = \beta\gamma\lambda_m\left|\dot{\mathbf{e}}\right|^{\gamma-1}\mathbf{K}_1 \tag{46}$$

and

$$\mathbf{H}_2 = \operatorname{diag}(h_{21}, h_{22}, \ldots h_{26}) = \beta\gamma\lambda_m\left|\dot{\mathbf{e}}\right|^{\gamma-1}\mathbf{K}_2 \tag{47}$$

are positive diagonal matrices anywhere except at sliding mode. After straightforward manipulations of (45), one can get

$$\dot{\mathrm{V}} \le -2h_1\mathrm{V} - 2^{(\rho+1)/2} - h_2\mathrm{V}^{(\rho+1)/2} \tag{48}$$

According to the finite-time stability (39), it is clear that (48) ensures that the tracking error $\mathbf{e}$ converges to zero along TSM in finite time. However, in practice, the vector of dynamic parameters χ is not exactly known since it changes with payloads. Also, external disturbances exist and must be taken into account in controller design. Therefore, an adaptive control scheme is recommended to cope with load changes and at the same time to guarantee robustness of performances. The main contribution of this paper is to extend the non-singular TSM control scheme explained above such that the dynamic parameters of CDPM are updated online via an adaptation mechanism to ensure convergence.

3.3 Adaptive TSM Control Scheme

In this section, a non-singular adaptive TSM is developed for CDPM based on the classical formulations presented in Sects. 3.1 and 3.2. The dynamic parameters are assumed to be unknown and have to be estimated. The main results of the proposed non-singular adaptive TSM control scheme are summarised through the following theorem.

Theorem 1 *Consider a CDPM described by* (23) *and* (26) *and satisfying* (27), (28), *and* (29) *and assume that the vector of dynamic parameters* χ *is unknown. If the non-singular TSM* (36) *is chosen and the following control scheme is applied:*

$$\mathbf{F}_{ext} = \Phi(\mathbf{x}, \dot{\mathbf{x}}, \mathbf{v})\hat{\chi} + u_s \tag{49}$$

where $\hat{\chi}$ is an estimate of χ, and

$$\mathrm{v} = \ddot{\mathbf{x}}_d - \frac{1}{\beta\gamma}\left|\dot{\mathbf{e}}\right|^{2-\gamma}\operatorname{sign}(\dot{\mathbf{e}}), \tag{50}$$

$$u_s = -\mathbf{K}_1\mathbf{s} - \mathbf{K}_2\left|\mathbf{s}\right|^{\rho}\operatorname{sign}(\mathbf{s}), \tag{51}$$

and

$$\dot{\hat{\chi}} = \beta\gamma\Lambda\Phi^T(\mathbf{x}, \dot{\mathbf{x}}, \mathbf{v})\left|\dot{\mathrm{e}}\right|^{1-\gamma}\mathbf{s} \tag{52}$$

where Λ is a positive-definite diagonal matrix with appropriate dimensions, then the closed-loop system reaches a neighborhood of TSM in finite time and the tracking error reaches the neighborhood of the origin also in finite time.

Proof Let $\tilde{\chi} = \chi - \hat{\chi}$ be dynamic parameters estimation error. Let us choose a Lyapunov function candidate as follows

$$\mathrm{V} = \frac{1}{2}\mathbf{s}^{\mathrm{T}}\mathbf{s} + \frac{1}{2}\tilde{\chi}^T\boldsymbol{\Lambda}^{-1}\tilde{\chi} \tag{53}$$

The time derivative of V is given by

$$\dot{\mathrm{V}} = \mathbf{s}^{\mathrm{T}}\dot{\mathbf{s}} - \tilde{\chi}^T\boldsymbol{\Lambda}^{-1}\dot{\hat{\chi}} \tag{54}$$

The time derivative of the terminal sliding **s** can be rewritten in the following form

$$\dot{\mathbf{s}} = \beta\gamma\left|\dot{\mathbf{e}}\right|^{\gamma-1}\mathbf{A}_{eq}^{-1}(\mathbf{A}_{eq}(\beta^{-1}\gamma^{-1}\left|\dot{\mathbf{e}}\right|^{2-\gamma}\operatorname{sign}(\dot{\mathbf{e}})) + \mathbf{A}_{eq}\ddot{\mathbf{e}}) \tag{55}$$

where $\mathbf{A}_{eq}$, $\mathbf{C}_{eq}$, $\mathbf{Q}_{eq}$, and $\mathbf{F}_{eq}$ are the inertia matrix, Coriolis and centrifugal matrix, gravity forces, and friction forces, respectively, where the arguments are suppressed for simplicity. The second-order time derivative of tracking error $\ddot{\mathbf{e}}$ can be written as

$$\mathbf{A}_{eq}\ddot{\mathbf{e}} = -\mathbf{A}_{eq}\ddot{\mathbf{x}}_d - \mathbf{C}_{eq}\ddot{\mathbf{x}} - \mathbf{Q}_{eq} - \mathbf{F}_{eq} + \mathbf{F}_{ext} + \mathbf{F}_d \tag{56}$$

Substituting (56) into (55) gives

$$\dot{\mathbf{s}} = \beta\gamma\left|\dot{\mathbf{e}}\right|^{\gamma-1}\mathbf{A}_{eq}^{-1}(\mathbf{A}_{eq}(\beta^{-1}\gamma^{-1}\left|\dot{\mathbf{e}}\right|^{2-\gamma}\operatorname{sign}(\dot{\mathbf{e}}) - \ddot{\mathbf{x}}_d) - \mathbf{C}_{eq}\ddot{\mathbf{x}} - \mathbf{Q}_{eq} - \mathbf{F}_{eq} + \mathbf{F}_{ext} + \mathbf{F}_d) \tag{57}$$

By definition of **v** in (50), Eq. (57) can rewritten as

$$\dot{\mathbf{s}} = \beta\gamma\left|\dot{\mathbf{e}}\right|^{\gamma-1}\mathbf{A}_{eq}^{-1}(-\boldsymbol{\Phi}(\mathbf{x}, \dot{\mathbf{x}}, \mathbf{v})\chi + \mathbf{F}_{ext} + \mathbf{F}_d) \tag{58}$$

Substituting the control law (51) and (49), then (58) can be simplified to

Fig. 2 CoGiRo CDPM prototype developed by CNRS-LIRMM and Tecnalia

$$\dot{\mathbf{s}} = \beta\gamma \left|\dot{\mathbf{e}}\right|^{\gamma-1} \mathbf{A}_{eq}^{-1}(\mathbf{\Phi}(\mathbf{x}, \dot{\mathbf{x}}, \mathbf{v})\tilde{\chi} - \mathbf{K}_1\mathbf{s} - \mathbf{K}_2 \left|\mathbf{s}\right|^{\rho} \operatorname{sign}(\mathbf{s}) + \mathbf{F}_d) \tag{59}$$

Substituting (59) into $\dot{\mathrm{V}}$ in (60) and substituting with the adaptive law (52), one can easily get

$$\dot{\mathrm{V}} \leq \beta\gamma\mathbf{s}^{\mathrm{T}} \left|\dot{\mathbf{e}}\right|^{\gamma-1} \mathbf{A}_{eq}^{-1}(-\mathbf{K}_1\mathbf{s} - \mathbf{K}_2 \left|\mathbf{s}\right|^{\rho} \operatorname{sign}(\mathbf{s}) + \mathbf{F}_d) \tag{60}$$

Using the properties (28) and (29), then one can simplify (60) as

$$\dot{\mathrm{V}} \leq \beta\gamma\mathbf{s}^{\mathrm{T}} \left|\dot{\mathbf{e}}\right|^{\gamma-1} \lambda_m(-\mathbf{K}_1\mathbf{s} - \mathbf{K}_2 \left|\mathbf{s}\right|^{\rho} \operatorname{sign}(\mathbf{s}) + \zeta) \tag{61}$$

Following the same line of proof as in [20], we can see that the system can reach neighborhood sliding-mode in finite-time which is defined by

$$\|\mathbf{s}\| \leq \delta = \min(\delta_1, \delta_2) \tag{62}$$

where $\delta_1 = \lambda_m\zeta/h_1$, $\delta_2 = (\lambda_m\zeta/h_2)^{(1/\rho)}$. In addition, once the sliding variable is within the region δ, the tracking error can reach the neighborhood of the origin in finite time as well. This completes the proof.

4 Simulation Study: Control of CoGiRo

4.1 CoGiRo Prototype

The CoGiRo prototype shown in Fig. 2 is a redundantly-actuated cable-suspended CDPM developed by CNRS-LIRMM and Tecnalia. CoGiRo occupies a space char-

Table 1 Cable attachment points

	Base points					Platform points	
	x_b	y_b	z_b		x_p	y_p	z_p
a_1	−7.5	−5.5	6	b_1	0.5	−0.5	0
a_2	−7.5	−5.5	6	b_2	−0.5	0.5	1
a_3	−7.5	5.5	6	b_3	−0.5	−0.5	0
a_4	−7.5	5.5	6	b_4	0.5	0.5	1
a_5	7.5	5.5	6	b_5	−0.5	0.5	0
a_6	7.5	5.5	6	b_6	0.5	−0.5	1
a_7	7.5	−5.5	6	b_7	0.5	0.5	0
a_8	7.5	−5.5	6	b_8	−0.5	−0.5	1

units are in m

Table 2 Platform inertial parameters

Parameter	XX	YY	ZZ	XY	XZ	YZ	MX	MY	MZ	M
χ_{PNL}	62.00	60.00	22.00	−0.32	−0.02	4.50	0.00	−7.50	0.00	79.00
χ_{PL}	132.00	135.00	45.00	0.00	−3.00	−2.00	−28.00	−7.50	0.00	200.00

[a] χ_{PNL} and χ_{PL} denote the platform inertial parameters without and with load, respectively. Units are in kg. m^2, kg. m, and kg

Table 3 Winch dynamic parameters

Parameter	I_a	F_v	F_c	r
Value	0.0311	0.0200	3.0000	0.0225

units are in kg. m^2, N. m. rad^{-1}.sec, N. m, m

acterised by a 15 m in length, 11 m in width, and 6 m in height giving a potential workspace of 677 m^3. The mobile platform is a cube with 1 m side length with a total mass of 79 kg. CoGiRo has 8-actuators with a 6-DOF moving platform i.e. it has an actuation redundancy of 2. CoGiRo is capable of manipulating payloads of 300 kg over the entire workspace and up to 500 kg if the task to be performed is not too close to the boundaries of the workspace. The lower limit of cable tension is $\mathbf{t}_{min} = 0$ N and the upper limit is $\mathbf{t}_{max} = 5000$ N. The base and platform cable attachment points of CoGiRo are given in Table 1 in units of m. The inertial parameters χ_{PNL} of the moving platform from the CAD model of CoGiRO are given in Table 2. The inertial parameters χ_{PL} of the platform with a typical payload previously used in an experimental study are used to assess the robustness of the proposed control scheme in this simulation study. The dynamic parameters of CoGiRo winches are shown in Table 3.

Table 4 Interpolated trajectory point sequence

Point	x	y	z	ϕ	θ	ψ
$\mathbf{x}_0$	0.000	0.000	1.296	0.000	0.000	0.000
$\mathbf{x}_1$	0.000	0.000	2.296	0.000	0.000	0.000
$\mathbf{x}_2$	−3.850	1.200	2.296	0.000	0.000	−45.840
$\mathbf{x}_3$	−3.850	1.200	1.307	0.000	0.000	−45.840
$\mathbf{x}_4$	−3.000	2.000	1.307	0.000	0.000	−45.840
$\mathbf{x}_5$	−3.000	2.000	1.796	0.000	0.000	−45.840
$\mathbf{x}_6$	4.000	−1.000	1.796	0.000	0.000	11.460
$\mathbf{x}_7$	4.000	−1.000	1.307	0.000	0.000	11.460
$\mathbf{x}_8$	4.300	−2.000	1.307	0.000	0.000	11.460
$\mathbf{x}_9$	4.300	−2.000	2.296	0.000	0.000	11.460
$\mathbf{x}_{10}$	0.000	0.000	2.296	0.000	0.000	0.000
$\mathbf{x}_{11}$	0.000	0.000	1.296	0.000	0.000	0.000

Table 5 Maximum velocity and acceleration along the trajectory

DOF	x	y	z	ϕ	θ	ψ
Maximum velocity $\mathbf{K}_v$ [m, rad/sec]	1.0000	1.0000	1.0000	1.5708	1.5708	1.5708
Maximum acceleration $\mathbf{K}_a$ [m, rad/sec^2]	0.1000	0.1000	0.1000	0.1571	0.1571	0.1571

Table 6 Adaptive TSM controller parameters

Parameter	Value
β	1.00
γ	1.50
ρ	0.33
$\mathbf{K}_1$	diag{70.00, 70.00, 105.00, 105.00, 105.00, 70.00}
$\mathbf{K}_2$	diag{25.00, 25.00, 37.50, 25.00, 25.00, 30.00}
$\Lambda_{\mathbf{I}_p}$	diag{1000.00, 1000.00, 1000.00, 1000.00, 1000.00, 1000.00}
$\Lambda_{\mathbf{MS}}$	diag{200.00, 200.00, 200.00}
$\Lambda_{\mathbf{I}_a}$	diag{0.10, 0.10, 0.10, 0.10, 0.10, 0.10, 0.10, 0.10}
$\Lambda_{\mathbf{F}_v}$	diag{0.01, 0.01, 0.01, 0.01, 0.01, 0.01, 0.01, 0.01}
$\Lambda_{\mathbf{F}_c}$	diag{0.10, 0.10, 0.10, 0.10, 0.10, 0.10, 0.10, 0.10}

4.2 Simulation Results

CoGiRo is supposed to perform pick-and-place tasks. In this simulation study, a trajectory of 11 mobile platform poses corresponding to a pick-and-place task is given in Table 4. It has been generated by a 5th-order polynomial interpolation subjected to the velocity and acceleration constraints given in Table 5. In order to

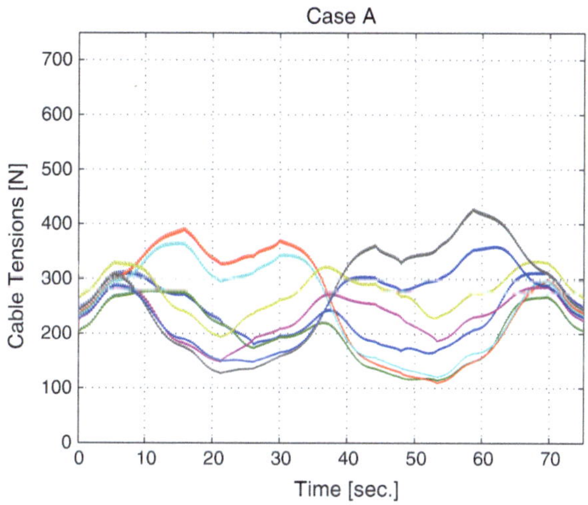

Fig. 3 Cable tensions: Case (A)

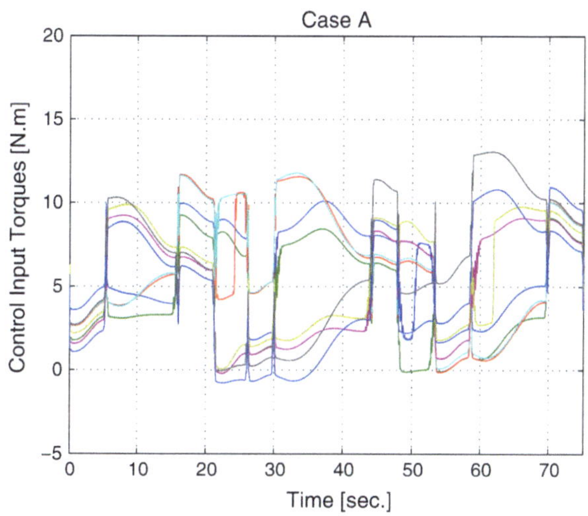

Fig. 4 Control input torques: Case (A)

assess the performance of the proposed adaptive TSM control scheme, we consider two simulation cases. In *Case A*, the adaptive TSM controller is simulated considering only 20 % of parametric uncertainty where in *Case B*, we consider loading the platform at configuration $\mathbf{x}_3$ and releasing the load at configuration $\mathbf{x}_6$ and, at the same time, the adaptive TSM control algorithm is initialised with 20 % of parametric uncertainty for the no-load platform inertial parameters χ_{PNL}. The parameters shown in Table 6 have been used in simulation of both cases where Λ_i is the adaptation gains corresponding to the parameters of i. Figures 3, 4, 5, and 6 show the simulation results

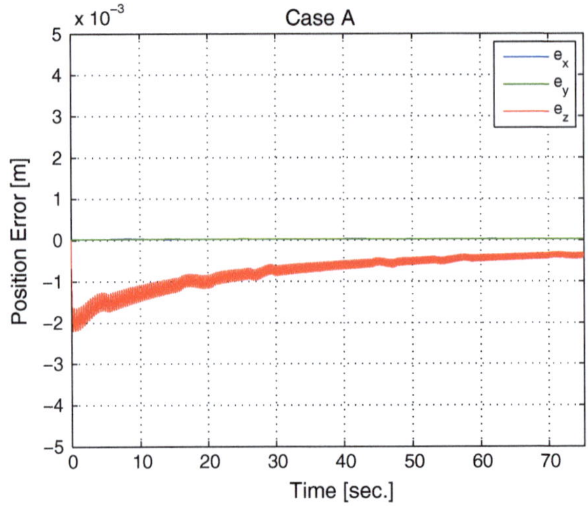

Fig. 5 Position error: Case (A)

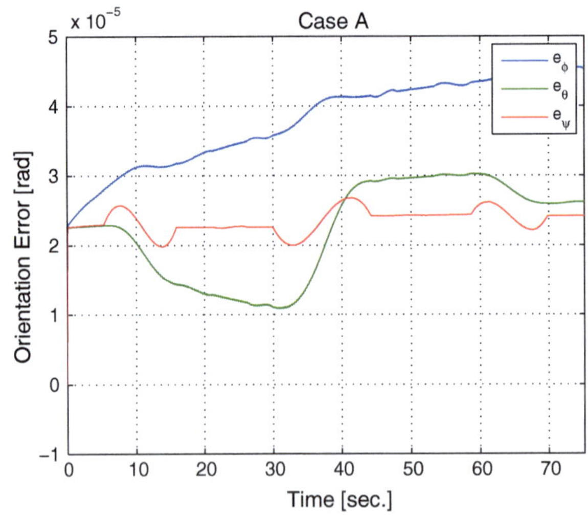

Fig. 6 Orientation error: Case (A)

of *Case A*. It can be seen that the proposed adaptive TSM control scheme is able to achieve good tracking performances even with parametric uncertainty. Figures 7, 8, 9, and 10 show the simulation results of *Case B*. It can be seen that at the loading instant, both cable tensions and control input torques increase instantaneously. However, the adaptive TSM is able to restore the tracking error in a short period of time and achieves a good tracking performance over the whole trajectory.

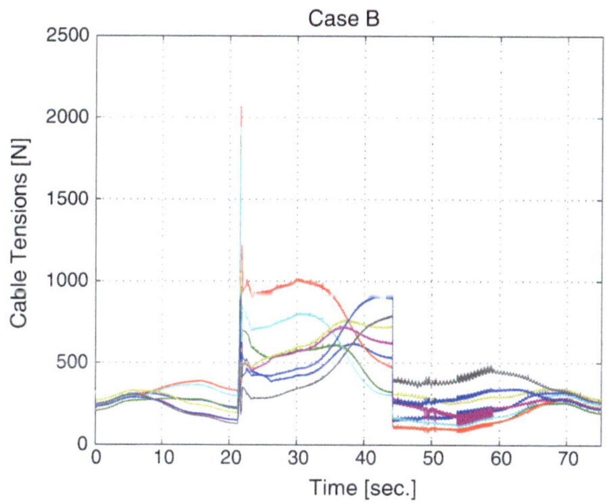

Fig. 7 Cable tensions: Case (B)

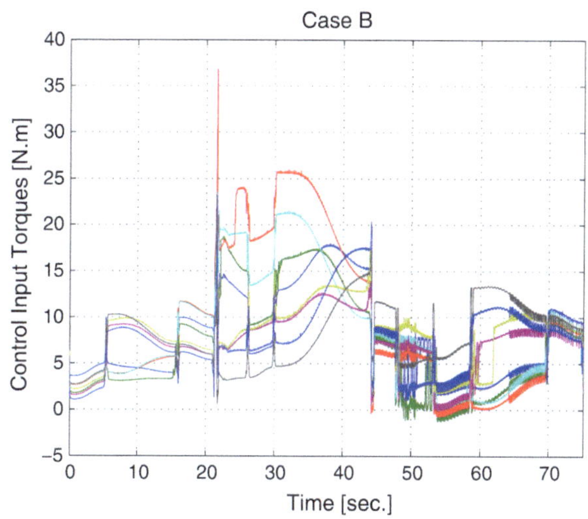

Fig. 8 Control input torques: Case (B)

5 Conclusion

The main contribution of this paper is to extend an early developed non-singular TSM control by synthesizing an adaptive law for CDPM dynamic parameters. The main motivation of using TSM for CDPMs is its finite-time stability which generally results in high tracking performances. Stability of conventional TSM for robotic manipulators are based on the assumption that the dynamics are partially known while the uncertain part is bounded. This assumption is true, however, it may results

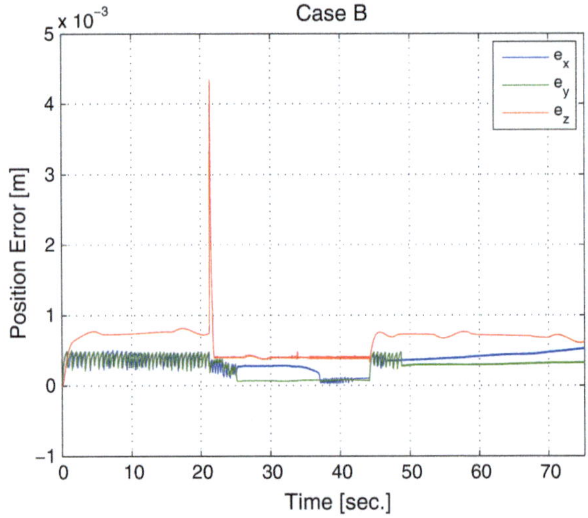

Fig. 9 Position error: Case (B)

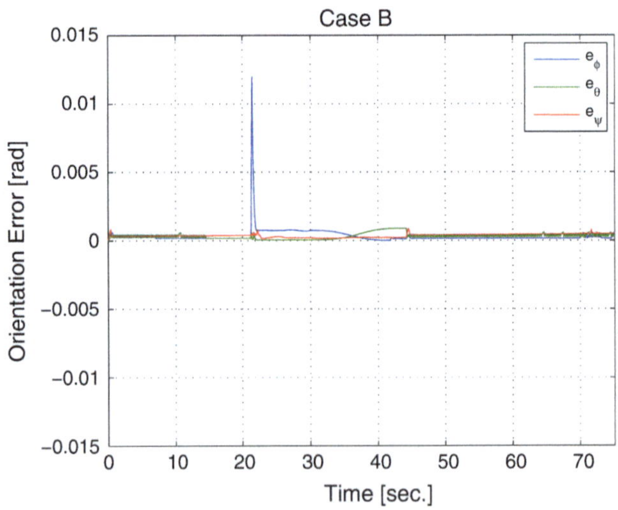

Fig. 10 Orientation error: Case (B)

in large control input torques especially when heavy payloads are to be manipulated. To tackle this problem, an adaptive mechanism is introduced to estimate online the dynamic parameters. The overall dynamics of CDPM has been reformulated in this paper in a regression form to facilitate the design of the adaptive control. Redundancy resolution is then solved based on the limits on control input torques corresponding to the tension limits. The adaptive TSM control scheme has been applied in simulation to a large cable-suspended CDPM-CoGiRo. CoGiRo is supposed to perform pick-

and-place of heavy payloads. The simulations show the effectiveness of the proposed scheme for large values of parametric uncertainties and loading conditions. However, the real benefit of the proposed adaptive TSM control will appear through experimental studies on CoGiRo. Our future work will be dedicated to these experimental validations.

Acknowledgments The financial support of the Région Languedoc-Roussillon (grant 140218) is greatly acknowledged.

References

1. Kawamura S, Kino H, Won C (2000) High-speed manipulation by using parallel wire-driven robots. Robotica 18:13–21
2. Shiang W-J, Cannon D, Gorman J (1999) Dynamic analysis of cable array robotic crane. In: Proceedings of the IEEE international conference on robotics and automation, ICRA1999, Detroit, Michigan, USA, pp 2495–2500
3. Lamaury J (2013) Contribution à la commande des robots parallèles à câble à redondance d'actionnement. Ph. D. thesis, Université Montpellier II
4. Lamaury J, Gouttefarde M, Tempier O (2012) Design and control strategies of a redundant suspended cable-driven parallel robot. latest advances in robot kinematics, ARK 2012. Springer Innsbruck, pp 1–9
5. Lamaury J, Gouttefarde M (2012) A tension distribution method with improved computational efficiency. In: Proceedings of the first international conference on cable-driven parallel robots, Stuttgart, Germany, pp 71–85
6. Fang S, Franitza D, Torlo M, Bekes F, Hiller M (2004) Motion control of a tendon-based parallel manipulator using optimal tension distribution. IEEE/ASME Trans Mechatro 9:561–568
7. Bruckmann T, Pott A, Hiller M, Franitza D (2006) A Modular controller for redundantly actuated tendon-based stewart platforms. In: The first 1st European conference on mechanism science, EuCoMes 2006, Obergurgl, Austria
8. Oh S-R, Agrawal SK (2005) A reference governor-based controller for a cable robot under input constraints. IEEE Trans Control Syst Technol 13:639–645
9. Pott A, Bruckmann T, Mikelsons L (2009) Closed-form force distribution for parallel wire robots. In: Proceedings of the 5th international workshop on computational kinematics, pp 25–34
10. Mikelsons L, Bruckmann T, Hiller M, Schramm D (2008) A real-time capable force calculation algorithm for redundant tendon-based parallel manipulators. In: IEEE international conference on robotics and automation, ICRA 2008, pp 3869–3874
11. Gholami P, Aref M, Taghirad HD (2008) On The control of the KNTU CDRPM: a cable driven redundant parallel manipulator. In: IEEE international conference on intelligent robots and systems, IROS 2008, pp 2404–2409
12. Khosravi MA, Taghirad HD (2014) Robust PID control of fully-constrained cable driven parallel robots. Mechatronics 24(2):87–97
13. Kino H, Yahiro T, Fumiaki T, Morizono T (2007) Robust PD control using adaptative compensation for completely restrained parallel-wire driven robots. IEEE Trans Rob 23(4):803–812
14. Williams RL II, Gallina P, Vadia J (2003) Planar translational cable-direct-driven robots. J Rob Syst 20(3):107–120
15. Lamaury J, Gouttefarde M, Chemori A, Herve PE (2013) Dual-space adaptive control of redundantly actuated cable-driven parallel robots. International Conference on Intelligent Robots and Systems (IROS), 2013 IEEE/RSJ, pp. 4879–4886 (2013).

16. Wijesoma SW, Richard RJ (1990) Robust trajectory following of robots using computed torque structure with VSS. Int J Control 52(4):935–962
17. Oh S-R, Agrawal SK (2004) Nonlinear sliding mode control and feasible workspace analysis for a cable suspended robot with input constraints and disturbances. In: Proceedings of the American control conference, Boston, MA, USA, pp 4631–4636
18. Man Z, Paplinski AP, Wu HR (1994) A robust MIMO terminal sliding mode control scheme for rigid robotic manipulators. IEEE Trans Autom Control 39(12):2464–2469
19. Feng Y, Yu X, Man Z (2002) Non-singular terminal sliding mode control of rigid manipulators. Automatica 38(12):2159–2167
20. Yu S, Yu X, Shirinzadeh B, Man Z (2005) Continuous finite-time control for robotic manipulators with terminal sliding mode. Automatica 41(11):1957–1964
21. Feng Y, Yu X, Han F (2013) On nonsingular terminal sliding-mode control of nonlinear systems. Automatica 49(6):1715–1722
22. Khalil W, Dombre E (2004) Modeling, identification and control of robots. Elsevier Science, Philadelphia
23. Mondal S, Mahanta C (2014) Adaptive second order terminal sliding mode controller for robotic manipulators. J Franklin Inst 351(4):2356–2377
24. Bruckmann T, Mikelsons L, Brandt T, Hiller M, Schramm D (2008) Wire robots part I: Kinematics, Analysis & Design. In Parallel Manipulators - New Developments, pp. 109–132 (2008).
25. Bruckmann T, Mikelsons L, Brandt T, Hiller M, Schramm D (2008) Wire robots part II: dynamics, control and application. in parallel manipulators—new developments, pp 133–152.
26. Diao X, Ma Ou (2009) Vibration analysis of cable-driven parallel manipulators. Multibody Syst Dyn 21:347–360
27. Craig J (1988) Adaptive control of mechanical manipulators. Addison-Wesley, Reading
28. Slotine J-JE, Li W (1991) Applied nonlinear control. Prentice Hall, Englewood Cliffs

Haptic Interaction with a Cable-Driven Parallel Robot Using Admittance Control

Wei Yang Ho, Werner Kraus, Alexander Mangold and Andreas Pott

Abstract Haptic interfaces are a possible solution for the intuitive operation of machines. In this paper, we present the implementation of an admittance control for a cable-driven parallel robot. The user can grasp the platform and push/pull it through the workspace. The dynamic behaviour (of the virtual system) can be parameterized regarding mass, stiffness and damping. With a system identification of the real robot we determine the maximum dynamic performance. It shows, that the robot can simulate (virtual) systems with a bandwidth up to 13.3 Hz.

Keywords Cable robot · Parallel robot · Haptic display · User interface

1 Introduction

In cable-driven parallel robots lightweight cables are used to move a platform within a huge workspace. The low moved mass enables for high dynamic with small drives. This makes cable robots also interesting as large workspace haptic device [1].

Already in 1993, Kawamura proposed a cable-driven system with impedance control as master device for teleoperation [2]. Under the name SPIDAR several string-based force displays were developed as haptic devices for virtual reality applications [3]. In [4] a cable-driven haptic interface with an impact generating winch for the

An erratum to this chapter is available at DOI 10.1007/978-3-319-09489-2_23

W. Yang Ho (✉) · W. Kraus · A. Mangold · A. Pott
Fraunhofer Institute for Manufacturing Engineering and Automation IPA,
Stuttgart, Germany
e-mail: WeiYang.Ho@ipa.fraunhofer.de

W. Kraus
e-mail: wek@ipa.fhg.de

A. Mangold
e-mail: mangold.alexander@googlemail.com

A. Pott
e-mail: asp@ipa.fhg.de

A. Pott and T. Bruckmann (eds.), *Cable-Driven Parallel Robots*,
Mechanisms and Machine Science 32, DOI 10.1007/978-3-319-09489-2_14

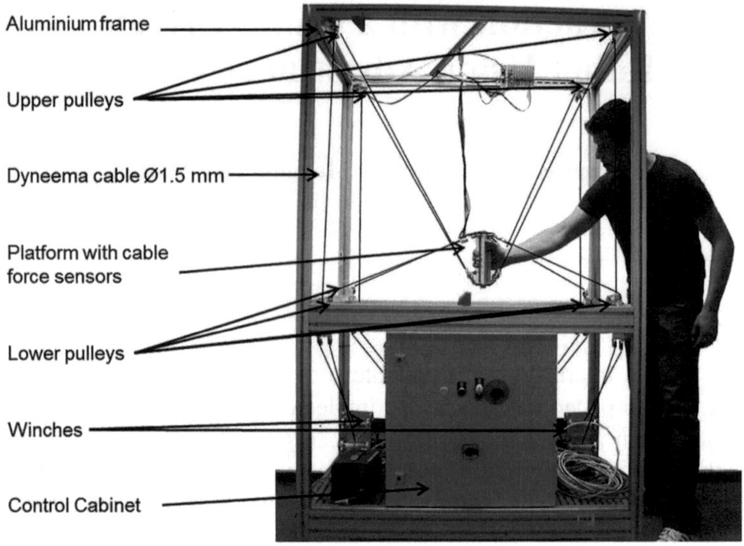

Fig. 1 Cable-driven parallel robot Mini-IPAnema 3

simulation of rigid impacts is proposed. A planar haptic device with four cables with a focus on the manipulability is investigated in [5, 6]. The concept of a spatial cable-driven haptic interface is proposed in [7]. Exhaustive work in the simulation of a virtual reality with a cable-driven locomotion interface including a hybrid control scheme is presented in [8]. The use of cable-driven system for gait rehabilitation was investigated with the STRINGMAN system in [9]. In [10] an impedance control of a serial multi body cable-driven mechanism is proposed.

The control of a haptic system can be realized either by impedance or admittance control. Both of them allow to control the movement of an actuator according to the interaction between the user and the actuator. On the basis of a spring mass damper system the applied force corresponds with an acceleration, a velocity and a position. In case of the impedance control, the input is the actual position and the output is an corresponding force. This force is then the input for a force controlled actuator. The admittance control on the other hand uses the actual applied force as an input and outputs a position setpoint. We implement an admittance control as it builds up on the established position control loop of the servo drive (Fig. 1).

In this paper, we present the development of control algorithms for an interactive haptic feedback. An admittance control to achieve a given system dynamic, represented by a spring mass damper system, is implemented. To guide the user within the workspace, a switchable stiffness is introduced. The maximum feasible dynamic of the simulated system is evaluated according to the transfer function of the robot. The haptic interface can be used to teach positions or as haptic device for the simulation of virtual systems. Other applications are force assistant in lifting or mounting processes.

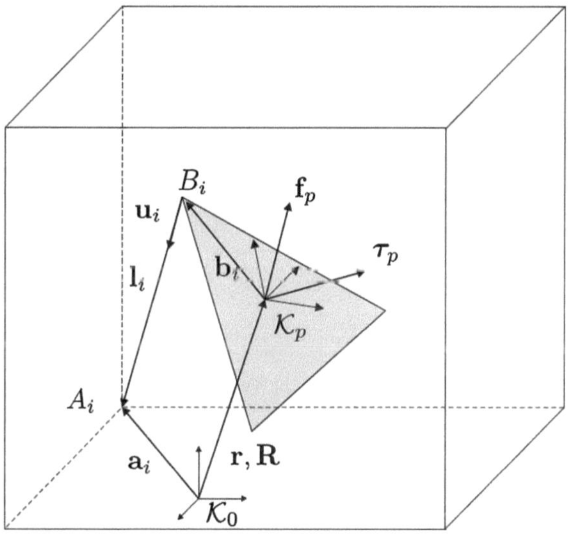

Fig. 2 Robot kinematic

The paper is organized as follows: The kinematic model of the cable robot is summarized in Sect. 2. The design of the admittance control and virtual workspace are described in Sect. 3. The integration in the control framework and the test setup is pictured in Sect. 4. Experimental results are presented and discussed in Sect. 5. Finally, conclusions and an outlook on future work is given in Sect. 6.

2 Robot Model

For completeness, we briefly review the robot model. The geometry of the robot is described by the proximal anchor points on the robot base $\mathbf{A}_i$ and the distal anchor points on the end-effector $\mathbf{B}_i$. The index i denotes the cable number and m is the absolute number of cables. By applying a vector loop as shown in Fig. 2 the cable vector $\mathbf{l}_i$ follows as

$$\mathbf{l}_i = \mathbf{a}_i - \mathbf{r} - \mathbf{R}\,\mathbf{b}_i, \tag{1}$$

while $\mathbf{r}$ is the platform position vector and rotation matrix $\mathbf{R}$ describes the orientation.

The structure equation with the structure matrix $\mathbf{A}^{\mathrm{T}}$ describes the force and torque equilibrium at the end-effector for a given cable

$$\underbrace{\begin{bmatrix} \mathbf{u}_1 & \cdots & \mathbf{u}_m \\ \mathbf{b}_1 \times \mathbf{u}_1 & \cdots & \mathbf{b}_m \times \mathbf{u}_m \end{bmatrix}}_{\mathbf{A}^{\mathrm{T}}(\mathbf{r},\mathbf{R})} \underbrace{\begin{bmatrix} f_1 \\ \vdots \\ f_m \end{bmatrix}}_{\mathbf{f}} = -\underbrace{\begin{bmatrix} \mathbf{f}_p \\ \boldsymbol{\tau}_p \end{bmatrix}}_{\mathbf{w}}, \tag{2}$$

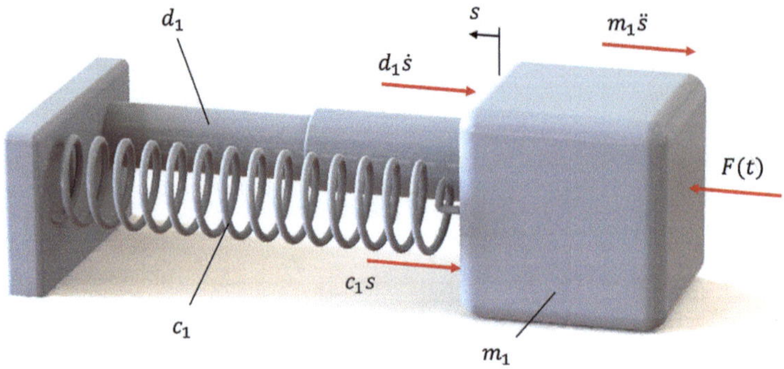

Fig. 3 Model of a 1D spring mass damper system with external force $F(t)$

with $\mathbf{u}_i = \frac{\mathbf{l}_i}{\|\mathbf{l}_i\|}$. The wrench $\mathbf{w}$ consists of external forces $\mathbf{f}_\mathrm{p}$ and torques $\boldsymbol{\tau}_\mathrm{p}$ and also includes the gravity g.

3 Admittance Control

3.1 Model of the Virtual System Behavior

The admittance control allows for the simulation of a virtual system behavior. The characteristic can be chosen nearly arbitrary, e.g. without gravity, viscous friction or very high stiffness. The task of the admittance control algorithm and its functions is to calculate a pose out of an applied wrench. The dynamic behavior of a simple mechanical system can be described with a model of a spring mass damper system. An one dimensional system, as shown in Fig. 3, consists of a stiffness c_1, a mass m_1 and a damping coefficient d_1 and is described by the following differential equation

$$m_1\ddot{s} + d_1\dot{s} + c_1 s = F(t), \tag{3}$$

with the position s of the body and its time derivatives $\dot{s}$ and $\ddot{s}$. For a given external force the position, velocity and acceleration of a body can be calculated by time-integration according to Eq. 3. To describe a system with 6 degrees of freedom (DOF) Eq. 3 is extended to

$$\underbrace{\mathbf{I}\,\ddot{\mathbf{r}}_\mathrm{set}}_{\textit{inertia force vector}} + \underbrace{\mathbf{D}\dot{\mathbf{r}}_\mathrm{set}}_{\textit{damping force vector}} + \underbrace{\mathbf{C}\mathbf{r}_\mathrm{set}}_{\textit{spring force vector}} = \underbrace{\mathbf{w}(t)}_{\textit{wrench}}, \tag{4}$$

with the commanded pose $\mathbf{r}_\mathrm{set} \in \mathbf{R}^{6\times 1}$ and its time derivatives $\dot{\mathbf{r}}_\mathrm{set}$ and $\ddot{\mathbf{r}}_\mathrm{set}$. The pose $\mathbf{r}$ consists of the position in x, y and z-direction and the Euler angles α, β and γ.

The external forces and torques are the elements of the wrench $\mathbf{w}(t) \in \mathbf{R}^{6\times 1}$. The matrices are

$$\mathbf{I} = diag([m_x \; m_y \; m_z \; I_\alpha \; I_\beta \; I_\gamma]), \tag{5}$$

$$\mathbf{D} = diag([d_x \; d_y \; d_z \; d_\alpha \; d_\beta \; d_\gamma]) \quad \text{and} \tag{6}$$

$$\mathbf{C} = diag([c_x \; c_y \; c_z \; c_\alpha \; c_\beta \; c_\gamma]). \tag{7}$$

The model allows for an implementation of the admittance control in both translational and rotational space. In the following, we focus on the translational workspace and with a fixed orientation.

3.2 Model of the Virtual Workspace

If the Eq. 4 would be used in the algorithm without any modification, it would be possible for the user to bring the platform to a certain pose by applying a *wrench*. But the platform will move to its origin as soon as the *wrench* is removed because of the spring force-term of the Eq. 4. To allow for positioning the platform we have to toggle the spring force on and off on demand. For this, we introduce a virtual workspace, in which the elements of the stiffness matrix $\mathbf{C}$ are a function $\mathbf{C} = f(\mathbf{r}_{\text{set}})$ of the pose. Additionally, this function limits the reachable workspace by applying a virtual spring force at the border. This allows the user to feel the end of the workspace area and additionally guides him back into the allowed workspace. For this purpose, the virtual workspace is divided into three zones, which are visualized in Fig. 4:

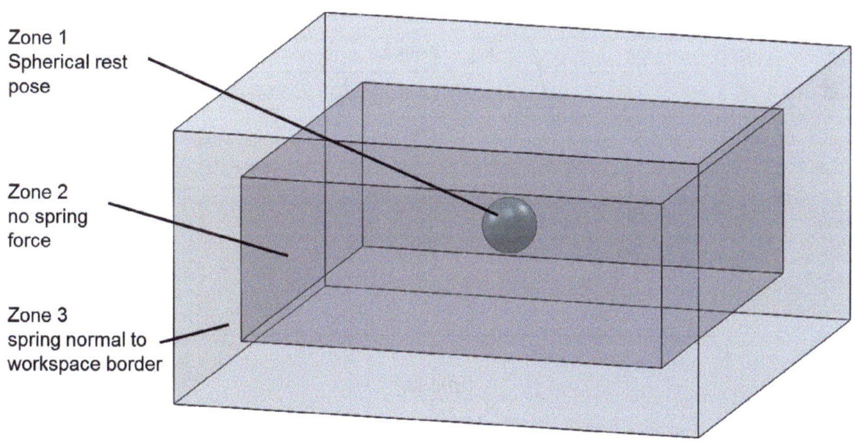

Fig. 4 Zones of the virtual workspace

Zone 1 (spherical rest pose) This zone is a sphere with the radius $R1$ an its midpoint corresponds to the home position of the robot. As long as the midpoint of the platform is inside zone 1, a spring force-vector is pointing to the origin of the sphere. Therefore, a spring force-vector pulls the platform to the origin. This works as a rest pose for the platform. Basically, every point within the workspace can be chosen as midpoint of the sphere. One variant is to adopt the actual pose as center of the sphere after defined time of standstill to fix the platform in this position.

Zone 2 (no spring force) In zone 2 the spring is disabled and only the damping and inertia of the virtual system are activated. If the external force is removed, the platform is slowed down by the damping and remains in the desired position.

Zone 3 (Spring force-vector normal to the borders of the workspace) If the platform is Zone 3, the spring force-vector which is normal to the borders of the workspace is toggled on. This force limits the workspace.

The Eq. 4 and the virtual workspace are implemented in Matlab/Simulink. In the implementation, the spring force $\mathbf{F}_F$ is calculated according to the above description and applied on Eq. 4.

3.3 Simulation of the Virtual System

To show the behavior of the virtual system, an external force in x-direction is generated and applied to the system. The force is designed to lead the platform in all three zones to show the different modes of the system.The applied force and the resulting position of the platform is visualized in Fig. 5 and described below:

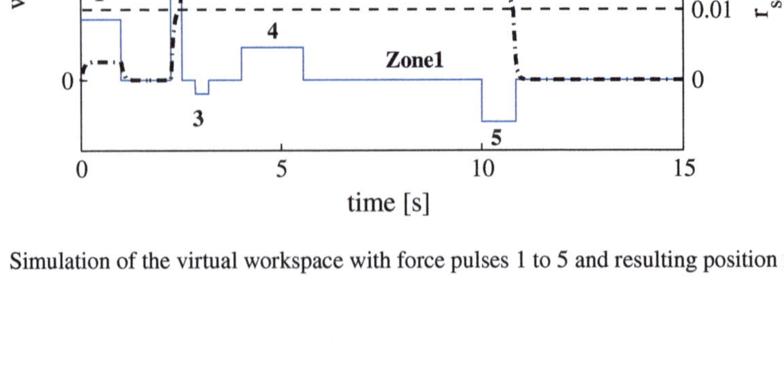

Fig. 5 Simulation of the virtual workspace with force pulses 1 to 5 and resulting position $\mathbf{r}_{set,x}$

Force pulse 1 In the time range 0 s $< t <$ 1 s a force with the magnitude of 2.6 N is applied to the platform. When the platform reaches a distance of 0.0026 m to the origin, a force equilibrium with the spring force is satisfied. So the applied force is not strong enough to guide the platform out of the Zone 1. As soon as the force is removed, the platform is moving back to the origin.

Force pulse 2 The second force pulse is applied to the platform in a time range of 2.25 s $< t <$ 2.52 s with a magnitude of 10.2 N. This time, the force is strong enough to overcome the counter-wise spring force of the Zone 1. As soon as the platform crosses the border to Zone 2, the spring force is set to zero. After removing the external force, the platform slows according to the system damping and holds the position.

Force pulse 3 At the time 2.85 s $< t <$ 3.2 s the platform is in the Zone 2. in this area of the virtual workspace is no spring force active. As it is shown in the figure, the platform moves according to the applied external force. As soon as the external force becomes zero, the platform is decelerated by the damping.

Force pulse 4 In this section the platform is moving trough the Zone 2 with constant velocity. At a time of 4.9 s the platform reaches the border of the working area (0.03 m). It is visible that the platform is slowed down by the counter force of the Zone 3. After the external force is removed, the platform is guided back into the allowed workspace described by Zone 2 and remains at the borderline.

Force pulse 5 The last force pulse is guiding the platform towards the origin. It is visible that the spring force of Zone 1 is catching the platform as soon as it reaches the area of the sphere, describing the border of the Zone 1.

4 Controller Implementation

4.1 Controller Design

In the following, the control structure is described, which is visualized in Fig. 6. The admittance control is based on an indirect force measurement over the cable force sensors. For the transformation from joint to Cartesian space, the structure matrix $\mathbf{A}^{\mathrm{T}}$ describing the actual cable directions is used. The wrench $\mathbf{w}$ is derived by

$$\mathbf{w} = -\mathbf{A}^{\mathrm{T}}\mathbf{f}_{\mathrm{is}}, \tag{8}$$

where $\mathbf{f}_{\mathrm{is}}$ corresponds to the measured cable forces.

The wrench includes the gravitational force of the platform and also measurement errors. At startup of the admittance control, the actual wrench $\mathbf{w}_{\mathrm{init}}$ is stored. Acceleration of the platform leads also to inertia forces of the real platform which are measured by the force sensor. Therefore, the inertia forces $F_i = m_{\mathrm{p}}\ddot{\mathbf{r}}_{\mathrm{set}}$ derived from

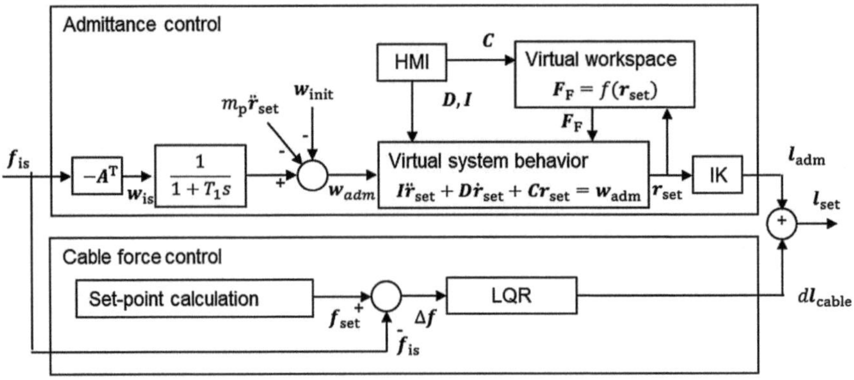

Fig. 6 Structure of the implementation with admittance and cable force controller

the platform mass m_p and the commanded acceleration $\ddot{\mathbf{r}}_{set}$ have to be excluded. The mass m_p corresponds to the real mass of the platform and is derived with $m_p = \frac{w_{init}}{g}$.

The input signal of the admittance controls $\mathbf{w}_{adm}$ yields

$$\mathbf{w}_{adm} = \mathbf{w}_{is} - \mathbf{w}_{init} - m_p \ddot{\mathbf{r}}_{set}, \tag{9}$$

which is smoothed with a PT_1-Filter with a time constant of $T_1 = 3\,\text{ms}$.

The position setpoint of the admittance control $\mathbf{r}_{set}$ is transformed from Cartesian to joint space with the inverse kinematics (IK) according to Eq. 1 and delivers $\mathbf{l}_{adm}$.

For the operation of the admittance control the stability of the platform is important. Therefore, a cable force control is applied to keep the cables under tension. For the determination of the setpoint cable forces $\mathbf{l}_{set}$, the closed-form solution presentend in [11] is applied. The setpoint cable forces are derived for the gravitational forces of the platform. Forces due to acceleration of the platform or the interaction with the user are not taken into account. The output of the cable force controller is a change in cable length $d\mathbf{l}_{cable}$. The output of the admittance and cable force controller are summed up by

$$\mathbf{l}_{set} = \mathbf{l}_{adm} + d\mathbf{l}_{cable}, \tag{10}$$

to the cable length setpoint $\mathbf{l}_{set}$ which is commanded to the servo drives.

4.2 Implementation

The control algorithms are implemented on the cable robot Mini-IPAnema 3 using eight cables and an aluminium platform as shown in Fig. 1. The space of the robot's frame is 1.1 x 0.8 x 1.0 m. The actuators are 200 W servo drives of type Beckhoff AM3121. Without an additional gear box, the drum with a diameter of 20 mm is connected. The theoretical rated cable force neglecting friction amounts to 65 N.

The robot control is realized on a Windows PC with Beckhoff TwinCAT 3.1 CNC at a cycle time of 1 ms. The field bus protocol is EtherCAT. For the force measurement each cable is equipped with a cable force sensor at the platform of type Futek LRM200 with a measurement range of 111 N. The analog output signal of the force sensors is digitized in A/D-converters and sent via the field bus to the control. The cable type is LIROS D-Pro 01505-0150 based on Dyneema SK 75 fibre (Polyethylene) with a diameter of 1.5 mm.

5 Experimental Investigation

5.1 Feasible Bandwidth

The maximum dynamic which can be simulated is limited by the system dynamic of the robot. The robot's dynamic is characterized by the bandwidth of the actuator unit. To get insight into the feasible bandwidth, we establish the transfer function $G_{\mathrm{p}}(s)$ describing the relation between setpoint position value l_{set} and the encoder feedback value l_{fb}.

The servo motor with the winch is a typical spring mass damper system. Therefore, we assume a second order system

$$G_{\mathrm{p}}(s) = K_{\mathrm{p}} \frac{1}{1 + 2\zeta T_{\mathrm{w}} s + (T_{\mathrm{w}} s)^2}, \tag{11}$$

for the transmission behavior. For the parameter identification based on a noise response we use the Matlab System Identification Toolbox. The resulting parameters are $T_{\mathrm{w}} = 0.0139$ and $\zeta = 0.587$. The measurement is reflected with a goodness fit of 98.5 % by the identified transfer function. The deviation of 1.5 % arises from sensor noise which is not modeled.

The bode diagram for $G_{\mathrm{p}}(s)$ representing the real robot dynamic properties is shown in Fig. 7. The transfer function $G_{\mathrm{p}}(s)$ reaches the -3 dB margin at a frequency of 13.3 Hz, which corresponds to the bandwidth of the haptic interface.

5.2 Eigenfrequency

In the following we investigate the natural oscillations for given spring mass damper systems. For this, we set the parameters $c_{\mathrm{x}}, d_{\mathrm{x}}$ and m_{x} to certain values. The theoretical eigenfrequency $f_{\mathrm{d,theo}}$ of a damped system is determined by

$$f_{\mathrm{d,theo}} = \frac{1}{2\pi} \sqrt{\frac{c_{\mathrm{x}}}{m_{\mathrm{x}}} - \left(\frac{d_{\mathrm{x}}}{2m_{\mathrm{x}}}\right)^2}. \tag{12}$$

Table 1 Verification of the dynamic parameters of the simulated system

Measurement	c_x [N/m]	d_x [Ns/m]	m_x [kg]	$f_{d,theo}$ [Hz]	$f_{d,exp}$ [Hz]
1	31.9	5	20	0.2	0.2025
2	197.5	5	20	0.5	0.504
3	395.4	5	10	1	0.997
4	9870.2	5	10	5	5.124

For the experiment, the user moves the platform by hand out of the home position and releases it. Then, the resulting platform oscillations were measured. The parameters of c_x, d_x and m_x and the theoretical eigenfrequency $f_{d,theo}$ according to Eq. 12 and measured eigenfrequency $f_{d,exp}$ are listed in Table 1.

For the parametrization of the admittance control one has to ensure that the virtual system dynamic lies under the robot dynamic. The eigenfrequency $f_{d,theo}$ according to Eq. 12 has to be smaller than the robot's bandwidth of 13.3 Hz. Simulating systems with a higher dynamic will lead to instabilities.

5.3 Operation

For the experimental evaluation of the system, the applied wrench in x-direction w_x and the resulting movement in x-direction $r_{set,x}$ is plotted in Fig. 8. In this experiment, an user grasps the platform and shifts it to several positions.

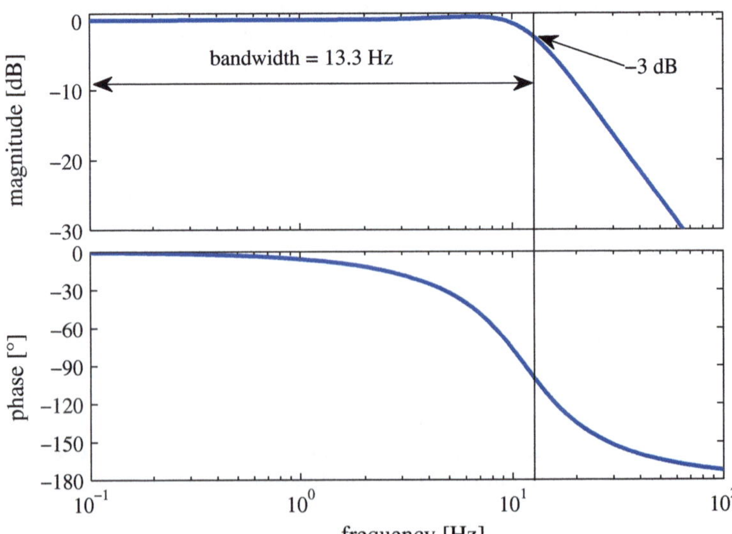

Fig. 7 Bode diagram of the transfer function $G_p(s)$

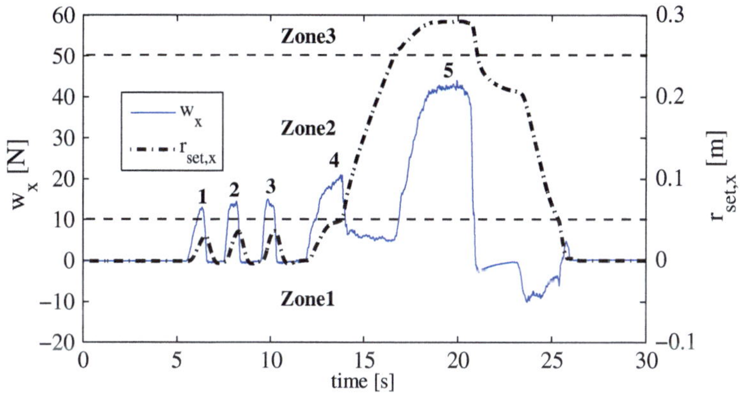

Fig. 8 Experiment: Measured user-applied force $\mathbf{w}_x$ and resulting position $\mathbf{r}_{set,x}$

The first three force pulses are not strong enough to bring the platform out of zone 1 and the platform is dragged back to the origin by the spring force. The third force pulse is strong enough to bring the platform to the zone 2, where the spring is turned off. The sudden missing of the stiffness increases the slope of the position and leads to a drop of the applied force w_x. The user's impression at this point is comparable to the feeling of dragging two magnets apart. Then, the applied force stays in a constant range while the slope of the position can be also described as constant in this area. This shows the active damping in zone 2. After the platform reaches the border of the workspace, the w_x value rises with a small delay. This delay is caused by the virtual inertia of the system. Due to that, the platform enters the zone 3 without additional force from the user. After the virtual kinetic energy is transformed into spring energy and energy dissipated by the virtual damper, the user has to increase the force to move the platform further into the zone 3. As it is visible, the user presses the platform into the zone 3 by a constant force. The position of the platform is not changing because the spring force of the zone 3 and the applied force by the user are equal. After the user releases the force, the platform is moving back to zone 2. The user feels a guiding force in normal direction to the border plane between zone 3 and zone 2. Back in zone 2 the platform is moved towards the origin. Finally, the spring force of the zone 1 catches the platform and drags it to the origin. As it can be seen from Fig. 8, the system behaves identical to the idealized one in Fig. 5. The $r_{set,x}$ is very responsive to the user input and system behavior can be controlled very precisely.

6 Conclusion and Outlook

We demonstrated the realization of an admittance control for a parallel cable robot. A switchable stiffness was introduced to guide the user with force feedback within the workspace. To estimate the maximum dynamic, the bandwidth of the investigated cable robot was determined to 13.3 Hz, which enables a very responsive system which can simulate in a very high range.

For the industrial application, the user inevitably is in contact with the robot. For the industrial deployment, a risk assessment and measures for a safe operation have to be performed. The algorithm can be developed further to a force assist in a lifting aid.

References

1. Gosselin F, Andriot C, Savall J, Martin J (2008) Large workspace haptic devices for human-scale interaction: a survey. In: Proceedings of Eurohaptics
2. Kawamura S, Ito K (1993) A new type of master robot for teleoperation using a radial wire drive system. In: Proceedings of IEEE/RSJ international conference on intelligent robots and systems
3. Sato M (2008) Development of string-based force display: spidar. In: Proceedings of the international conference on virtual systems and multimedia
4. Billette G, Gosselin C (2009) Producing rigid contacts in cable-driven haptic interfaces using impact generating reels. In: Proceedings of IEEE international conference on robotics and automation, pp 307–312
5. Gallina P, Rosati G, Rossi A (2001) 3-d.o.f. wire driven planar haptic interface. J Intell Rob Syst 32:23–36
6. Gallina P, Rosati G (2002) Manipulability of a planar wire driven haptic device. Mech Mach Theory 37:215–228
7. Williams II RL (1998) Cable suspended haptic interface. Int J Virtual Reality 3(3): 13–21
8. Otis MJD, Mokhtari M, Du Tremblay C, Laurendeau D, Rainville FMD, Gosselin CM (2008) Hybrid control with multi-contact interactions for 6dof haptic foot platform on a cable-driven locomotion interface. In: Proceedings of Haptic interfaces for virtual environment and teleoperator systems, pp 161–168
9. Surdilovic D, Zhang J, Bernhardt R (2007) String-man: wire-robot technology for safe, flexible and human-friendly gait rehabilitation. In: Proceedings of IEEE 10th international conference on rehabilitation robotics (ICORR), pp 446–453
10. Rezazadeh S, Behzadipour S (2007) Impedance control of cable-driven mechanisms. In: Proceedings of ASME international design engineering technical conferences and computers and information in engineering conference
11. Pott A, Bruckmann T, Mikelsons L (2009) Closed-form force distribution for parallel wire robots. In: Proceedings of 5th international workshop on computational kinematics 2009. Springer, Duisburg, Germany, pp 25–34, 6–8 May 2009

A Kinematic Vision-Based Position Control of a 6-DoF Cable-Driven Parallel Robot

Ryad Chellal, Loïc Cuvillon and Edouard Laroche

Abstract This paper introduced an original setup for the validation of vision-based position control methodologies dedicated to 6-DoF Cable-Driven Parallel Robots. The cable robot is an INCA 6D with eight cables initially developed by Haption as a haptic interface, equipped with a motion-capture system Bonita developed by Vicon to measure the end-effector pose. In addition to the description of this setup, this paper reports simulation and experiment results obtained with an original control scheme based on a cascaded control architecture in two parts. First is the position control to ensure an accurate end-effector positioning, which includes two nested closed-loops: an external vision-loop based on the pose measurement that drives the motors, equipped of inner speed-loop previously designed in order to control each speed in a decoupled fashion when rejecting the inherent non-linear behaviour of the cables. Second is the tension distribution to maintain the cables under feasible tensions, the paper makes an extensive review of the available methods and presents an algorithm inspired from one of them extended to the dynamic control.

1 Introduction

Comparing with rigid link-actuated manipulators, Cable-Driven Parallel Robot (CDPR) manipulators benefit from large workspaces, high speed motions and modular geometries. However, it appears that their control is a more complex issue as the cables must remain under tensions at any time [1]. To solve this issue, the proposed approaches in the literature can be classified in two main categories. In the off-line solutions, a path planning step is used prior to any motion in order to design a reference trajectory which guarantees that the cables will remain under tension

R. Chellal (✉) · L. Cuvillon · E. Laroche
ICube Laboratory of Engineering, Computer and Image Sciences,
University of Strasbourg and CNRS, 67412 Illkirch, France
e-mail: ryad.chellal@unistra.fr

A. Pott and T. Bruckmann (eds.), *Cable-Driven Parallel Robots*,
Mechanisms and Machine Science 32, DOI 10.1007/978-3-319-09489-2_15

during the predefined motion [2, 3], but assumes that a perfect control of the robot is available. In the on-line solutions, an algorithm of tension distribution (also known as redundancy resolution or force calculation) is used to maintain the cable tensions inside a predefined feasible workspace during the motion [4, 5]. This is a typical solution for redundant manipulators, where there are more cables than Degrees of Freedom (DoF) of the end-effector, and it is the solution considered in this work.

Concerning the position control of CDPR manipulators, most of the proposed methods rely on the joint position measurements. According to the coordinate space chosen to solve this control problem, there are two alternatives. In the first one, the controllers are designed in the joint space coordinates. Using the Inverse Position Kinematic Model (IPKM), the reference end-effector pose is converted in reference joint positions which are then controlled by a feedback loop. Some related works are the joint space PD controller proposed by Kawamura et al. applied to the SEGESTA robot [6] and later to the KNTU robot by Gholami et al. [7], and the joint space PID controller for the redundant suspended ReelAx8 prototype presented by Lamaury et al. [8]. In the second one, the controllers are designed in the task space coordinates. Assuming that the Direct Position Kinematic Model (DPKM) is available, the end-effector pose is calculated from the joint position measurements and a feedback control allows to track a reference pose. Gholami et al. evaluated such a task space PD controller and compared it to the previous approach [7]. However, for parallel manipulators, the direct kinematics is difficult to obtain (see for instance Carricato and Merlet for an intensive study on the matter [9]).

In the previously mentioned control schemes, the modelling errors and the deformations of the cables directly result in errors on the end-effector pose. One solution for improving the accuracy is then to use some exteroceptive sensors in order to obtain a direct measurement of the end-effector pose. Some preliminary works using cameras have been proposed by Dallej et al. for controlling the redundant suspended ReelAx8 robot [10] or the large-dimension CoGiRo robot [11].

The purpose of this paper is twofold. First, an original evaluation setup is introduced, based on an INCA 6D robot with eight cables developed by Haption, equipped with a motion-capture system Bonita developed by Vicon used for the measurement of the end-effector pose. Second, results from simulation and experiment are reported, based on an original control scheme relying on a cascaded control architecture in two parts. First is the position control to ensure an accurate end-effector positioning, which includes two nested closed-loops: an external vision-loop based on the pose measurement that drives the motors, equipped of inner speed-loop previously designed in order to control each speed in a decoupled fashion when rejecting the inherent non-linear behaviour of the cables. Second is the tension distribution to maintain the cables under feasible tensions, the paper makes an extensive review of the available methods and presents an algorithm inspired from one of them extended to the dynamic control.

The paper is organised as follows: Sect. 2 describes the setup (the INCA robot and the Bonita motion-capture system). In Sect. 3 the model of the robot is developed. In Sect. 4 the control strategies that are considered for evaluation are presented. The simulation and experiment results are presented in Sect. 5.

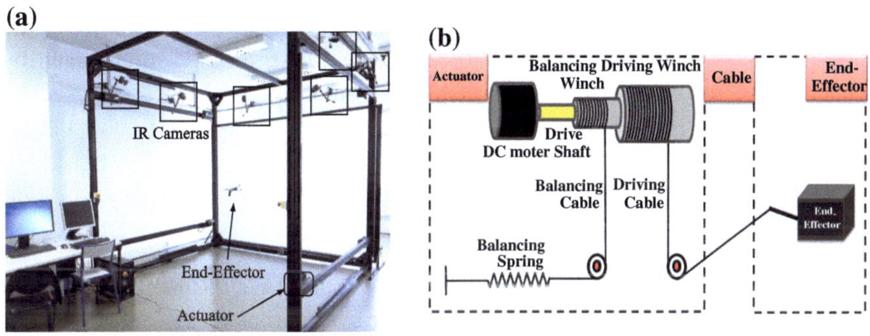

Fig. 1 The INCA robot **a** The INCA robot equipped with the 6 Bonita IR cameras. **b** Actuation scheme

2 System Description

2.1 *INCA Robot*

The INCA robot [12] developed by Haption has a cubic configuration of 3 m by side, and uses eight driving cables to move the end-effector and eight balancing cables to ensure pretension in the driving cables when the motors are non-powered (Fig. 1b). Each actuator is placed on one of the eight vertices of the workspace (Fig. 1a) and is composed of a DC motor with a current-loop controller, which drives both the driving and balancing winches (Fig. 1b) to store, wind and unwind the cables.

A measurement of the motor positions and currents are respectively achieved by incremental optical encoders and current proprioceptive sensors.

2.2 *Bonita Motion-Capture System*

The Bonita motion-capture system used to measure the pose of the INCA end-effector is composed of six IR cameras (Fig. 1a) and a tracker software running on a Windows PC, all from the Vicon company. Each camera has its own emitting source and delivers a grayscale image with VGA resolution. Assuming that this stereo system has been previously calibrated, the pose of the INCA end-effector fitted with five retro-reflective fixed markers can be tracked by the software.

The temporal and spatial performances of the pose reconstruction are critical for the robot control and are evaluated:

- the global latency of the system (delay between the start of the image acquisition and the availability of the pose measurement) is 10.7 ± 0.7 ms at a 200 Hz camera frame-rate. This latency of roughly twice the acquisition period is the sum of one period of image acquisition and one period for the pose reconstruction.

- the accuracy of the pose is 1.7 ± 0.4 mm and evaluated by the RMS error in the IR images between the maker positions and their expected positions given by back-projection of the end-effector geometry with the reconstructed pose.

3 Modelling of the 6-DoF CDPRs

The model is derived from a generic model of the m-DoF CDPR manipulators with n cables, which is augmented with the pretension system (balancing cables, winches and springs) specific to the INCA prototype.

Given the size of the INCA robot and its cables of 1 mm diameter, the considered model assumes that the cables are of negligible mass (straight) and of infinite stiffness (inextensible). The model is briefly presented in this section, details can be found in [12].

3.1 Kinematics Modelling

To design the vision-loop control law, the Inverse Velocity Kinematic Model (IVKM) can be calculated by differentiating the Inverse Position Kinematic Model (IPKM) with respect to the time, and is given by the Jacobian matrix J_θ that relates the end-effector velocity $V_e = [v_e^T \; w_e^T]^T$ (including the linear v_e and angular w_e velocities) to the motor velocities vector $\dot{\theta} = [\dot{\theta}_1 \cdots \dot{\theta}_n]^T$ such as:

$$\dot{\theta} = J_\theta(X_e)\; V_e \tag{1}$$

where J_θ can be easily calculated for parallel manipulators from the end-effector pose $X_e = [P_e^T \; \Phi_e^T]^T$ (including the position P_e and orientation Φ_e) as given in [12].

Considering the chosen representation of the orientation based on a system of three pure rotations Roll-Pitch-Yaw of angles $\Phi_e = [\phi_r, \phi_p, \phi_y]^T$ respectively around the principal axes (X_o, Y_o, Z_o), the IVKM can be rewritten in terms of the time variation of the end-effector pose $\dot{X}_e$ such as:

$$\dot{\theta} = J_\theta(X_e)\; J_\Phi(\Phi_e)\; \dot{X}_e \tag{2}$$

where $J_\Phi(\Phi_e) = \text{blocdiag}\,(I_3, J_{rpy}(\Phi_e))$, in which the Jacobian J_{rpy} relates the angular velocity w_e to the speed of the chosen orientation $\dot{\Phi}_e$ as in [12]:

$$w_e = J_{rpy}(\Phi_e)\; \dot{\Phi}_e \tag{3}$$

3.2 Dynamics Modelling

To determine the joint speed-loop control law, and for the purpose of simulations, a Direct Dynamic Model (DDM) has been developed. The DDM of the whole system projected in the task space coordinates can be written in the following form:

$$M(X_e)\,\ddot{X}_e + C(X_e, \dot{X}_e)\,\dot{X}_e + K(X_e) + G(X_e) = F_{ev} \tag{4}$$

under the following n constraints bounding the cable tensions vector $T = [T_1 \cdots T_n]^T$:

$$T_{\min} \leq T(I_m, \theta, \dot{\theta}, \ddot{\theta}) \leq T_{\max} \tag{5}$$

where the details of the inertia matrix M, the vector of forces and moments of Coriolis and centrifugal forces C, the vector of stiffness K and the vector of forces and moments of gravity G are given in [12]. The input vector of virtual wrenches $F_{ev} = [f_{ev}^T\; m_{ev}^T]^T$ meaning the external forces f_{ev} and moments m_{ev} acting on the end-effector resulting from the applied motor currents vector $I_m = [I_{m1} \ldots I_{mn}]^T$ is:

$$F_{ev} = W_I(X_e)\, I_m \tag{6}$$

with the wrench matrix $W_I(X_e) = -J_\theta^T(X_e)\, K_{em}$, and the diagonal matrix of the motor torque constants $K_{em} = \text{diag}(k_{em1}, \ldots, k_{emn})$.

The cable tensions vector T can be estimated using the actuators dynamics by:

$$T(I_m, \theta, \dot{\theta}, \ddot{\theta}) = T_o + R_{pm}^{-1}\, [K_{em}\, I_m - J_{eq}\, \ddot{\theta} - F_{veq}\, \dot{\theta} - F_{seq}\, \text{sign}(\dot{\theta}) - K_{eq}\, \theta] \tag{7}$$

where the details of the diagonal matrices of driving winches radius R_{pm}, equivalent inertia moment J_{eq}, equivalent viscous F_{veq} and Coulomb F_{seq} friction coefficients, balancing springs equivalent rotational stiffness K_{eq}, and the vector of initial cables pretension T_o are given in [12].

4 Kinematic Vision-Based Position Control

4.1 Overview

The cascaded control architecture considered here in Fig. 2 is implemented in two parts detailed later in this section:

1. The position control in itself, a kinematic position-based vision control (also known as Position-Based Visual Servoing (PBVS) or 3D Visual Servoing) which includes two nested closed-loops:

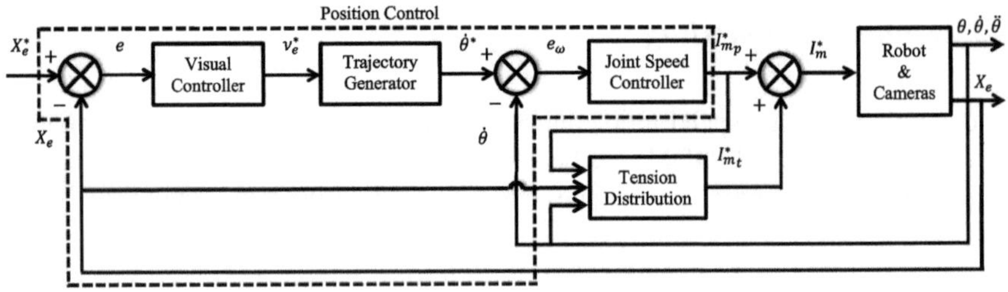

Fig. 2 Kinematic vision-based position control scheme with the tension distribution part

- the external vision-loop in the task space coordinates aims to ensure an accurate end-effector positioning from the end-effector pose error.
- the internal speed-loop in the joint speed coordinates aims to ensure a decoupled control of the motor speeds requested by the external loop while rejecting the inherent non-linear behaviour of the cables.

2. The tension distribution, that aims to maintain the cables under a tension that respects interval constraints.

4.2 Position Control

4.2.1 Vision-Loop

For the system with measured pose X_e and controlled velocity $v_e = \dot{X}_e$, an exponential convergence of the actual pose X_e towards the reference X_e^* is ensured by the proportional control law:

$$v_e^* = K\,(X_e^* - X_e) \tag{8}$$

where the proportional gain matrix chosen under the shape $K = \text{blocdiag}(k_P\, I_3, k_\Phi\, I_3)$ to tune the position and orientation, should be positive to adjust the response time of a stable control, and also diagonal to realise a decoupling between the position P_e and orientation Φ_e components of end-effector pose X_e.

The vision-loop control signal v_e^* being expressed in the task speed coordinates, is then converted into the joint speed coordinates $\dot\theta^*$ using the IVKM (2).

4.2.2 Speed-Loop

The original proposed methodology relies on joint speed control. For the controller tuning, only one actuator control is considered, the others being considered as tension disturbances. The corresponding model is presented in Fig. 3, it includes a first-order

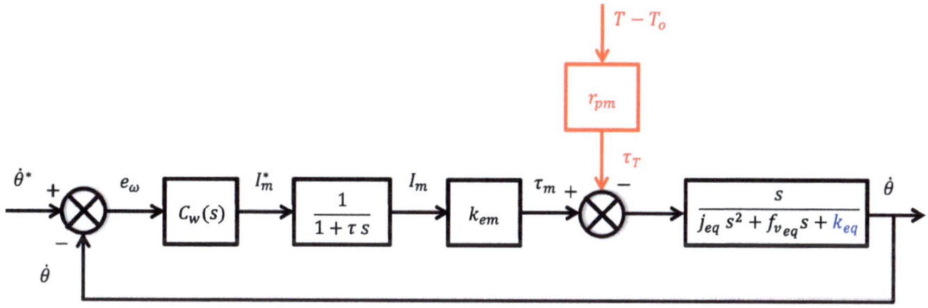

Fig. 3 Speed-loop control scheme of one DC motor subject to the effects of its driving and balancing cables

model of the current-loop dynamics with the time constant τ, the motor torque constant k_{em} and a second-order model of the actuator dynamics represented by the inertia j_{eq}, the viscous friction $f_{v_{eq}}$ and the rotational stiffness k_{eq} due to the balancing spring. These parameters have been estimated experimentally in a previous work [12]. The cable tension T can be considered as a disturbance acting on the motor current control signal I_m^*, and then should be rejected or compensated by the joint speed controller C_w.

A PI controller $C_w(s)$ is generally sufficient for controlling the speed of a DC motor, the integral term being necessary in order to reject the disturbances:

$$C_w(s) = K_p \left(1 + \frac{1}{T_i\, s}\right) \tag{9}$$

It was tuned in order to reach a high bandwidth. Therefore, a design model was established by considering a high frequency approximation of the transfer from the motor torque τ_m to the motor velocity $\dot{\theta}$, resulting in the control model $G_c(s)$ which suits to the controller design using the symmetrical optimum method (detailed in Sect. 5.2) for an efficient disturbances rejection [13]:

$$G_c(s) = \frac{k_{em}}{j_{eq}\, s\,(1 + \tau\, s)} \tag{10}$$

4.3 Tension Distribution

4.3.1 State of the Art

In order to ensure that the vector of cable tensions T remains inside the feasible tensions workspace $[T_{\min}\ T_{\max}]$, the speed-loop control signal of motor currents vector $I_{m\,p}^*$ leading to the virtual wrench F_{ev} according to (6), is modified to solve the system of algebraic equations (6) under the inequality constraints (5). Due the robot redundancy, the system of equations (6) is under-determined and then has an infinity of solutions (assuming that W_I has full rank $r = n - m$) that can be determined by resolving the quadratic optimisation problem of the objective function E given by:

$$E = \frac{1}{2}\left(I_m^* - I_{obj}\right)^T \left(I_m^* - I_{obj}\right) + \left(W_I\left(I_m^* - I_{mp}^*\right)\right)^T \lambda \tag{11}$$

where the variable objective currents vector I_{obj} corresponding to an objective tensions vector T_{obj}, and the vector $\lambda \in \mathbb{R}^m$ is the Lagrange multiplier associated to the equality constraints $W_I\ (I_m^* - I_{mp}^*) = 0$.

The set of solutions of (6) evolves in an r-dimensional subspace that can be written as following:

$$I_m^* = I_{mp}^* + I_{mn}^* \tag{12}$$

where:

- the particular solution I_{mp}^* of minimum-norm aims to control the end-effector pose, and it is the previous PI speed-loop control signal.
- the homogeneous solution I_{mn}^* of W_I null-space aims to satisfy the tension constraints, while maintaining the end-effector on its actual pose:

$$I_{mn}^* = [I_n - W_I^+\ W_I]\ I_{\text{obj}} \tag{13}$$

Let us review the literature dedicated to the determination of the objective tensions vector T_{obj} that maintains the cable tensions vector T within a feasible tensions workspace, ensuring the constraints (5). The available works can be classified in two categories. The first approach opted for iterative algorithms, so that efficient constrained optimisation methods can be used such as Linear Programming Methods (LPM) [1, 14], but the cable tensions continuity is not guaranteed. Other optimisation methods are also used such as Non-Linear Programming Methods (NLPM) in the particular case of Quadratic Programming Methods (QPM) [1, 15], and the general NLPM with the gradient descent method to resolve the problem in a quadratic formulation [7]. These quadratic methods guarantee the tensions continuity but have a non-predictable runtime. However, these algorithms are not suitable for the constraints of real-time control. Hence, the second approach relies on non-iterative algorithms to handle the real-time control constraints. For instance, Mikelsons et al. proved that the Center of Gravity (CoG) of the feasible tensions distribution workspace (the set of solutions of equations (6) satisfying the tension constraints (5)) is a solution that ensures the tensions continuity [16]. Recently, some other works have been proposed by Borgstrom [17], and by Lamaury and Gouttefarde which optimised the CoG method to the case of two degrees of redundancy [18].

4.3.2 Considered Algorithm

Unlike this previous second category of non-iterative algorithms with an evolutionary criterion, this work is inspired from an non-evolutionary algorithm proposed by Lafourcade [5] more appropriate to satisfy our real-time constraints (less than 1 ms), it is adapted to our case also considering a constant vector T_{obj} but variable I_{obj} due to the actuators dynamics of (7).

The proposed algorithm consists in: (1) selecting T_{obj} inside the feasible tensions workspace and calculating I_{obj} by inverting (7), (2) resolving the optimisation problem (11) without the tension inequalities constraints (5), (3) if some tension inequality constraints are violated then the q concerned inequalities selected among them are transformed into current equality constraints (not more than r tension inequality constraints can be saturated simultaneously) and included into the optimisation problem. All combinations of 1 to r violated tension inequality constrained are considered until one solution is met satisfying all the non-saturated tension inequalities constraints. If no solution is found, the vector I_{obj} can be scaled by a scalar factor to not modify the trajectory and the process could be repeated. If the problem has no solution, the reference trajectory should be modified (not detailed herein).

When saturating a combination of q violated tension constraints, the control signal of motor currents vector I_m^* can be obtained by resolving the new quadratic optimisation problem of the objective function E_{sat} given by:

$$E_{sat} = E + (S^T\ I_m^* - I_{\text{sat}})^T\ \mu \tag{14}$$

where the selection matrix $S = [s_1 \dots s_q] \in \mathbb{R}^{n\times q}$ concatenates the vectors s_k of the canonical basis of $\mathbb{R}^n$ to select the combinations of the violated tension constraints to be saturated, and the vector $\mu \in \mathbb{R}^q$ is the Lagrange multiplier associated to the current equality constraints $S^T\ I_m^* - I_{\text{sat}} = 0$, such as the vector $I_{\text{sat}} \in \mathbb{R}^q$ is the currents vector that corresponds to the saturated tensions vector by inverting (7).

The set of solutions of the optimisation problem (14) can be written as:

$$I_m^* = I_{m\,p}^* + I_{m\,t}^* \tag{15}$$

where:

$$I_{m\,t}^* = I_{mn}^* + \left[W_I^+\ W_I^{\text{sat}} - S\right]\ [S^T\ W_I^+ W_I^{\text{sat}} - I_q]^{-1}\ \Delta_{\text{sat}}$$

in which the resulting saturated wrench matrix is $W_I^{\text{sat}} = W_I\ S$, and the vector of the excessive motor currents $\Delta_{\text{sat}} = I_{\text{sat}} - S^T\ (I_{mp}^* + I_{mn}^*)$ is the image of the excess in cable tensions evaluated by (7).

5 Results

5.1 Evaluation Task

In order to validate the vision-based position control scheme of Sect. 4, a set of end-effector trajectories are tested in simulation and experimentally. The end-effector being located on the measured initial pose X_{eo}, it is moved as follows:

- Step trajectories: the end-effector reference trajectories are three pure translations of magnitudes $+0.3$ m, and rotations of angles $+15°$ along/around respectively X_o, Y_o and Z_o axes.
- Tracking trajectories: the end-effector reference trajectory is a circle belonging to the plane $z = 0$ m, located at the center of the workspace and of radius 0.3 m.

5.2 Controllers Design

This part details the design of the controllers used:

- for the visual controller, the proportional gain matrix K is chosen as $k_P = k_\Phi = 6\,\text{rad}^{-1}$ to get a time response of 500 ms on each component of the pose X_e.
- for the joint speed controller, the gains K_p and T_i are tuned using the symmetrical optimum method as follows:
  - first, the integration time constant T_i is chosen as $T_i = a\,\tau$, where the coefficient a allows as to tune the phase margin $\Delta\phi$ so that $a = \tan^2(\frac{\Delta\phi}{2} + \frac{\pi}{4})$.
  - then, the proportional gain K_p is calculated as $K_p = j_{eq}/(\tau\sqrt{a})$, so that the phase margin corresponds to the maximal phase of the corrected system $\phi_{\max} = \Delta\phi - \pi$, which occurs at the pulsation $w_{\max} = 1/(\tau\sqrt{a})$.

  In our case, by selecting $a = 20$ (leading to $\Delta\phi \simeq 65°$), we obtain: $K_p = 0.1107$ A·s/rad and $T_i = 26$ ms. The achieved bandwidth for the joint speed control is 172 rad/s.

5.3 Obtained Performances

For the step trajectories, the time responses of each controlled component of the end-effector pose are given in Fig. 4 for the simulated and experimental cases. The response time is of 500 ms and the precision is less than 1 mm for the translations and less than 1° for the rotations.

The circular tracking trajectories of the end-effector are shown in Fig. 5 for the simulation and experimental cases. The corresponding time responses of the speed and current of the motor 1 are respectively given in the Figs. 6a and b.

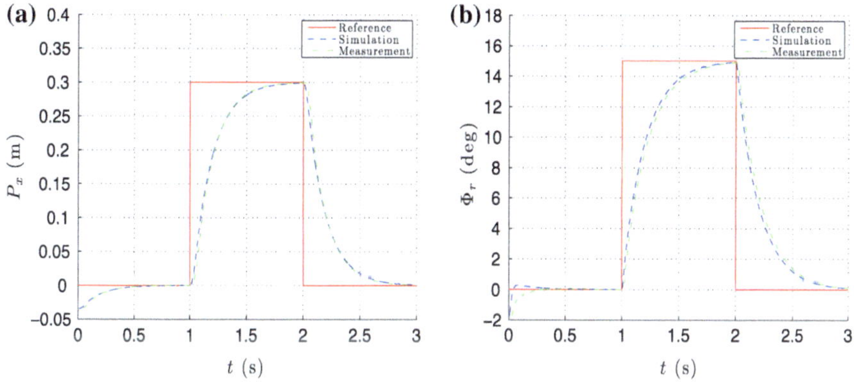

Fig. 4 End-effector pose for the step trajectories. **a** Position along X_o. **b** Orientation around X_o

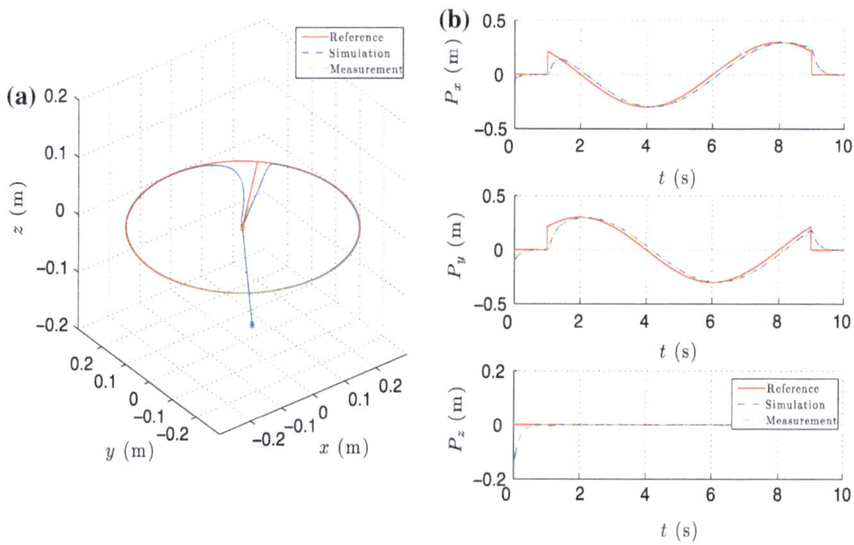

Fig. 5 End-effector pose for the circular tracking trajectory. **a** Spacial trajectory. **b** Temporal trajectory

The results provided from the test trajectories clearly show that the reference signals are nicely tracked, with the previous response time and a very little error, showing a very close matching between the simulations and the experiments.

The feasible tensions workspace is defined by the boundaries $T_{\min} = 1.48$ N and $T_{\max} = 18.52$ N, which have been calculated based on the static model of the actuators, considering the current limits of the motors [0 3] A, and the unwinding cable length limits [0 4.82] m. The objective tension T_{obj} has been dimensioned as $T_{\text{obj}} = 10\ I_{8\times1}$ N. The objective current I_{obj} is updated at each sample time using

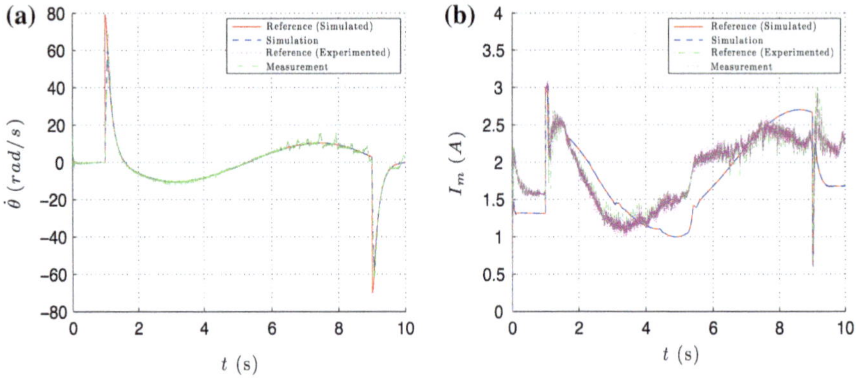

Fig. 6 Motor 1 responses for the circular tracking trajectory. **a** Motor 1 speed. **b** Motor 1 current

Fig. 7 Cable tensions for the circular tracking trajectory estimated using the actuators dynamics

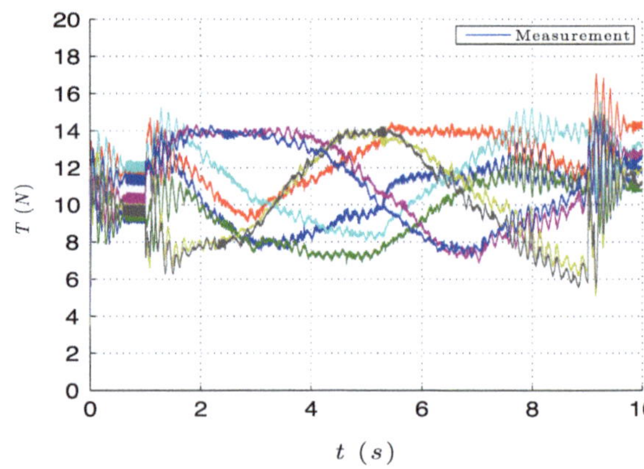

the dynamic model of the actuators from the objective tension T_{obj}. It can be seen on Fig. 7 that the eight cables are maintained under feasible tensions that are continuous during the circular motion of the end-effector.

6 Conclusion

This paper reports simulation and experiment results of the vision-based position control of a 6-DoF redundant CDPR INCA developed by Haption, equipped with the Bonita motion-capture system developed by Vicon to measure the end-effector pose. The control laws allow to track the pose reference trajectories with a response time of 500 ms and a precision less than 1 mm in translation and less than 1° in rotation, while maintaining the cables under feasible tensions. The corresponding bandwidth of approximatively 6 Hz would allow to achieve pick-and-place tasks at a quite high pace. Deeper investigations will be necessary in order to evaluate the limitations of the control laws in terms of reachable bandwidth and robustness.

References

1. Oh SR, Agrawal S (2005) Cable suspended planar robots with redundant cables: controllers with positive tensions. IEEE Trans Robot 21(3):457–465
2. Gosselin C (2012) Global planning of dynamically feasible trajectories for three-DoF spatial cable-suspended parallel robots. In: Cable-driven parallel robots. Springer, Berlin
3. Trevisani A (2012) Experimental validation of a trajectory planning method avoiding cable slackness and excessive tension in underconstrained translational planar cable-driven robots. In cable-driven parallel robots. Springer, Berlin
4. Ming A, Higuchi T (1994) Study on multiple degree-of-freedom positioning mechanism using wire-concept, design and control (part 1). Int J Jpn Soc Precis 28(2):131–138
5. Lafourcade P (2004) Etude des manipulateurs parallèles à câbles, conception d'une suspension active pour soufflerie. Ph. D thesis, ENSAE School
6. Kawamura S, Kino H, Won C (2000) High-speed manipulation by using parallel wire-driven robots. Robotica 18:13–21
7. Gholami P, Aref MM, Taghirad HD (2008) On the control of the KNTU CDRPM: A cable driven parallel manipulator. In: IEEE/RSJ international conference intelligent robots and systems (IROS), pp 2404–2409, September 2008
8. Lamaury J, Gouttefarde M, Michelin M, Tempier O (2012) Design and control strategies of a redundant suspended cable-driven parallel robot. In: Lenarcic J, Husty M (eds) Advances in robot kinematics (ARK). Springer, Berlin, pp 237–244
9. Carricato M, Merlet JP (2011) Direct geometrico-static problem of under-constrained cable-driven parallel robots with three cables. In: International conference robotics and automation (ICRA), pp 3011–3017, May 2011
10. Dallej T, Gouttefarde M, Andreff N, Michelin M, Martinet P (2011) Towards vision-based control of cable-driven parallel robots. In: IEEE/RSJ international conference intelligent robots and systems (IROS), pp 2855–2860, September 2011
11. Dallej T, Gouttefarde M, Andreff N, Dahmouche R, Martinet P (2012) Vision-based modeling and control of large-dimension cable-driven parallel robots. In: IEEE/RSJ international conference intelligent robots and systems (IROS), pp 1581–1586, October 2012
12. Chellal R, Laroche E, Cuvillon L, Gangloff J (2012) An identification methodology for 6-DoF cable-driven parallel robots parameters—application to the INCA 6D robot. In: Bruckmann T, Pott A (eds) First international conference on cable-driven parallel robots. Cable-driven parallel robots. Springer, pp 301–317, September 2012
13. Kessler C (1958) Das symmetrische optimum. In: Regelungstechnik, vol 6, pp 395–400
14. Yu K, Lee LF, Krovi VN (2009) Simultaneous trajectory tracking and stiffness control of cable actuated parallel manipulator. In: International conference design engineering technical and computers and information in engineering, 2009
15. Vafaei A, Aref MM, Taghirad HD (2010) Integrated controller for an over constrained cable driven parallel manipulator: KNTU CDRPM. In: IEEE international conference on robotics and automation (ICRA), pp 650–655, May 2010
16. Mikelsons L, Bruckmann T, Hiller M, Schramm D (2008) A real-time capable force algorithm for redundant tendon-based parallel manipulators. In: IEEE international conference robotics and automation (ICRA), pp 3869–3874, May 2008
17. Borgstrom PH, Jordan BL, Sukhatme GS, Batalin MA, Kaiser WJ (2009) Rapid computation of optimally safe tension distributions for parallel cable-driven robots. IEEE Trans Robot 25(6):1271–1281
18. Lamaury J, Gouttefarde M (2012) A tension distribution method with improved computational efficiency. In: Bruckmann T, Pott A (eds) First internatioanl conference on cable-driven parallel robots. Cable-driven parallel robots. Springer, pp 71–85, September 2012

Analysis of a Real-Time Capable Cable Force Computation Method

Katharina Müller, Christopher Reichert and Tobias Bruckmann

Abstract This paper deals with the problem of computing minimum feasible cable force distributions for redundantly actuated cable-driven parallel robots. In this context, the known Closed Form Method as well as its limitations are presented and potential improvements are identified. This finally leads to an approach called Improved Puncture Method. Then the methods are analyzed regarding their covered workspace, the resulting cable force distributions and the needed computation time. Finally the Improved Puncture Method was implemented into the augmented PD controller and run on the SEGESTA prototype.

Keywords Cable-driven parallel robot · Force distribution · Threefold redundancy · Geometrical approach · Improved puncture method

1 Introduction

Cable-driven parallel robots (CDPRs) belong to the class of parallel robots. In contrast to serial robots where joints and links (e.g. struts) are arranged in series, for parallel robots the actuators are directly connected to the end-effector. That leads to better load spreading, higher stiffness and higher precision. On the other hand, classical parallel robots usually have a relatively small workspace compared to serial manipulators.

This can be counteracted by using cables instead of stiff struts. The cables are connected to the end-effector in a parallel topology and wound up by a motor-driven

K. Müller (✉) · C. Reichert · T. Bruckmann
Chair of Mechatronics, University Duisburg-Essen, Essen, Germany
e-mail: mueller@imech.de

C. Reichert
e-mail: reichert@imech.de

T. Bruckmann
e-mail: tobias.bruckmann@uni-due.de

A. Pott and T. Bruckmann (eds.), *Cable-Driven Parallel Robots*,
Mechanisms and Machine Science 32, DOI 10.1007/978-3-319-09489-2_16

winch. Winding up cables can be done very fast. Thus, the end effector can reach high velocities and accelerations [1, 5, 8].

CDPRs can be classified by their degree of redundancy r [9, 12]. This can be determined by $r = m - n$ where m is the number of cables and n is the degree of freedom. Because cables can only transmit tensile forces, to fully restrain a CDPR the number of cables must exceed the degree of freedom at least by one. In the case of minimum redundancy ($r = 1$) the CDPR is called Completely Restrained Parallel Manipulator (CRPM). The class Incompletely Restrained Parallel Manipulator (IRPM) denotes the case of $r < 1$ and the class Redundantly Restrained Parallel Manipulator (RRPM) describes the case of $r > 1$ [2].

2 Problem Definition

2.1 Mathematical Definition

Figure 1 shows the forces (cable forces and external forces) and torques which effect the end effector.

For the static equilibrium, both the sum of the forces and the sum of the torques have to be zero:

$$\sum_{i=1}^{m} \mathbf{f}_\mathrm{i} + \mathbf{f}_\mathrm{p} = \mathbf{0} \text{ and } \sum_{\mathbf{i=1}}^{\mathbf{m}} \mathbf{p}_\mathrm{i} \times \mathbf{f}_\mathbf{i} + \boldsymbol{\tau}_\mathrm{p} = \mathbf{0}. \tag{1}$$

In short form this can be written as:

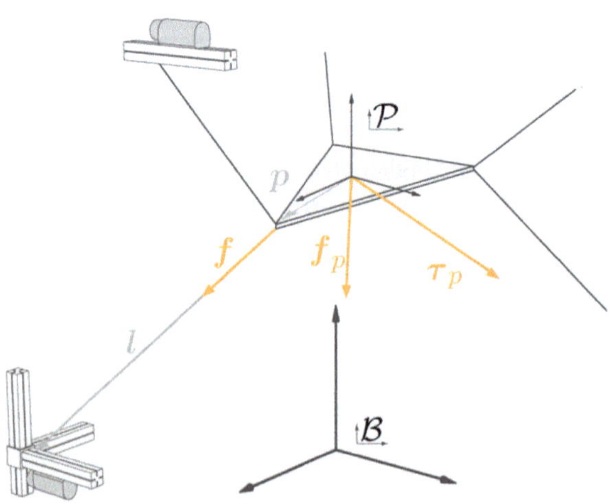

Fig. 1 Wrench applied onto the platform. *Source* [3]

$$\mathbf{A}^{\mathrm{T}}\mathbf{f} + \mathbf{w} = \mathbf{0} \tag{2}$$

$$\Leftrightarrow \quad \mathbf{f} = \underbrace{-\mathbf{A}^{+\mathrm{T}}\mathbf{w}}_{\mathbf{f}_0} + \mathbf{H}\boldsymbol{\lambda}. \tag{3}$$

Here $\mathbf{A}^{\mathrm{T}} \in \mathbb{R}^{\mathrm{n}\times\mathrm{m}}$ is the structure matrix, $\mathbf{f} \in \mathbb{R}^{\mathrm{m}}$ is the cable force vector, $\mathbf{w} \in \mathbb{R}^{\mathrm{n}}$ is the vector of the external forces and torques (e.g. caused by inertia of process loads), $\mathbf{A}^{+\mathrm{T}} \in \mathbb{R}^{\mathrm{m}\times\mathrm{n}}$ is the Moore–Penrose pseudoinverse of the structure matrix and $\mathbf{H} \in \mathbb{R}^{\mathrm{n}\times\mathrm{r}}$ is the kernel (or null space) of the structure matrix.

For a CDPR, Eq. (3) is additionally bounded by a maximum and a minimum permissible cable force. To avoid slackness the minimum forces have to be greater than zero, because of the fact that cables can only transmit tensile forces. For technical reasons, the cable forces are also bounded above. This yields in the following inequality:

$$\mathbf{0} < \mathbf{f}_{\mathrm{min}} \leq \mathbf{f} \leq \mathbf{f}_{\mathrm{max}}. \tag{4}$$

Cable forces which fulfill Eq. (3) as well as Eq. (4) are called feasible cable forces.

$\mathbf{f}_0$ is a minimum solution, solving the system of equations described in Eq. (2) with the smallest Euclidean norm. Because the minimum solution usually lies outside the admissible solution space defined by the force limits defined in Eq. (4), it needs to be moved by choosing an appropriate value for $\mathbf{H}\boldsymbol{\lambda}$.

For CRPMs and RRPMs this system of equations is under-determined. Accordingly, there usually exists more than one solution.

An additional demand on the computed force distributions is that they have to be continuous while the end effector moves along a continuous trajectory. This is crucial for control purposes.

2.2 Solution Approaches

To solve the presented problem, different approaches have been developed. In [10] Pott provides a list of most of the known approaches.

Generally, there are two different ways to calculate the feasible cable forces: Either this problem can be formulated as an optimization problem or it can be interpreted and solved geometrically. Normally, optimization problems are solved iteratively. For this reason, the needed computation time usually cannot be exactly predicted and therefore real-time capability cannot be guaranteed which again is crucial for control purposes.

Because of their non-iterative character, geometrical approaches are a promising alternative for real-time applications. In [7] basic geometrical approaches are briefly presented. Two of them, the Closed Form Method and the Puncture Method, will be analyzed in the next sections.

3 Improved Puncture Method

The Improved Puncture Method (IPM) is a further development of the Puncture Method (PM), which to the authors' best knowledge was first presented in [7]. It may be used in combination with the improved Closed Form Algorithm presented by Pott in [10]. It belongs to the geometrical approaches to solve the presented problem, meaning that the given equations are interpreted geometrically.

As shown in Fig. 2 for $m = 3$ cables and $n = 1$ degree of freedom, Eq. (4) can be interpreted as an m-dimensional hypercube $\mathscr{C}$, whereas solutions of Eq. (3) lie in the r-dimensional solution space $\mathscr{S}$. Feasible cable forces, which fulfill both equalities (Eqs. 3 and 4) lie inside of the intersection $\mathscr{F}$.

Using the **Closed Form Method** (CFM) to find one feasible cable force distribution, two steps have to be performed:

- First the center point of the hypercube $\mathscr{C}$ is projected onto the solution space $\mathscr{S}$.
- Then it has to be checked whether the projected center point lies within the intersection $\mathscr{F}$ [11].

This method has two main advantages, both regarding the computation time: As experiments showed, it needs very low average computation time. Additionally, the worst-case computation time can be strictly bounded.

Not all computation methods cover the same workspace (the space that can be reached by the end effector). Some may fail to find a solution where others identify a feasible force distribution. For the CFM the covered workspace is relatively small compared to those of other methods. Besides, the level of the computed cable force distributions is quite high.[1]

For practical applications it is often desirable to have rather low cable force tensions. One method to achieve this is the **Puncture Method** (PM). Here two points on the solution space $\mathscr{S}$ are needed:

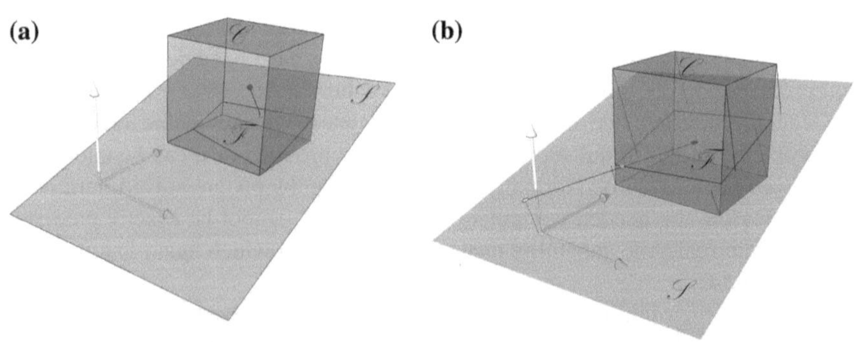

Fig. 2 Schematic representation of the CFM and the IPM for $m = 3$ cables and $n = 1$ degree of freedom. **a** Closed form method. **b** Improved puncture method. *Source* [3]

[1] Research results regarding this topic can be found in [6, 7].

- One point must lie inside of the hypercube (e.g. the solution of the Closed Form Method).
- One point must lie outside of the hypercube, preferably near the origin. Therefore, the projection of the origin onto the solution space lends itself to being used.
- When these two points are connected by a line, the desired cable force distribution lies at the intersection of this line with one of the minimum cable force boundaries.

Research results show that this method (based on the Closed Form Method) still is very fast, while the computed cable force distributions lie on a much lower tension level, which in fact is very similar to the solutions of Quadratic Programming. Nevertheless, the covered workspace is still as small as the one of the Closed Form Method.[2]

Addressing this problem, in [10] Pott presents an extension to the CFM to enlarge the covered workspace. In this paper it is referred to as **Improved Closed Form Method** (ICFM).

The added algorithm sets in when the solution of the CFM lies outside of the cable force boundaries:

- The cable force which is farthest away of the upper or lower cable force boundary is picked and fixed to its maximum $f_{\max}$ or minimum $f_{\min}$ respectively. For the remaining cable forces, the CFM is executed again.
- If the new solution now lies inside of the cable force boundaries, it is the sought solution and the algorithm is stopped.
- Otherwise, the next cable force with the largest distance to the cable force boundaries gets fixed to a constant value as long as the remaining number of unfixed cable forces is larger then the degree of freedom n of the robot.

Consequently, the added algorithm can be performed r times at most. This means that the worst-case computation time of the improved Closed Form Method can still be strictly bounded.

Now these two approaches, the ICFM and the PM, get combined into the **Improved Puncture Method** (IPM). Thus, the objective is to obtain a method which, on the one hand, is very fast with a strictly bounded worst-case computation time and which is therefore real-time capable. On the other hand, the resulting cable force distributions shall lie on a preferably low tension level.

4 Analysis for a CDPR with a Threefold Redundancy

To check, whether the IPM meets the declared objectives, it was exemplarily analyzed for a spatial parallel robot with six degrees of freedom and nine cables (as illustrated in Fig. 3).

[2] See [6, 7] for more information.

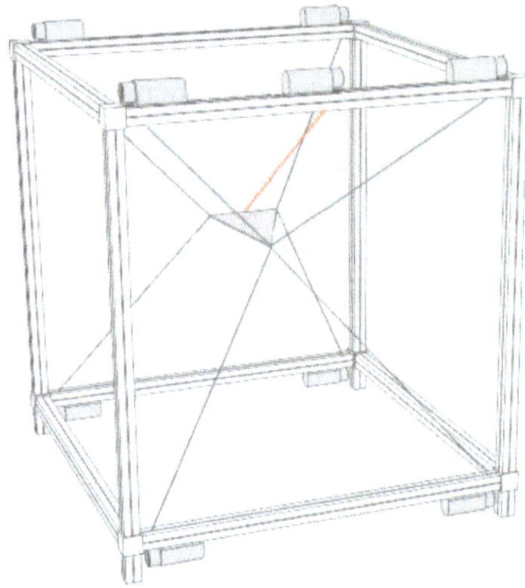

Fig. 3 Schematic representations of a RRPM with $m = 9$ and $n = 6$. *Source* [2]

4.1 Workspace

To measure the covered workspace, the available space inside of the robot frame (size 1.48 m × 1.36 m × 1.00 m) was discretized and the distance between the grid points was set to 0.02 m. Then it was checked for which grid points a feasible cable force distributions could be found. The cable force boundaries were defined as $f_{\min} = 10\,\text{N}$ and $f_{\max} = 100\,\text{N}$.

In Fig. 4a the workspace covered by the CFM is represented. For 36,648 points a feasible force distribution could be found.

When the PM is based on the basic CFM, its workspace is the same.

By using the ICFM (getting the IPM), the covered workspace could be considerably enlarged. It contains 64,362 points and is pictured in Fig. 6b.

4.2 Resulting Cable Force Distributions

Next the resulting cable forces are analyzed. Figure 5a shows the trajectory for which the resulting force distributions where computed with the different methods. It was supposed that the end effector (EE) moves along the spiral path from the lowest to the highest end. Again the cable force boundaries were defined as $f_{\min} = 10\,\text{N}$ and $f_{\max} = 100\,\text{N}$.

Figure 5b shows the cable force distributions that result from Quadratic Programming. In this context it shall only be used as a reference for comparison.

The cable forces resulting from the CFM (Fig. 5c) are on a comparatively high level. Here no kind of minimization takes place. Besides, in the area between about

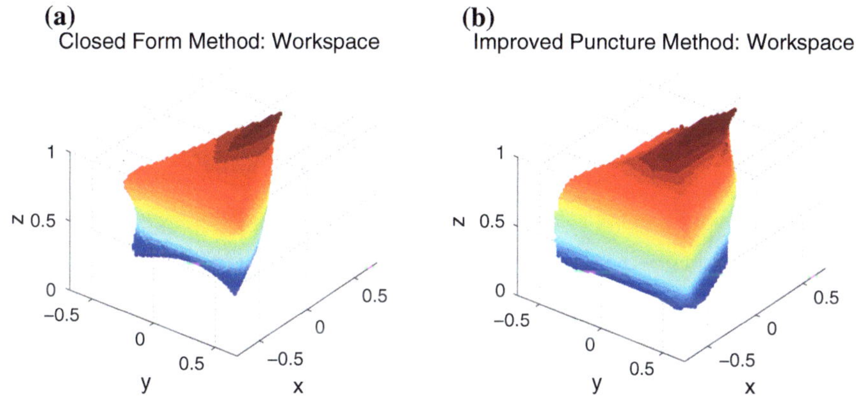

Fig. 4 Covered workspace. **a** Closed Form Method. **b** Improved Puncture Method

90 and 96 % of the trajectory and at its end no feasible cable force distributions can be found by this method. For this reason, the graph shows gaps in these ranges.

The resulting cable forces of the PM based on the CFM are on a much lower level, which is quite similar to the one of Quadratic Programming, but not identical. However, the graph still has gaps in the critical areas. Although, they are slightly smaller.

The graph of Fig. 5e which belongs to the ICFM shows no gaps anymore. Instead, the critical cable forces were fixed to the upper or lower limit. Anyhow, the values of the cable forces lie on a comparatively high level.

For this reason, the PM can be applied again. The results are represented in Fig. 5f. As expected the cable forces could be reduced again in the uncritical area where the end effector moves inside of the CFM's workspace. But when the end effector leaves it and the extended algorithm of the ICFM sets in the IPM does not show the desired minimizing effect. The resulting cable forces do not differ distinctly from the ones of the ICFM in the range of trajectory process between around 90 and 96 %.

4.3 Computation Time

The computation time that needed by the different methods was exemplarily tested by running each method on a real time system using `TwinCAT3`® by BECKHOFF on an Intel® Core2™ Duo CPU T9400 processor.

The results for the CFM as well as for the IPM are shown in Fig. 6.

Both methods need very little computation time. The average computation time of the CFM is 22.25 µs. The computation time is quite constant and the marginal oscillation to a maximum of around 3 µs can be led back to system latency.

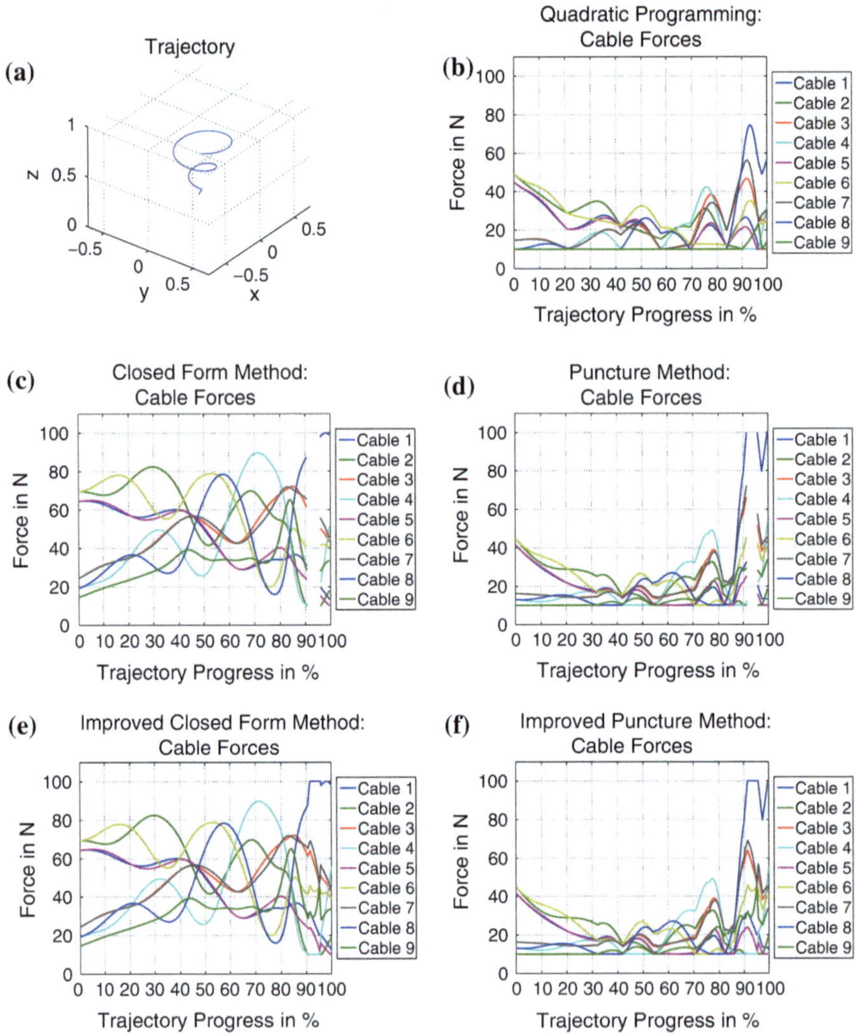

Fig. 5 Resulting cable force distributions. **a** Spiral trajectory. **b** Quadratic Programming. **c** Closed Form Method. **d** Puncture Method. **e** Improved Closed Form Method. **f** Improved Puncture Method

The IPM needs more than twice that much computation time, but that is still very fast. However, it is striking, that the computation time increases up to 104.8 μs in the critical area at a trajectory process of around 90 till 95 % where the CFM does not find feasible solutions and where for the IPM the extended algorithm sets in.

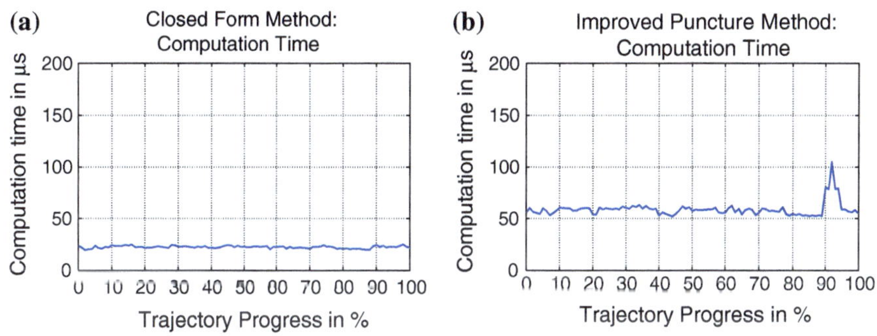

Fig. 6 Computation time. **a** Closed Form Method. **b** Improved Puncture Method

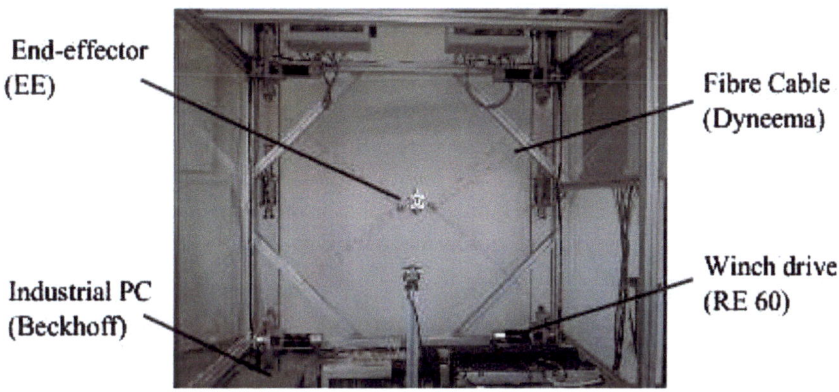

Fig. 7 SEGESTA-prototype of the chair of mechatronics at the University of Duisburg-Essen

4.4 Real-Time Control Application

The presented Improved Puncture Method was implemented on the 6-DOF SEGESTA prototype shown in Fig. 7. The CDPR is controlled by means of eight BLDC-Motors (Maxon RE 60) linked to the winches with a nominal torque generation of $c_{nom} = 0.85\,\mathrm{N\,m}$. Thus $m = 8$, i.e. the CDPR has a twofold redundancy. The mass of the EE is approximately 0.3 kg. To measure the forces strain-gauge beam arrangements (Megatron KM302) are integrated into the winches.

The applied augmented PD (APD) control scheme [4] is based on a joint space PD controller to regulate the cable length regarding of the desired EE pose. A non-linear feed forward control law using the inverse dynamics solution is integrated to compensate the platform dynamics by means of an operational space force vector described by v_{ff}. To compensate the winch dynamics like inertia and friction forces a joint space force vector f_{ff} is in-cooperated into the control scheme. The augmented PD control law is given by

Table 1 Coordinates of the chosen trajectory

Path	0	1	2	3	4	5	6	7	8
x	0	0.2	0.2	−0.2	−0.2	−0.2	−0.2	0.2	0.2
y	0	−0.2	0.2	0.2	−0.2	−0.2	0.2	0.2	−0.2
z	0.5	0.4	0.4	0.4	0.4	0.6	0.6	0.6	0.6

$$f_m = D\ddot{q}_d + f_c sign(\dot{q}) + f_v\dot{q} + f_{TD} \tag{5}$$

where f_m is the commanded actuator input force, f_{TD} are the desired cable forces based on a force distribution algorithm as presented in the previous sections and f_c and f_v are the diagonal matrices of Coulomb and viscous coefficients, respectively.

Within this control approach the input vector for the force distribution algorithm is

$$v_c = v_{PD} + v_{ff} \tag{6}$$

where $v_{PD} = A^T \left(K_p e + K_D \dot{e}\right)$ defines the controller output. K_p and K_D are positive definite diagonal matrices, e is the cable length error vector and $\dot{e}$ is the cable velocity error vector. Based on the inverse kinematics solution the desired cable length q_d will be computed based on a given EE pose. Here the EE pose refers to the position of the

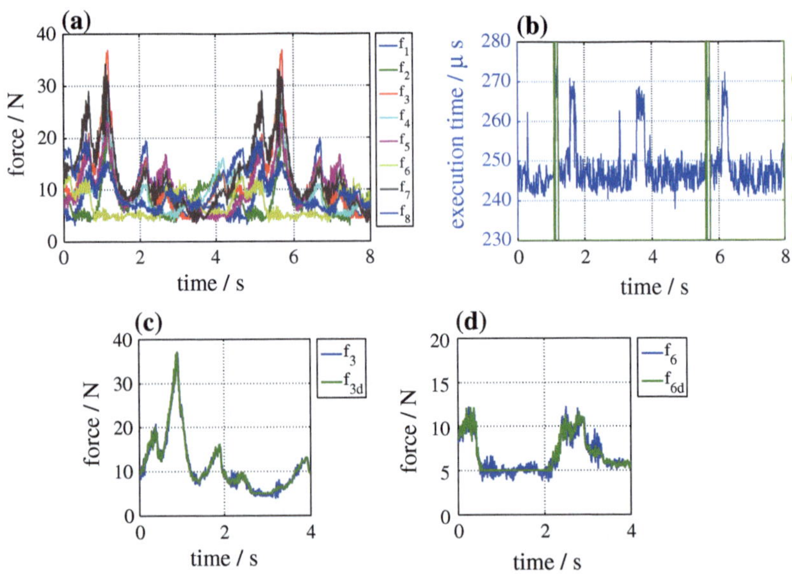

Fig. 8 Measured data. **a** Measured cable forces. **b** Controller computation time. **c** Forces of third cable. **d** Forces of sixth cable

EE and the orientation of the EE with respect to the given base frame. The desired cable velocity $\dot{q}_d = A\dot{x}_d$ will be computed by the help of the structure matrix.

The augmented PD controller was implemented on the SEGESTA prototype shown in Fig. 7. The EE was controlled along the path shown in Table 1 and the cable force boundaries were set to $f_{min} = 5\,\text{N}$ and $f_{max} = 50\,\text{N}$.

Figure 8a shows the measured cable forces while in Fig. 8b the controller's computation time is illustrated. When k is 1 the extended algorithm of the IPM sets in. It can be seen that the computation time of the controller increases by about 25 μs then. The remaining jumps are caused by other parts of the controller.

Figure 8c and d show the desired and the measured forces of cable 3 and cable 6 which were exemplarily selected.

5 Conclusion

This paper dealt with the problem of computing minimal cable force distributions for RRPMs for real-time applications. In this context, the CFM and its limits were presented as well as potential improvements which finally led to an algorithm called Improved Puncture Method. The methods were analyzed regarding their covered workspace, the resulting cable force distributions and the needed computation time.

In conclusion, it can be stated that the Improved Puncture Method still is very fast and that it covers a much larger workspace. A non-negligible drawback, however, is that the minimization only works properly as long as the end-effector moves within the workspace that is covered by the basic Closed Form Method. In the outer areas which are only covered because of the extended algorithm, the Puncture Method does not have the desired minimizing effect.

Acknowledgments The research leading to these results has received funding from the European Community's Seventh Framework Programme under grant agreement No. NMP2-SL-2011-285404 (CableBOT).

References

1. Bruckmann T (2010) Auslegung und Betrieb redundanter paralleler Seilroboter. Ph. d. dissertation, Universität Duisburg-Essen
2. Bruckmann T, Mikelsons L, Brandt T, Hiller M, Schramm D (2008) Wire robots Part I—kinematics, analysis and design. In: Lazinica A (ed) Parallel manipulators—new developments. I-Tech Education and Publishing, Vienna, pp 109–132
3. Bruckmann T, Reichert C (2013) Real-time determination of set point cable force distributions. European Robotics Forum, Lyon
4. Lamaury J, Gouttefarde M (2013) Control of a large redundantly actuated cable-suspended parallel robot. In: IEEE international conference on robotics and automation (ICRA) 2013, pp. 4659–4664. doi:10.1109/ICRA.2013.6631240

5. Landsberger SE (1984) Design and construction of a cable-controlled. Parallel Link Manipulator. Masterthesis, Massachusetts Institute of Technology
6. Müller K (2013) Methoden zur Berechnung der Seilkraftverteilungen höherredundanter paralleler Seilroboter. Master's thesis
7. Müller K, Reichert C, Bruckmann T Analysis of geometrical force calculation algorithms for cable-driven parallel robots with a threefold redundancy. In: 14th international symposium on advances in robot kinematics. (submitted)
8. Merlet JP (2007) Kinematics of the wire-driven parallel robot MARIONET using linear actuators. ANR SIROPA http://hal.archives-ouvertes.fr/hal-00172932
9. Ming A, Higuchi T (1994) Study on multiple degree-of-freedom positioning mechanism using wire (Part 1). Int J Jpn Soc Precis Eng 28:131–138
10. Pott A (2014) An improved force distribution algorithm for over-constrained cable-driven parallel robots. In: Thomas F, Perez Gracia A (eds) Computational kinematics. Mechanisms and machine science, vol 15. Springer, Netherlands, pp 139–146
11. Pott A, Bruckmann T, Mikelsons L (2009) Closed-form force distribution for parallel wire robots. In: Kecskemethy A, Müller A (eds) Computational kinematics. Springer, Berlin, pp 25–34
12. Verhoeven R (2004) Analysis of the workspace of tendon-based stewart platforms. Dissertation, Gerhard-Mercator-Universität -GH- Duisburg

First Experimental Testing of a Dynamic Minimum Tension Control (DMTC) for Cable Driven Parallel Robots

Saeed Abdolshah and Giulio Rosati

Abstract Cable tension distribution is an important issue in parallel cable-driven robots to obtain high efficiency and accuracy of motion. In this paper, a novel approach is introduced to optimize cable tension distribution of cable-driven parallel robots, which consists in modifying the minimum tension of the cables according to the dynamics of the system. This method has been compared to the traditional, fixed-minimum tension approach on a 2-cable, 1 DOF test bed with different settings of the controller. First experimental results showed that Dynamic Minimum Tension Control (DMTC) can be better than traditional approaches in terms of accuracy and energy consumption.

1 Introduction

Cable-Driven Parallel Robots have a short but growing history. They can be defined as closed loop mechanisms consisting of multiple actuated cables and an end-effector. There are some advantages such as very large workspace and high acceleration due to less mass and inertia, however inaccuracy and necessity of cables to be in tension can be regarded as disadvantages. Cable driven systems have been used widely in different applications. SkyCam [1], rehabilitation [2], and high speed manipulation [3] are the most important applications of such systems.

Recently, there have been major advances in different fields of cable driven robots related to statics, dynamics, control and design [4]. Performance study and cable tension distribution are some of the most significant subjects which were discussed by different researchers. In these studies, a maximum limitation of cable tension has been considered based on cable and actuators properties, and a minimum boundary is usually set to keep cables in tension to reduce control issues and vibrations. Several

S. Abdolshah · G. Rosati (✉)
Department of Management and Engineering (DTG), University of Padua, 35131 Padova, Italy
e-mail: giulio.rosati@unipd.it

S. Abdolshah
e-mail: saeed.abdolshah@studenti.unipd.it

A. Pott and T. Bruckmann (eds.), *Cable-Driven Parallel Robots*,
Mechanisms and Machine Science 32, DOI 10.1007/978-3-319-09489-2_17

algorithms were proposed to find the most efficient cable tension in this range, to apply the smallest torque values without changing the working condition.

Different solutions have been derived accroding to the level of redundancy of the system being controlled, but all of them employ a fixed value for the minimum allowable tension in the cables. The most common algorithm of CRPMs cable tension distribution consists in finding the set of cable tensions, among those ensuring the desired end-effector wrench, which has at least one cable at minimum tension [5].

In the case of overconstrained cable robots, many researchers investigated on estimation of optimal tension distribution using linear, and quadratic programming. Shiang et al. [6] proposed a standard linear programming to minimize sum of all tensions while maximizing tension on two longest cables in order to refuse any kind of slack on a four-cable array robotic crane. In [7] an analytical solution was suggested based on sum of tensions along cables as small as possible at every pose of platform without violating the controllable workspace condition. The minimum Tension was chosen referring to some experiments for an accurate path tracking in a completely constrained 6-DOF robot.

Pham et al. [8] proposed a recursive algorithm to check the existence of positive torques, and then optimized torque objective function with linear programming. In another study, tension distribution has been implemented by feedback linearization, however some modifications were suggested in case of mathematical constraints or negative force [9]. Necessity of finding a proper starting point and possibility of jump from one extreme point to another between successive computations are problems of LP, however with introducing an optimally safe tension distribution with a slack variable, fast generation of proper starting and optimal point were gained [10]. Moreover, quadratic programming methods for estimation of two-norm optimal tension distributions were applied [11, 12], and Lim et al. [13] proposed a gradient programming method and compared it to linear and quadratic programming.

Furthermore some other methods based on convex optimization for minimization of actuator forces were applied [14]. Minimization of p-norm [15], and L1-norm optimization [16] have been applied for tension distribution, however Mikelsons et al. [17] investigated on safe tension distribution using noniterative method to gain higher robot stifness. Also since the fingers of a grasp are unidirectional, some relationships between planar cable-driven robots and spatial antipodal grasp theorem were studied to achieve proper cable tension distribution [18].

Pott et al. [19] had proposed a closed form solution which has been modified in his recent study [20]. In addition, properties of different cable tension distribution methods were explained and compared regarding to some factors such as computational speed, workspace coverage, capability of real time responses, applicability for different redundancies and force margin. This comparision shows that each of common methods have some kinds of problems and are not capable of controlling all kind of cable driven robots efficiently.

In this paper, we propose a novel approach for cable tension calculation, based on the estimated dynamics of the end-effector, which sets a dynamically variable value for the minimum tension of cables according to the desired wrench. The algorithm has been tested on a 2-cable, 1 DOF prototype, and results were compared

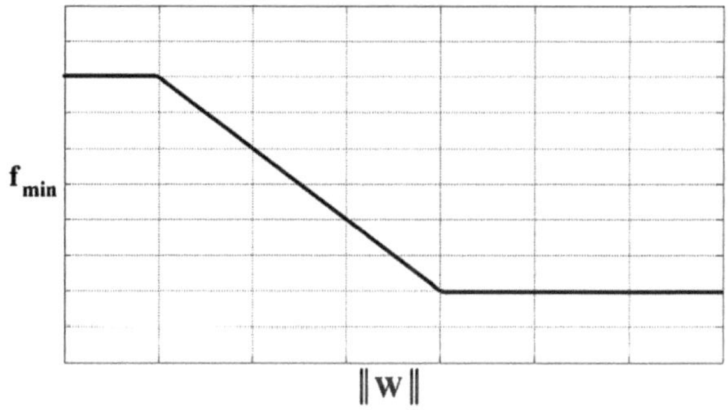

Fig. 1 Sample mapping function used for changing the minimum cable tension $f_{\min}$ according to the absolute desired wrench W

with traditional, fixed minimum tension methods. In Sect. 2, the new algorithm is described. Section 3 presents the experimental setup and the analysis of results.

2 Dynamic Minimum Tension Control

The main idea behind the DMTC control consists in dynamically calculating a suitable minimum value of the lower boundary $f_{\min}$ of cable tension, before applying a standard tension distribution algorithm. The main difference with already existing approaches lies in that the value of $f_{\min}$ is not fixed. On the contrary, it is changed on-the-fly according to the dynamics of the end-effector. More precisely, $f_{\min}$ is increased when the *desired end-effector wrench* W is small, and it is decreased when its absolute value gets greater.[1]

The main reason for adjusting $f_{\min}$ is that, on the one hand, higher values of $f_{\min}$ are preferrable, as they allow to avoid cable slacking and they usually yield higher positioning accuracy [7]; on the other hand, increasing $f_{\min}$ can lead to saturation of actuators, especially when the total wrench W is large. Based on such considerations, we propose to reduce $f_{\min}$ as long as the absolute value of W increases, and to restore higher minimum tensions when it gets close to zero, as qualitatively shown in Fig. 1. Such an approach can be applied to any cable-driven parallel robot, regardless the number of cables and the number of degrees of freedom, as the mapping function converts one scalar quantity (the absolute desired wrench) into another scalar quantity (the minimum tension of the cables). In the implementation presented in this paper, this is obtained using a look-up table (LUT).

It is well known that the static equilibrium of an n-DOF cable driven robot, controlled by m cables, can be expressed by the following linear equation system:

[1] For such systems with rotational and translational degrees of freedom, the units in the wrench vector W must be normalized before calculating its absolute value, e.g., by dividing the torques by the pulley radius.

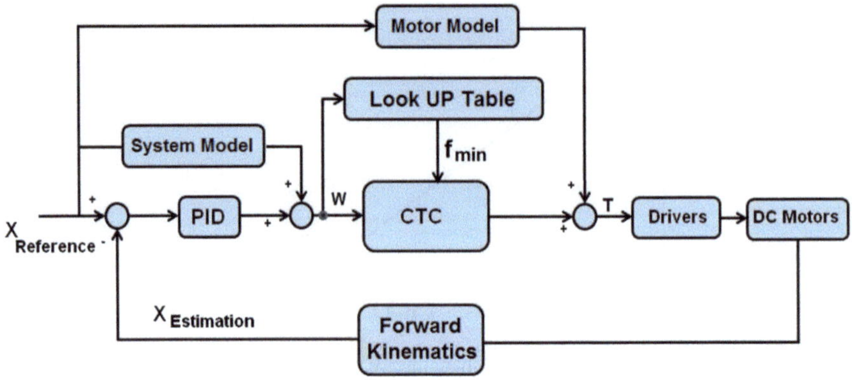

Fig. 2 Schematic of a position control using the Dynamic Minimum Tension Control (DMTC)

$$AT = W \tag{1}$$

where $A \in \mathbb{R}^{n\times m}$ is the structure matrix of the robot, W is the wrench applied to the end-effector, and T is the vector of cable tensions, which can be written as:

$$W = \begin{bmatrix} F \\ M \end{bmatrix} \in \mathbb{R}^n \qquad T = \begin{bmatrix} T_1 \\ . \\ . \\ . \\ T_m \end{bmatrix} \in \mathbb{R}^m \tag{2}$$

It has been demonstrated in [21] that, as long as the end-effector lies in the force-closure workspace, vector T can be calculated by taking the sum of one particular solution T_p of system (1) and one vector belonging to the kernel of the matrix A with all positive components, which we call T_k, whose norm increases with $f_{\min}$:

$$T = T_p + T_k = T_p + N\lambda(f_{\min}) \tag{3}$$

where $N \in \mathbb{R}^{m\times(m-n)}$ is a basis of the kernel of A; $\lambda \in \mathbb{R}^{m-n}$ contains the weights of the linear combination of the columns of N yielding T_k. By properly chosing the vector λ, the vector T will hold at least one cable at tension $f_{\min}$ and all others cables at $f_{\min}$ or greater. Clearly, by reducing $f_{\min}$ we will get a vector T satisfying system (1) with reduced norm. This is particularily clear, for example, in the case of planar point-mass CRPMs, where for a given direction of the desired force W, there will be at least one cable whose direction is opposite to that of W (in the sense that the dot product between the desired force W and the direction of the cable will be negative), so the reduction of its tension ($f_{\min}$) will help reducing the tension in all other cables.

Figure 2 presents a schematic of a position control using the DMTC approach. The output W of the controller, given by the sum of feedforward and feedback actions,

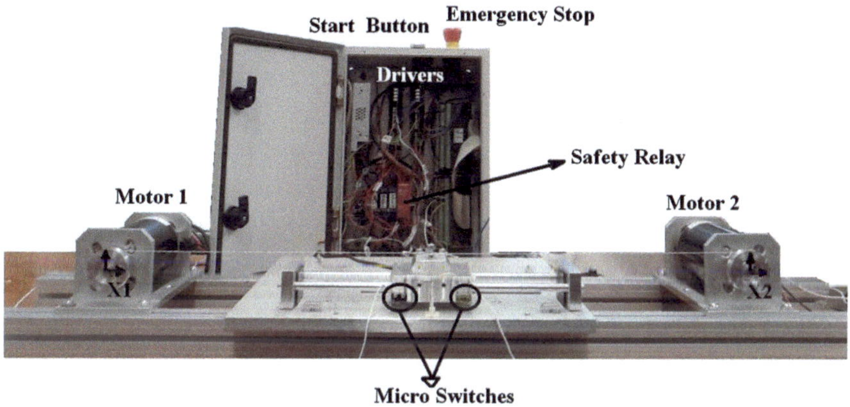

Fig. 3 Experimental setup for testing optimal force distribution

enters the DMTC block which calculates $f_{\min}$. Both W and $f_{\min}$ enter the Cable Tension Computation (CTC) block, which outputs reference to drivers. Position feedback by encoders closes the feedback loop. A similar approach, except from the DMTC, is that presented by Lamaury et al. in [22], where they proposed the PID position control in the Cartesian space of a 6-DOF cable-suspended parallel robot, based on forward kinematics and on a particular PID tuning procedure. Clearly, the DMTC approach can be applied also in the case of joint space position control, provided that also in this case the output of the controller is the end-effector wrench, which is obtained from the motor torques vector by using the pseudo-inverse of the Jacobian matrix.

3 Experimental Setup and Results

To test the new method, a 1 DOF cable-driven robot with two DC motors and a slider (end-effector) was built (Fig. 3). One linear and two rotational encoders were used to measure the position of the slider and of the motors. This plant was connected to a PC via a PCI Multifunction I/O Sensoray626 board to control the system by Matlab and Simulink RTWT.

In the feedforward loop, slider friction was modeled as a function of velocity, based on sum of Stribeck, Coulomb, and viscous components, which were estimated through experiments. In the feedback loop, the position of the end-effetor was estimated using rotational encoder readings, with additional compensation of cable elongation. The linear encoder was used only to measure the actual position of the end-effector for final assesment of controller performance.

The CTC block distributed the force in the two cables by simply imposing their tensions to $f_{\min}$ and $W + f_{\min}$, according to the direction of W. Then, it calculated

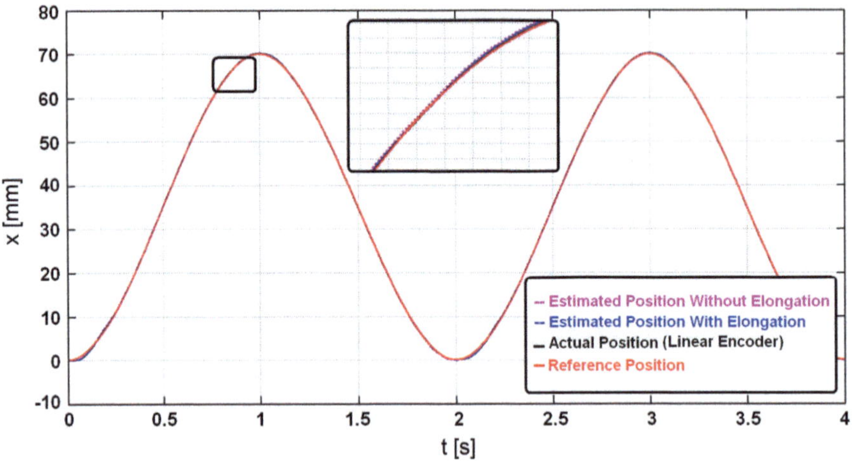

Fig. 4 Comparision of motion reference and position estimation of encoders

the current reference for the drivers in two ways: either proportionally to the desired cable tension (CTC1), or by calculating a position reference for each motor and yielding to the driver a PID control action to track this reference (CTC2). The CTC1 implements a very easy but not accurate controller, which totally neglects actuator dynamics. Such solution can be used only with very slow motion, when actuator dynamics is negligible. The CTC2 implements a more accurate controller, which in turn requires a model of cable elasticity to compute motor position reference and the tuning of a PID controller for each motor.

We used a periodic, third degree polinomial reference (see Fig. 4), with various frequencies (0.5, 1 and 2 Hz) and fixed amplitude (35 mm). Each reference was tested with both CTC algorithms and with four different minimum tensions: LUT, 1, 4 and 7 N. An upper bound of 30 N was used for cable tension. In LUT mode, the minimum tension varied between 1 and 7 N according to the absolute value of W.

To compare the different working conditions, the variance of position error and a power-related index were used. The former was calculated as the variance of the difference between reference and actual (linear encoder) positions. The latter was calculated as the average sum of the squared current references provided to the drivers. Measures were repeated during ten periods (i.e., twenty travels of the slider).

A Comparative Index i_C was also calculated, given by the following expression:

$$i_C = \frac{var_x}{var_{LUT}} * \frac{i_{P,x}}{i_{P,LUT}} \tag{4}$$

where var_x is the variance of error with $f_{\min} = x[N]$, $i_{P,x}$ is the power index in the same condition, whereas LUT values refer to LUT $f_{\min}$. We will consider a value of i_C greater than one as an indicator that the LUT method outperforms the fixed $f_{\min}$ method. In fact, if $i_C > 1$, the ratio between the performance parameters must be

Table 1 Comparison of LUT and fixed minimum tension with different references and CTC1

Frequency (Hz)	Minimum cable tension	Error variance	Power consumption	Comparative index
0.5	Look-Up Table	0.00376	2.77	—
	1 N	0.00498	1.58	0.756
	4 N	0.00399	3.21	1.230
	7 N	0.00358	5.34	1.836
1	Look-Up Table	0.00709	2.87	—
	1 N	0.00839	1.27	0.524
	4 N	0.00759	2.69	1.003
	7 N	0.00677	5.18	1.723
2	Look-Up Table	0.0537	3.67	—
	1 N	0.131	1.58	1.050
	4 N	0.0701	3.64	1.295
	7 N	—	—	—

more than one in at least one case, indicating superiority of LUT for such parameter; on the other hand, the other parameter may yield a ratio lower than one, but with a smaller difference.

Results are summarized in Tables 1 and 2. The LUT method tends to provide better results, as measured by the comparative index, considering a mix of accuracy and power consumption. This result holds especially when the LUT method is compared to a fixed maximum tension mode where f_{min} equals the maximum value provided by the LUT. In this case, the LUT method yields comparable accuracy with less power consumption with respect to the fixed mode. The greatest difference is obtained with CTC1 (simplest control).

When compared to fixed modes with reduced constant tension, the LUT mode performs better in terms of accuracy, but yields greater power consumption. In such cases, the comparative index may be close to one or even smaller than one. The latter case indicates that the relative (percentage) benefit in terms of accuracy is smaller than the increase in power consumption.

Figure 5 shows the plots of power consumption versus variance of position error for all testing conditions. Each plot refers to a specific CTC and frequency condition, and renders in blue the three fixed modes, in red the LUT mode. Blue points are connected with lines to highlight the trend. The plots in Fig. 5 indicate that an increase in accuracy is usually obtained through greater power consumption. They also show that the LUT method tends to stay below the trend of the fixed modes, although further testing is needed to verify if this holds also for different settings of the LUT.

Table 2 Comparison of LUT and fixed minimum tension with different references and CTC2

Frequency (Hz)	Minimum cable tension	Error variance	Power consumption	Comparative index
0.5	Look-Up Table	0.00173	1.09	—
	1 N	0.00196	1.02	1.060
	4 N	0.00142	1.38	1.039
	7 N	0.00125	2.12	1.406
1	Look-Up Table	0.00591	1.03	—
	1 N	0.00646	0.958	1.017
	4 N	0.00595	1.15	1.124
	7 N	0.00432	1.78	1.263
2	Look-Up Table	0.0674	1.69	—
	1 N	0.0789	1.45	1.004
	4 N	0.0737	1.61	1.042
	7 N	0.00624	2.26	1.238

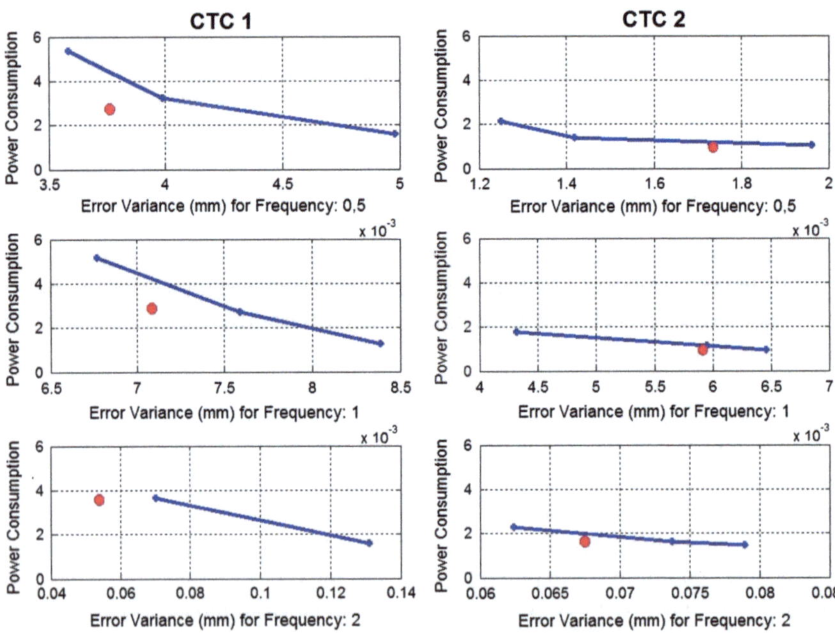

Fig. 5 Prediction of cable tension estimation for constant minimum value and look up table

4 Conclusion

In this paper, a dynamic minimum tension control for cable-driven robots was proposed and tested on a simplified scenario. This method is based on changing the minimum tension of cables according to the dynamics of the system. In particular, minimum cable tension is reduced when the end-effector wrench is large, with the aim of avoiding saturation of actuators and limiting power consumtion.

First experimental results show that this approach yields large benefits in terms of power consumption, although a reduction in accuracy is observed, especially with slow reference. When compared to fixed minimum tension modes with smaller minimum tension, the DMTC control performs comparably to the fixed modes, except when the fixed value becomes extremely small.

One limitation of this study is that the experimental setup included a single-dof system, so further testing is needed to verify if the DMTC concept can be applied succesfully in more complex contexts. Moreover, our system had a quite high friction, due to the coupling of the slider, which may have influenced results, and which is not usually present in cable-driven parallel robots. On the other hand, this setting allowed for precise measurement of end-effector motion. Finally, further investigation is needed to tune LUT parameters with the aim of obtaining the best trade-off between acuracy and power consumption.

Acknowledgments This work was supported by the Italian Ministry of University and Research (MIUR), call PRIN 2012, grant no. 20124SMZ88.

References

1. Taghirad HD, Nahon M (2008) Kinematic analysis of a macro-micro redundantly actuated parallel manipulator. Adv Robot 22(6–7):657–687
2. Rosati G, Gallina P, Masiero S (2007) Design, implementation and clinical tests of a wire-based robot for neurorehabilitation. IEEE Trans Neural Syst Rehabil Eng 15(4):560–569
3. Kawamura S, Kino H, Won C (2000) High-speed manipulation by using parallel wire-driven robots. Robotica 18:13–21
4. Rosati G, Zanotto D, Agrawal SK (2011) On the design of adaptive cable-driven systems. J Mech Robot Trans ASME 3(2):021004
5. Gallina P, Rosati G (2002) Manipulability of a planar wire driven haptic device. Mech Mach Theory 37(2):215–228
6. Shiang W-J, Cannon D, Gorman J (1999) Dynamic analysis of the cable array robotic crane. In: Proceedings of the IEEE international conference on robotics and automation, vol 4. pp 2495–2500
7. Fang S, Franitza D, Torlo M, Bekes F, Hiller M (2004) Motion control of a tendon-based parallel manipulator using optimal tension distribution. IEEE/ASME Trans Mechatron 9:561–568
8. Pham CB, Yang G, Yeo SH (2005) Dynamic analysis of cable-driven parallel mechanisms. In: Proceedings of IEEE/ASME international conference on advanced intelligent mechatronics. pp 612–617
9. Oh SR, Agrawal S (2005) Cable suspended planar robots with redundant cables: controllers with positive tensions. IEEE Trans Robot 21:457–465
10. Borgstrom P, Jordan B, Sukhatme G, Batalin M, Kaiser W (2009) Rapid computation of optimally safe tension distributions for parallel cable-driven robots. IEEE Trans Robot 25:1271–1281
11. Borgstrom PH, Borgstrom NP, Stealey MJ, Jordan B, Sukhatme GS, Batalin MA, Kaiser WJ (2009) Design and implementation of nims3d, a 3-d cabled robot for actuated sensing applications. Trans Rob 25:325–339

12. Barrette G, Gosselin CM (2005) Determination of the dynamic workspace of cable-driven planar parallel mechanisms. J Mech Des 127:242–248
13. Lim W, Yeo S, Yang G, Mustafa S, Zhang Z (2011) Tension optimization for cable-driven parallel manipulators using gradient projection. In: IEEE/ASME international conference on advanced intelligent mechatronics (AIM). pp 73–78
14. Hassan M, Khajepour A (2008) Optimization of actuator forces in cable-based parallel manipulators using convex analysis. IEEE Trans Robot 24:736–740
15. Gosselin C, Grenier M (2011) On the determination of the force distribution in overconstrained cable-driven parallel mechanisms. Meccanica 46(1):3–15
16. Snyman J, Hay A (2004) Analysis and optimization of a planar tendon-driven parallel manipulator. In: On advances in robot kinematics. Springer, Netherlands, pp 303–312
17. Mikelsons L, Bruckmann T, Hiller M, Schramm D (2008) A real-time capable force calculation algorithm for redundant tendon-based parallel manipulators. In: IEEE international conference on robotics and automation, ICRA 2008. pp 3869–3874
18. Voglewede P, Ebert-Uphoff I (2005) Application of the antipodal grasp theorem to cable-driven robots. IEEE Trans Robot 21:713–718
19. Pott A, Bruckmann T, Mikelsons L (2009) Closed-form force distribution for parallel wire robots. In: Computational kinematics. Springer, Berlin Heidelberg, pp. 25–34
20. Pott A (2014) An improved force distribution algorithm for over-constrained cable-driven parallel robots. In: Computational kinematics. Springer, Netherlands, pp 139–146
21. Roberts RG, Graham T, Lippitt T (1998) On the inverse kinematics, statics, and fault tolerance of cable-suspended robots. J Robot Syst 15(10):581–597
22. Lamaury J, Gouttefarde M (2013) Control of a large redundantly actuated cable-suspended parallel robot. In: IEEE international conference on robotics and automation (ICRA), 2013. pp 4659–4664

Modeling and Control of a Large-Span Redundant Surface Constrained Cable Robot with a Vision Sensor on the Platform

Amber R. Emmens, Stefan A.J. Spanjer and Just L. Herder

Abstract A type of cable driven redundant parallel manipulator is considered that is spanned across a surface, instead of suspended in space. A control strategy needs to be implemented that guarantees positive cable tensions and provides good performance in trajectory tracking. Therefore a workspace control method is designed based on a model of the robot, which includes a solution for the cable tension distribution. As it is hard to measure the absolute position of the end-effector, a new approach is presented where feature detection is used to update an approximated position. Experiments are carried out to compare performances with and without the addition of a vision system for a simple line motion. The error between the platform and the reference position is measured for both cases as well as the cable tensions. Mean absolute errors of 8.8 and 13 mm are obtained for the case with and without vision system respectively and positive cable tensions are measured for the complete motion. This indicates that the principle of including feature detection in the position measurement is a suitable control strategy.

1 Introduction

In parallel manipulators rigid links can be replaced by cables. The advantage is that light weight constructions are obtained that can span large workspaces for high-speed motions. A possible application for this type of robot could be as a window cleaning device, because the cleaning of large buildings with free-form architecture can be challenging with conventional equipment [15, 16]. A disadvantage of using

A.R. Emmens · S.A.J. Spanjer (✉) · J.L. Herder
University of Twente, Drienerlolaan 5, Enschede, The Netherlands
e-mail: s.a.j.spanjer@utwente.nl

A.R. Emmens
e-mail: a.r.emmens@utwente.nl

J.L. Herder
e-mail: j.l.herder@utwente.nl

A. Pott and T. Bruckmann (eds.), *Cable-Driven Parallel Robots*,
Mechanisms and Machine Science 32, DOI 10.1007/978-3-319-09489-2_18

cables is that cables can only pull and not push, so for proper control cables must be kept in tension at all times. To obtain a fully constrained system and to enhance the workspace, these robots typically have more cables than degrees of freedom. This redundancy and the non-linear coupled behaviour of the cables makes the control design of a cable robot challenging compared to parallel manipulators with rigid links.

In general, control methods are designed in joint space or workspace. In joint space cable lengths are measured and controlled to a desired cable length, which follows from the inverse kinematics of a desired trajectory. The tracking performance reduces, because elongation and coupled behaviour of the cables are neglected. Still this is a popular control strategy. The choice for a joint space controller is a matter of convenience, due to its relative easy implementation [9, 14, 18]. The change in cable length is often derived with an encoder on the servomotor [5, 8, 11, 17], the end-effector coordinates however, are not measured. Therefore in this paper a workspace controller is implemented. When the workspace is used the position and orientation of the platform have to be measured during the control process, which can be difficult and expensive. Ways to measure the position and orientation of the end-effector are with lasers [6, 17] or vision sensors [4, 10]. Both have disadvantages as lasers may induce problems with alignment and eye-safety, whilst vision sensors limit the workspace of the robot. In the field of mobile robots sometimes optical sensors are used as a replacement for rotary encoders to detect motion relative to the ground [1, 3]. The applied control strategy in this research is inspired by the latter and uses feature detection with a camera on the platform to obtain the position of the end-effector.

In this paper a type of non-suspended cable robot is considered, referred to as surface constrained cable robot, which is different from the cable robots mainly discussed in literature as it is spanned across a surface [15]. Therefore it suffers from friction and its degrees of freedom are limited. Furthermore the actuators are placed on the platform instead of at a fixed base. The objective of the paper is to present a suitable control strategy for a large-span surface constrained cable robot. For that purpose a workspace control strategy is designed using vision feedback and a solution for cable the tension distribution. During line motion experiments the absolute position of the end-effector is measured to obtain the error on the reference trajectory. Performances of the vision and non-vision based control strategies are compared. Furthermore a model is made for simulations, which can be compared with the real motion and cable tensions of the prototype.

The paper is structured as follows: first the modeling method is explained including the kinematics and dynamics of the platform as well as the control of the platform position and the cable tensions. The experimental setup and the implementation of the control strategies are presented. Results of experiments and simulations are discussed.

2 Method

A model of the cable robot is made that will be used for simulations. Based on this model a controller is designed for implementation. An experimental setup is used to compare different control strategies and to validate the model.

2.1 Model

2.1.1 Kinematics

Figure 1 shows the model of the robot. It has three degrees of freedom: two translations x ans y along the surface and a rotation θ parallel to the surface. Four cables are used to move the platform, so it has one redundant cable. The motors and drums to wind the cables are attached to the platform. It has two reference frames: frame A attached to the base and frame B attached to the platform. The attachment points of the cables at the base and frame are denoted A_i and B_i respectively, where i is the cable index. The cable vector l_i and the cable length L_i are

$$\mathbf{l_i}^A = \mathbf{A_i}^A - \mathbf{B_i}^A \tag{1}$$

$$L_i = \left\| (\mathbf{A_i}^A - \mathbf{B_i}^A) \right\| \tag{2}$$

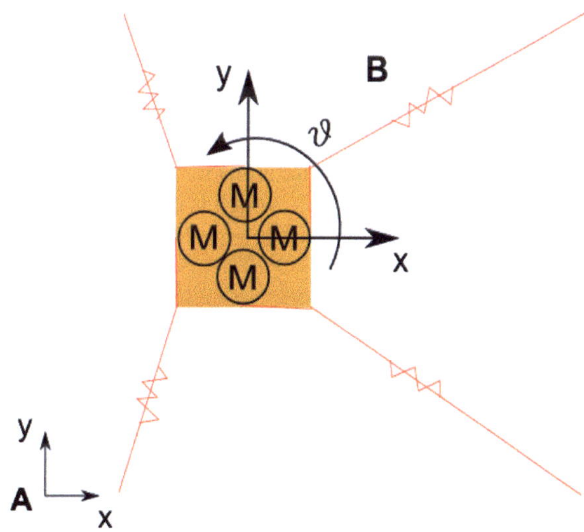

Fig. 1 Model of the cable robot on a flat surface

The transformation from a point Q in frame A to frame B is

$$\mathbf{Q}^B = R_A^B(\mathbf{Q}^A - \mathbf{p}_B^A) \tag{3}$$

where R_A^B is the 2×2 rotation matrix defined by θ and p_B^A the translation vector defined by the platform positions x_p^A and y_p^A.

2.1.2 Dynamics

The equations of motion of the actuators are

$$J_a\ddot{\theta}_i = T_i - R_b\omega_i - r_p t_i \tag{4}$$

where J_a is the moment of inertia of the actuator, θ_i the rotation of the motor, T_i the torsion delivered by the actuator, R_b a coefficient to compensate for internal viscous friction, ω the angular speed of the motor, r_p the radius of the pulley and t_i the tension of cable i. The cables are modeled as preloaded springs with stiffness k_i dependent on the non-stretched cable length L_{0i} and damping c due to some internal viscous friction.

$$t_i = k_i\Delta L_i + c\Delta\dot{L}_i \tag{5}$$

$$k_i = \frac{AE}{L_{0i}} \tag{6}$$

where A and E are the cross sectional area and elastic modulus of the cable respectively. ΔL_i is the cable elongation dependent on the amount of winded cable and the position of the platform. The total cable length L_i is

$$L_i = L_{0i} + \Delta L_i \tag{7}$$

Forces acting on the platform in the directions of the degrees of freedom are the cable induced forces F_k and the friction F_w, which is dependent on the normal force, the movement direction and a friction coefficient. The equations of motion of the platform are

$$\begin{Bmatrix} M\ddot{x} \\ M\ddot{y} \\ J\ddot{\theta} \end{Bmatrix} = \begin{Bmatrix} \mathbf{F_k} \\ M_k \end{Bmatrix} - \begin{Bmatrix} \mathbf{F_w} \\ M_w \end{Bmatrix} \tag{8}$$

For the cable induced forces the wrench matrix W is used.

$$\begin{Bmatrix} \mathbf{F_k} \\ M_k \end{Bmatrix} = W\mathbf{t} = \begin{bmatrix} \mathbf{d_1^A} & \dots & \mathbf{d_i}^A \\ \mathbf{b_1}^A \times \mathbf{d_1}^A & \dots & \mathbf{b_i}^A \times \mathbf{d_i}^A \end{bmatrix} \begin{Bmatrix} t_1 \\ t_2 \\ t_3 \\ t_4 \end{Bmatrix} \tag{9}$$

With d_i^A the direction of cable i and b_i^A the vector from B_i^B to the origin of frame B rotated into frame A.

$$\mathbf{d_i} = \frac{\mathbf{l_i}^A}{L_i} \tag{10}$$

$$\mathbf{b_i}^A = R_B^A \mathbf{B_i}^B \tag{11}$$

2.2 Control

Most cable robots are controlled in joint-space. This means that from the desired trajectory of the platform, the desired cable lengths are calculated and each cable is controlled individually. A problem with this way of controlling is that the motor torque is allowed to grow excessively when motors are pulling against each other. Therefore the choice is made for a workspace controller. A PID controller is used to control the error between the position of the platform and the reference position. From the desired position the desired cable tensions must be deduced. Therefore a solution for the redundancy must be obtained. Recap Eq. 9:

$$\mathbf{F_p} = W\mathbf{t} \tag{12}$$

where F_p is the desired resultant force on the platform induced by the cables. In case of redundancy W is not square. Therefore the inverse of W does not exist. When the desired platform forces are known a primary solution for the cable tensions t_p can be found by applying the Moore-Penrose pseudoinverse.

$$\mathbf{t_p} = W^{\dagger}\mathbf{F_p} \tag{13}$$

To this solution, solutions in the null-space of W can be added, which is the space of cable tensions that do not add to the resultant force on the platform.

$$\mathbf{t} = \mathbf{t_p} + \mathbf{t}_{\aleph} \tag{14}$$

$$W\mathbf{t}_{\aleph} = \mathbf{0} \tag{15}$$

$$\mathbf{t}_{\aleph} = (I - W^{+}W)\nu \tag{16}$$

where $t_{\aleph}$ is the vector with tensions from the null-space, I the identity matrix and ν is an arbitrary vector that can be chosen to satisfy positive cable tensions [7]. Tension distribution solutions are discussed in more detail in [2, 12, 13]. An overview of the control method is shown in Fig. 2.

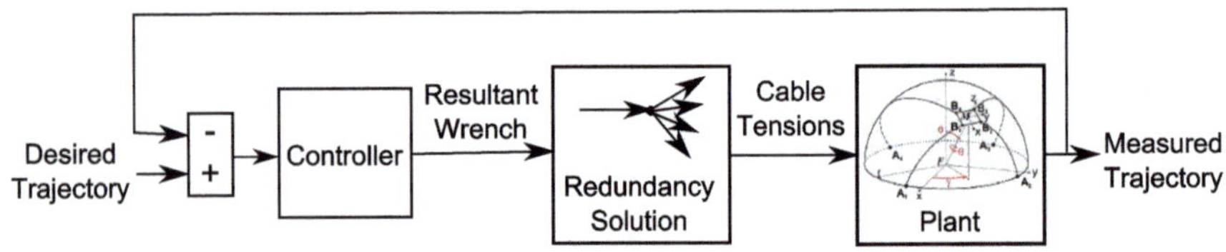

Fig. 2 Control scheme

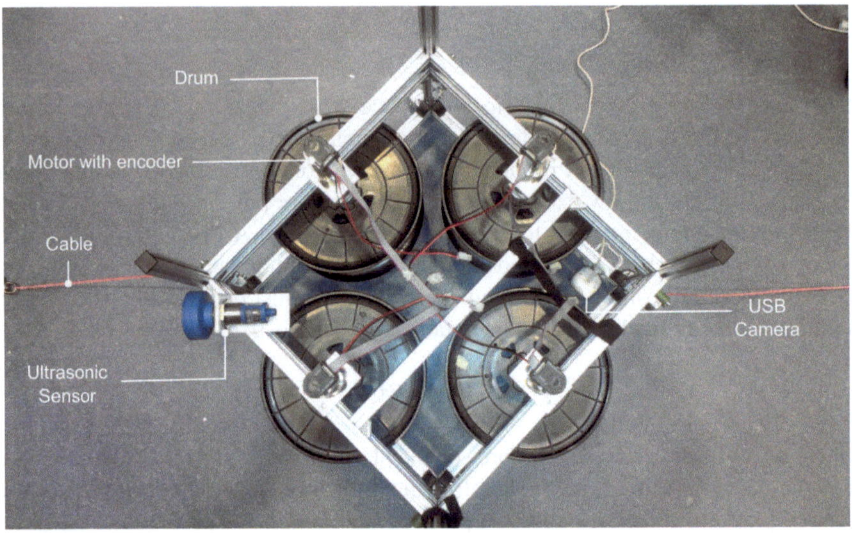

Fig. 3 Prototype of the cable robot

2.3 *Experimental Setup*

The prototype of the robot is shown in Fig. 3. It consist of four Maxon motors DCX35L GB KL 24V with encoders ENC30 HEDL 500IMP and planetary gearhead GPX42 156:1. Furthermore a Microsonic mic+ ultrasonic sensor is used to measure the absolute position and Vishay Celtron S-Type load cells to measure the cable tension. The cables are made of Liros D-Pro Dyneema rope with diameter 2.5 mm and winded on drums with a radius of 0.0557 m. The usb-camera used is a Microsoft Lifecam VX-800 with resolution 160×120 and frame rate $1/30$. The system is controlled using xPC and Simulink. In the test a line motion is made of two meters from left to right and back with a maximum velocity of 0.2 m/s using only two of the four motors. The choice is made to execute a one dimensional so that an ultrasonic sensor can be used to obtain the real trajectory of the robot and to simplify the initial vision control. The robot is subjected to a skew sine reference profile. Performances of two different measurement methods are compared: using the encoders

Table 1 Parameter values used for simulation and control

Parameter	Value
Suspension locations A_i^A	$\begin{bmatrix}0 & 4.54\end{bmatrix}$
Suspension locations B_i^B	$\begin{bmatrix}-0.34 & 0.34\end{bmatrix}$
Platform size	0.48×0.48 m
Platform mass M	14 kg
Cable elasticity E	$69 \cdot 10^9$ Pa
Cable diameter	2.5 mm
Drum radius r_p	0.0557 m
Gear reduction	156:1

on the servomotor and an optimization with the vision systems. Indicators of good performance are positivity of cable tensions and accurate trajectory tracking. The load cells are used to check whether the cable tensions remain positive and the mean distance between the reference and the real position of the platform is a measure for the trajectory tracking. A list with parameter values is shown in Table 1.

2.3.1 Control

The previously described control method is implemented on the prototype. A discretized PID controller of the form $C = K_p \frac{s\tau_z + 1}{s\tau_p + 1}(1 + \frac{1}{\tau_i s})$ is used with $K_p = 7.97 \times 10^3$, $\tau_z = 0.105$, $\tau_p = 0.0105$ and $\tau_i = 0.211$. The input of the PID controller is the tracking error on the position. Because the absolute position is not measured the motor encoders are used to approximate the position of the platform. The output of the controller is the desired resultant force on the platform. The desired cable tensions are found by using the pseudo-inverse of the wrench matrix and adding null-space solutions such that the minimum cable tension is 20 N. In the one dimensional case finding the null-space solution is quite straightforward, because increment of the tension in one cable leads to similar increment in the second cable. A velocity feed-forward is added to compensate for the large amount of friction in the system.

A second test is carried out with the addition of a vision system. A camera is plugged in the target pc and detects black lines on a white surface at known positions. First edges are detected in the input image using the Sobel method. Then local maxima are selected in the hough transform of the obtained black and white image that correspond to lines in the image. Because the locations of the lines are known as well as the location of the camera with respect to the platform, the position of the platform can be deduced. The difference between the approximated position based on the encoders and the position measured by the camera is added to the encoder based position with a rate limitation. Every time a new line is detected the difference is updated, else the last available difference is used. This way the tracking error

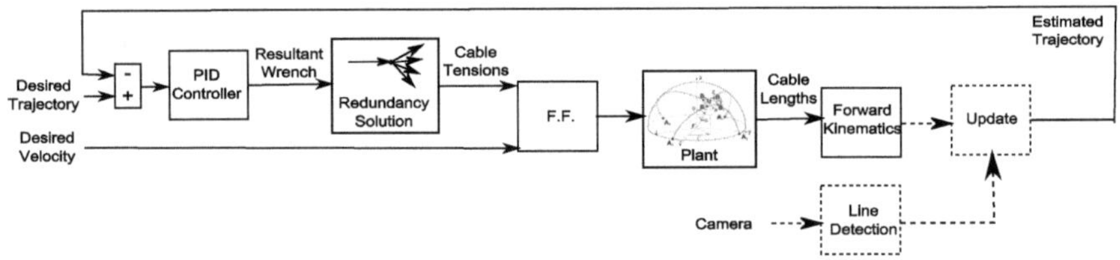

Fig. 4 Implemented control scheme

changes and the control action adapts to the new situation. The implemented control scheme is shown in Fig. 4.

3 Results

3.1 Experimental Results

The first tests are executed without vision feedback, so only the approximated position based on the encoder output is used. Figure 5 shows the tracking error between this position and the reference.

With the used control method the reference can be tracked with an accuracy of 2.5 mm. However, the position deduced from the encoders is not the real position of the platform. The real trajectory is measured with the ultrasonic sensor. The data obtained is smoothed with a moving average filter. Figure 7 shows that the real error

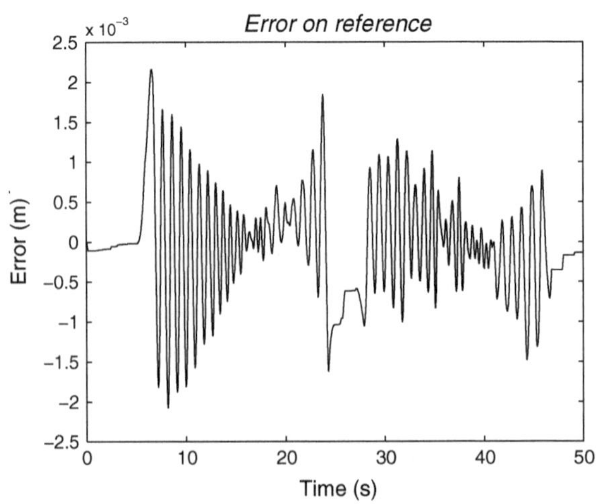

Fig. 5 Error between the reference position and the position based on the motor encoders

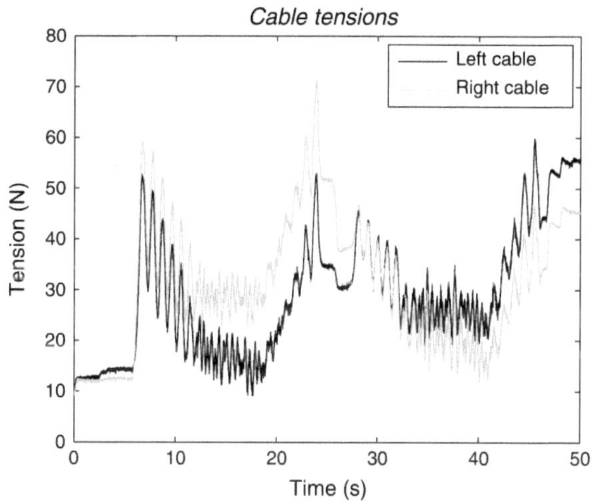

Fig. 6 Cable tensions

between the position of the platform and the reference rises up to 28 mm and has a mean of 13 mm during motion, much larger than the accuracy obtained previously.

The cable tensions during motion are shown in Fig. 6. They remain positive at all times and between the start up and stopping phase the tension in the redundant cable are near the desired minimum tension.

In the subsequent test a vision system is introduced to update the encoder measurements. Figure 7 also shows the difference between the reference position and the real position measured with the ultrasonic sensor during this test. The mean distance between both is 8.8 mm during motion.

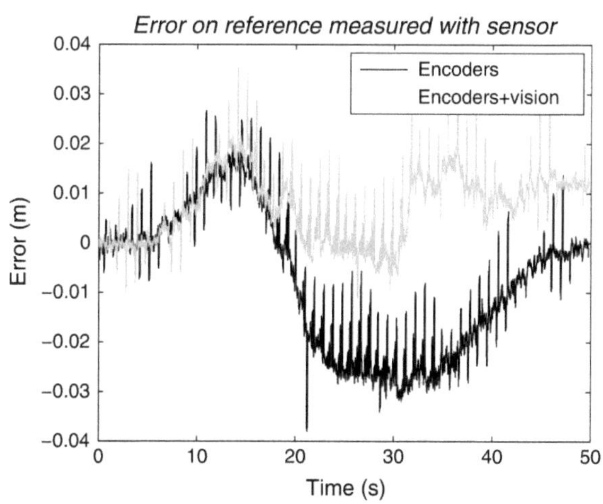

Fig. 7 Error between the reference and real trajectory for the case with and without vision system

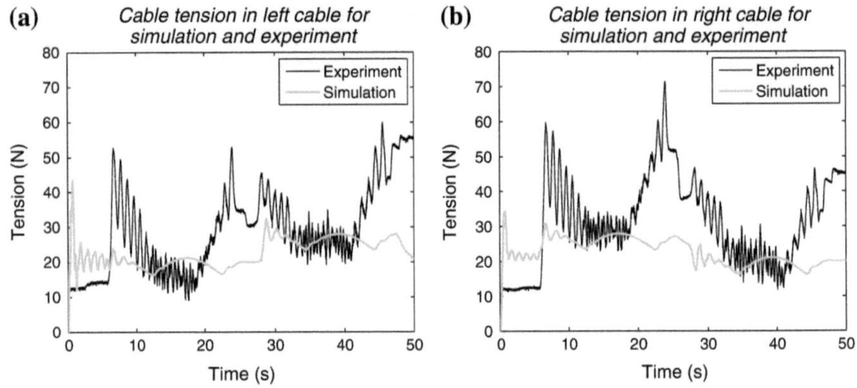

Fig. 8 Cable tensions obtained in simulation and experiment without vision system. **a** Cable tension in left cable for simulation and experiment. **b** Cable tension in right cable for simulation and experiment

3.2 Simulation Results

A line motion is simulated similar as in the experiment. The same control strategy is implemented and the plant is modeled according to the kinematics and dynamics described previously. A list of used values for several variables can be found in Table 1. Figure 8 shows the simulated cable tensions during motion compared to the experimental ones. The simulation results do not show the vibrations in the tension that occurred in the tests. A similar observation can be made for the tracking error in Fig. 9.

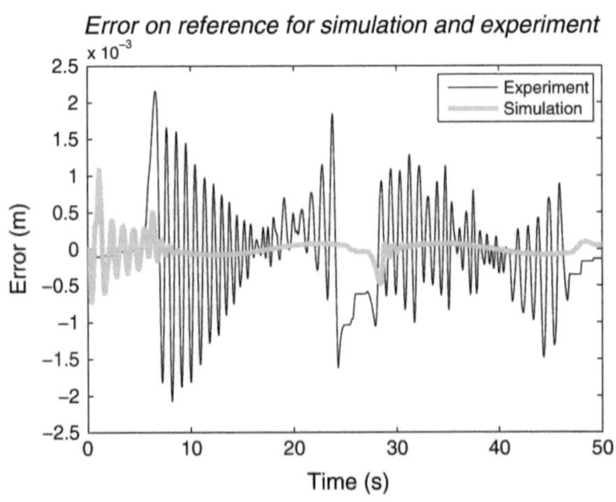

Fig. 9 Tracking error during simulation and the experiment without vision system

4 Discussion

Figures 8 and 9 show differences between the tests and the simulated model. For the cable tensions these may occur due to static friction, creating an overshoot when the robot starts moving.

The drum radius is assumed to be constant, whilst in reality it increases when cables are winded on top of each other resulting in a large error in the trajectory (Fig. 7). When the winding is controlled in such a way that drum radius is indeed constant this can be improved, though an error will still remain due to varying cable tensions and winding angles.

Figure 7 shows that the error between the reference and the real position of the platform reduces and the manipulator stops closer to the desired set-point when a vision system is added. This proves that using line detection to update the approximated position is a good solution to improve the performance. The vision system however is not working flawless yet. Inaccuracies were introduced by inconsistencies in the edge detection and low camera quality. It is expected that already a better performance can be obtained by improving the vision system and image processing. Furthermore other sensors could be used. In this case a camera was chosen as additional sensor to provide more accurate position measurements, but it might as well could have been a different sensor to detect features, such as a depth or heat sensor. It is shown the principle of feature detection to update an approximated position can be a good solution when the real position is hard to measure. In the ideal case features are extracted from the surroundings instead of detecting placed markers. Theretofore other feature detection systems may be more suitable for application and improve the performance.

5 Conclusion

A workspace control strategy is designed that provides positive cable tensions during a motion. When the plant is controlled using only the encoder based position measurement the mean absolute error between the reference and the real position of the end-effector is 13 mm measured with the ultrasonic sensor. When a vision system is added to update the approximated position this error is 8.8 mm, so a better performance is obtained with the latter control method. Simulation results show similarities with the experimental results, though differences occur due to uncertain parameters in the model such as friction and drum radius. The improvement of the performance shows that using a feature detection system fits in a suitable control strategy for cable robots that are spanned across a surface.

References

1. Abeykoon AMHS, Udawatta L, Dunuweera MS, Gunasekara RT, Fonseka M, Gunasekara SP (2011) Enhanced position sensing device for mobile robot applications using an optical sensor. In: 2011 IEEE international conference on mechatronics (ICM). pp 597–602. doi:10.1109/ICMECH.2011.5971186
2. Bruckmann T, Pott A, Hiller M (2006) Calculating force distributions for redundantly actuated tendon-based stewart platforms. In: Lennari J, Roth B (eds) Advances in robot kinematics, Springer, Netherlands, pp 403–412. doi:10.1007/978-1-4020-4941-5_44
3. Cimino M, Pagilla P (2011) Optimal location of mouse sensors on mobile robots for position sensing. Automatica 47(10):2267–2272
4. Dallej T, Gouttefarde M, Andreff N, Michelin M, Martinet P (2011) Towards vision-based control of cable-driven parallel robots. In: 2011 IEEE/RSJ international conference on intelligent robots and systems (IROS), pp 2855–2860, doi:10.1109/IROS.2011.6094591
5. De Luca A, Siciliano B, Zollo L (2005) Pd control with on-line gravity compensation for robots with elastic joints: theory and experiments. Automatica 41(10):1809–1819
6. Du J, Bao H, Cui C, Duan X (2012) Nonlinear pd control of a long-span cable-supporting manipulator in quasi-static motion. J Dyn Syst Meas Contr Trans ASME 134(1):11022–11030
7. Fahimi F (2009) Redundant manipulators. In: Autonomous robots, Springer, US, pp 1–36. doi:10.1007/978-0-387-09538-7_2
8. Fang S, Franitza D, Torlo M, Bekes F, Hiller M (2004) Motion control of a tendon-based parallel manipulator using optimal tension distribution. IEEE/ASME Trans Mechatron 9(3):561–568
9. Gholami P, Aref M, Taghirad H (2008) On the control of the kntu cdrpm: a cable driven redundant parallel manipulator. In: IEEE/RSJ international conference on intelligent robots and systems, IROS 2008, pp 2404–2409. doi:10.1109/IROS.2008.4650740
10. Kino H, Yahiro T, Takemura F, Morizono T (2007) Robust pd control using adaptive compensation for completely restrained parallel-wire driven robots: translational systems using the minimum mumber of wires under zero-gravity condition. IEEE Trans Rob 23(4):803–812
11. Lum GZ, Mustafa S, Lim HR, Lim WB, Yang G, Yeo S (2010) Design and motion control of a cable-driven dexterous robotic arm. In: 2010 IEEE conference on sustainable utilization and development in engineering and technology (STUDENT), pp 106–111. doi:10.1109/STUDENT.2010.5686997
12. Mikelsons L, Bruckmann T, Hiller M, Schramm D (2008) A real-time capable force calculation algorithm for redundant tendon-based parallel manipulators. In: IEEE international conference on robotics and automation, ICRA 2008, pp 3869–3874. doi:10.1109/ROBOT.2008.4543805
13. Oh SR, Agrawal S (2003) Cable-suspended planar parallel robots with redundant cables: controllers with positive cable tensions. In: Proceedings of IEEE international conference on robotics and automation, ICRA '03., vol 3. pp 3023–3028. doi:10.1109/ROBOT.2003.1242055
14. Su Y, Sun D, Ren L, Mills J (2006) Integration of saturated pi synchronous control and pd feedback for control of parallel manipulators. IEEE Trans Rob 22(1):202–207
15. Voss K, Wijk V, Herder J (2013) A cable-driven parallel mechanism for the interaction with hemispherical surfaces. In: Viadero F, Ceccarelli M (eds) New trends in mechanism and machine science, mechanisms and machine science, vol 7, Springer, Netherlands, pp 409–417. doi:10.1007/978-94-007-4902-3_44
16. Voss KHJ, Wijk V, Herder JL (2012) Investigation of a cable-driven parallel mechanism for interaction with a variety of surfaces, applied to the cleaning of free-form buildings. In: Husty M (ed) Lenari journal of latest advances in robot kinematics, Springer, Netherlands, pp 261–268
17. Zi B, Duan B, Du J, Bao H (2008) Dynamic modeling and active control of a cable-suspended parallel robot. Mechatronics 18(1):1–12
18. Zi B, Zhu ZC, Du JL (2011) Analysis and control of the cable-supporting system including actuator dynamics. Control Eng Pract 19(5):491–501

Part IV
Application

Cable Function Analysis for the Musculoskeletal Static Workspace of a Human Shoulder

Darwin Lau, Jonathan Eden, Saman K. Halgamuge and Denny Oetomo

Abstract The study of *cable function* allows the contribution of particular cables towards the generation of motion to be determined for cable-driven parallel manipulators (*CDPMs*). This study is fundamental in the understanding of the arrangement of cables for CDPMs and can be used within the design of optimal cable arrangements. In this paper, the analysis of cable function for the musculoskeletal static workspace of a human shoulder is performed. Considering the muscles within the shoulder as state dependent force generators, the set of muscles required in sustaining the gravity force is determined for each workspace pose. As a result, the set of poses that each muscle is responsible for (*muscle function*) can be computationally determined. By comparing the results to the muscle function from biomechanics studies, it is shown that the results from the proposed cable function analysis are consistent with that reported in the literature of human studies.

1 Introduction

Cable-driven parallel manipulators (*CDPMs*) have been widely studied in recent years due to their distinctive advantages: reduced end-effector weight and inertia compared to serial and traditional parallel rigid link mechanisms [1], potentially large reachable workspace [2] and high reconfigurability [3]. Moreover, CDPMs have been studied as bio-inspired systems [4–6] due to their anthropomorphic nature. Cables

D. Lau (✉) · J. Eden · S.K. Halgamuge · D. Oetomo
Department of Mechanical Engineering, University of Melbourne, Victoria 3010, Australia
e-mail: laudt@unimelb.edu

J. Eden
e-mail: j.eden2@student.unimelb.edu.au

S.K. Halgamuge
e-mail: saman@unimelb.edu.au

D. Oetomo
e-mail: doetomo@unimelb.edu.au

A. Pott and T. Bruckmann (eds.), *Cable-Driven Parallel Robots*,
Mechanisms and Machine Science 32, DOI 10.1007/978-3-319-09489-2_19

and rigid links of multilink cable-driven manipulators (*MCDMs*) can be regarded as structurally analogous to the muscles and bones of musculoskeletal systems, respectively. Furthermore, both cables and muscles can only provide unilateral actuation (*positive cable force*).

Workspace analysis has been widely studied on CDPMs to provide the operational region for a particular arrangement of cables [7–10]. This can also be applied to musculoskeletal systems, however, the difference in the actuation dynamics between the cables and muscles must be considered. In [11], the impact of considering the *state-dependent force generator* property in physiological muscles on the static workspace of the human shoulder was studied. By comparing the static workspace with that generated using the *ideal force generator* cable model, it was observed that the state dependent active and passive muscle characteristics of a physiological muscle significantly impacted the static workspace. The musculoskeletal static workspace was compared to the range of motion reported by human benchmarks, and it was shown that the inclusion of the physiological muscle model resulted in a workspace more realistic to that of a human shoulder.

Workspace analysis is one approach that allows the impact of different arrangements in cable attachment locations to be observed [3, 12]. However, these approaches require that the workspace has to be regenerated each time in order to study the impact of different cable arrangements. As a result, the determination of the contribution of individual cables to the workspace is computationally expensive and has not been investigated in previous CDPM studies. The role of individual cables in generating motion or workspace (*cable function*) is beneficial in understanding the design of cable arrangement for a CDPM. This could be used in the optimisation of the attachment locations of the cables in a CDPM. For musculoskeletal systems, the study of *muscle function* can be used in a range of applications in rehabilitation robotics and biomechanics. For example, in the treatment of upper limb poststroke rehabilitation, the knowledge of muscle function would allow diagnosis of impairment and eventually lead to a more targeted rehabilitation treatment.

In biomechanics, muscle function has been studied for a range of human movements, such as the walking and running patterns of the gait [13], or in the motion of the shoulder [14–18]. In these studies, muscle function was determined by performing inverse dynamics on different trajectory motions. The limitations of this approach include: the determined muscle function results are limited to the selected trajectory, and the accuracy of the results is dependent on the choice of objective function used to resolve the muscle actuation redundancy.

In this paper, the analysis of muscle function to the musculoskeletal static workspace of the human shoulder is performed. The set of muscles that contribute to a particular static workspace pose can be generated by computing the maximum mass that the shoulder can withstand in that pose. This can be determined by computing the intersection of the ray of the gravity wrench vector and the surface of the wrench zonotype. The proposed approach allows the cable function for any state dependent force generator of CDPMs to be determined. By studying the muscle function of the human shoulder, it is shown that the results obtained are consistent with that reported in the literature of human studies. Compared with studying the muscle function for a specific trajectory, the proposed method allows the muscle function for the entire workspace to be determined without the need to solve for the inverse dynamics.

The remainder of the paper is organised as follows: Sect. 2 describes the analysis of the workspace of the human shoulder. Section 3 proposes the method to determine the cable function for any particular pose within the static workspace. Section 4 presents the study of muscle function over the static workspace of the human shoulder. Section 5 concludes the paper and presents areas of future work.

2 Musculoskeletal Static Workspace of the Human Shoulder

In this section, the physiological muscle model used in the cable function analysis and the generation of the musculoskeletal static workspace for the human shoulder are presented.

2.1 Hill-type Muscle Model

One widely accepted model of the physiological muscle is the *modified Hill-type model* [19] consisting of tendon and muscle elements connected in series. The combined muscle-tendon length l^{mt} can be expressed with respect to the tendon length l^t and the muscle length l^m. The relationship between the force that can be produced and the muscle-tendon length can be described by a set of generic force relationships and muscle specific properties [20]: peak isometric muscle force F_0^m, optimal muscle fibre length l_0^m, optimal muscle fibre pennation angle α_0 and tendon slack length l_s^t.

The tendon behaves as a passive non-linear elastic element. One model for the generic tendon force-strain relationship [20] can be analytically expressed as

$$\hat{F}^t(\varepsilon) = \begin{cases} 0 & \varepsilon < 0 \\ 0.10377\left(e^{91\varepsilon} - 1\right) & 0 \leq \varepsilon < 0.01516 \\ 37.526\varepsilon - 0.26029 & 0.01516 \leq \varepsilon < 0.1 \end{cases}, \quad (1)$$

where $\hat{F}^t$ and ε are the normalised tendon force and tendon strain, respectively. Tendon strain is defined by $\varepsilon = (l^t - l_s^t)/l_s^t$ and normalised tendon force is $\hat{F}^t = F^t/F_0^m$. The normalised muscle force $\hat{F}^m = F^m/F_0^m$ can be expressed as

$$\hat{F}^m(\eta) = \hat{F}_a^m(\eta)a(t) + \hat{F}_p^m(\eta), \quad (2)$$

where $\eta = l^m/l_0^m$ is the normalised muscle length and $0 \leq a(t) \leq 1$ is the activation level of the muscle at time t. The active muscle force $\hat{F}_a^m(\eta)$ relationship [20] and passive muscle force $\hat{F}_p^m(\eta)$ relationship [21] can be expressed as

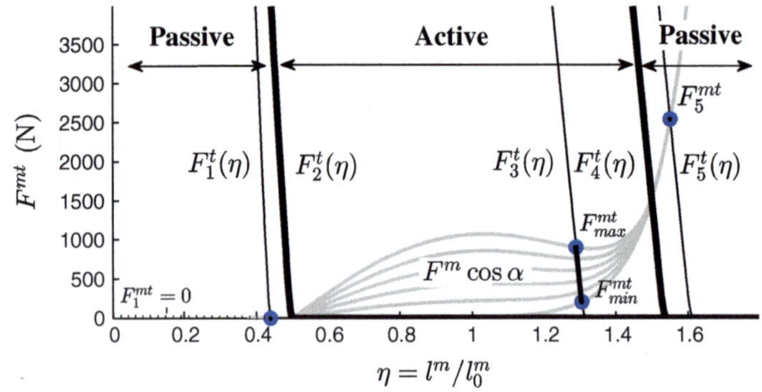

Fig. 1 Muscle-tendon forces for a range of l^{mt}, showing the scenarios in which the muscle is active and passive

$$\hat{F}_a^m(\eta) = \begin{cases} 1 - \left(\frac{\eta - 1}{0.5}\right)^2 & 0.5 < \eta < 1.5 \\ 0 & \text{otherwise} \end{cases} \tag{3}$$

$$\hat{F}_p^m(\eta) = \eta^3 e^{8\eta - 12.9} . \tag{4}$$

In [11], it was shown that in static equilibrium the forces that can be produced by the muscle-tendon complex can be determined at any pose. Figure 1 shows the solution muscle-tendon forces for various muscle-tendon lengths l^{mt}. Curves F_1^t and F_5^t represent scenarios when the muscle is passive, with muscle-tendon forces of $F_1^{mt} = 0$ and F_5^{mt}, respectively. For curve F_3^t, the muscle-tendon complex is active and can have a range of muscle force $F^{mt} \in [F_{min}^{mt}, F_{max}^{mt}]$. The bold lines F_2^t and F_4^t represent the minimum and maximum muscle-tendon lengths for which the muscle is active, respectively.

2.2 Shoulder Workspace

Using the Hill-type muscle model presented in Sect. 2.1 as a state dependent force generator, the static workspace for the human shoulder was studied [11]. As shown in Fig. 2a, the human shoulder consists of the *humerus* bone (end-effector) that is connected to the *scapular-clavicle* bone (base) through the *glenohumeral* joint. Accurate kinematics of the human shoulder were obtained from the well accepted OpenSim shoulder model developed by Holzbaur et al. [22], as shown in Fig. 2b. OpenSim is a widely accepted simulation platform in the biomechanics community used in performing analysis on musculoskeletal systems [23], for example, to study the muscle lengths and forces for a particular trajectory of motion.

The *glenohumeral* joint possesses three degrees-of-freedom and the pose of the system can be represented by $\mathbf{q} = \begin{bmatrix} \alpha & \beta & \gamma \end{bmatrix}^T$, where α, β, and γ are the xyz-Euler

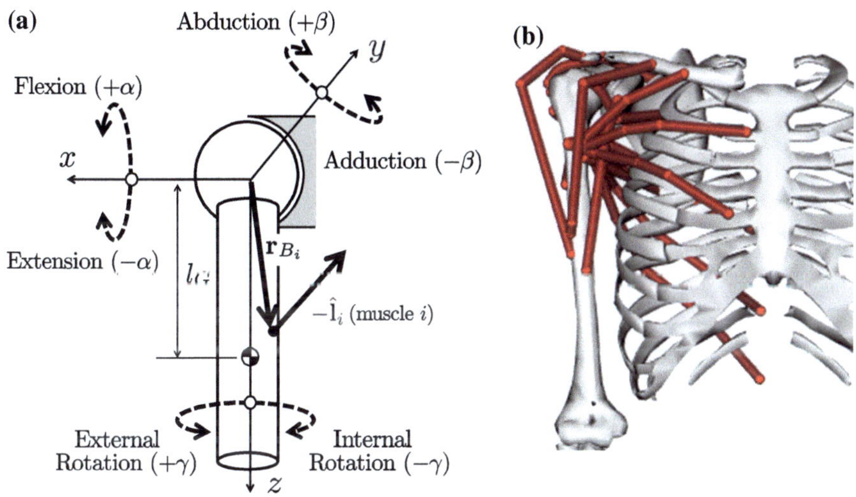

Fig. 2 Shoulder model consisting of the humerus bone and the glenohumeral joint. The rotations α, β and γ in (**a**) represent pure rotations about the x, y and z axes, respectively. The muscle geometry shown in (**b**) were obtained from the OpenSim shoulder model [22]

angles of the *glenohumeral* joint. Figure 2a shows the physiological interpretations for rotations in α, β and γ.

The mass, inertia and location of the centre of gravity for the average human *humerus* were obtained from [24]. The muscles for the shoulder model consists of $m = 15$ muscle sections that are identified as the main contributors to shoulder motion [22]: *deltoid* (*anterior, middle, posterior*), *supraspinatus, infraspinatus, subscapularis, teres minor, teres major, pectoralis major* (*clavicular, sternal, ribs*), *latissimus dorsi* (*thoracic, lumbar, iliac*) and *coracobrachialis*. The muscle properties F_0^m, l_0^m, l_s^t and α_0 for each muscle sub-region were obtained from [22]. For the shoulder system, the transpose of the Jacobian matrix can be expressed as

$$L^T = \left[\mathbf{r}_{B_1} \times \hat{\mathbf{l}}_1 \ \mathbf{r}_{B_2} \times \hat{\mathbf{l}}_2 \cdots \mathbf{r}_{B_{15}} \times \hat{\mathbf{l}}_{15} \right]. \tag{5}$$

As shown in Fig. 2a, the vectors $\mathbf{r}_{B_i}$ and $\hat{\mathbf{l}}_i$ represent the insertion location to the humerus and the direction vector of the insertion, respectively, for muscle i. The equations of motion for the system can be represented as

$$M(\mathbf{q})\ddot{\mathbf{q}} + \mathbf{C}(\dot{\mathbf{q}}, \mathbf{q}) + \mathbf{G}(\mathbf{q}) = -L(\mathbf{q})^T \mathbf{f}, \tag{6}$$

where M, $\mathbf{C}$, $\mathbf{G}$ correspond to the mass inertia matrix, centrifugal and Coriolis force vector, and gravitational vector, respectively. The musculoskeletal static workspace can be defined as

$$SW^* = \left\{ \mathbf{q} : \mathbf{G}(\mathbf{q}) + L_p^T \mathbf{f}_p = -L_a^T \mathbf{f}_a, \exists \mathbf{f}_a \in [\, \underline{\mathbf{f}_a}, \overline{\mathbf{f}_a}] \right\}. \tag{7}$$

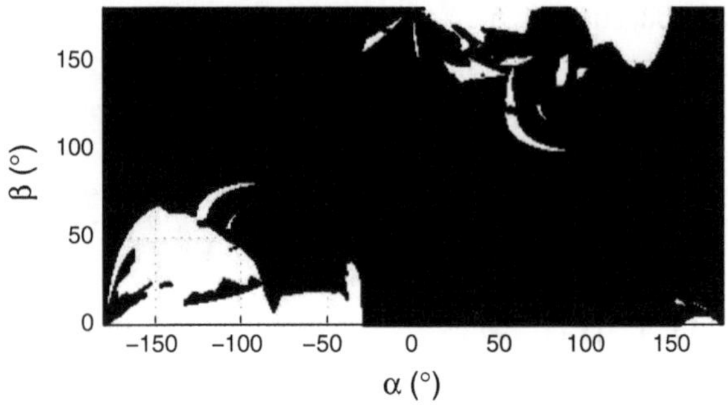

Fig. 3 The α-β cross section of the musculoskeletal static workspace for the human shoulder for zero rotation ($\gamma = 0°$)

The vector $\mathbf{f}_p$ and matrix L_p^T represent passive muscle forces and the transpose of the passive muscle Jacobian matrix, respectively. Similarly, the vector $\mathbf{f}_a$ and matrix L_a^T represent the active muscle forces and transpose of the active muscle Jacobian matrix, respectively. The passive muscle forces $\mathbf{f}_p$ and active muscle force ranges $\mathbf{f}_a \in [\underline{\mathbf{f}_a}, \overline{\mathbf{f}_a}]$ can be solved using the approach presented in [11]. Figure 3 shows the musculoskeletal static workspace defined by (7) for the human shoulder [11].

3 Gravity Force Cable Function at a Pose

One approach to study the cable function of a musculoskeletal system is to consider the set of cables required in generating the maximum weight that the system can sustain. First, consider the set of wrenches $\mathscr{W}_{avail}(\mathbf{q})$ that can be produced by a state dependent force generator at a given pose as

$$\mathscr{W}_{avail}(\mathbf{q}) = \left\{\mathbf{w} : \mathbf{w} = -L^T(\mathbf{q})\mathbf{f},\ \forall \mathbf{f} \in [\underline{\mathbf{f}}, \overline{\mathbf{f}}]\right\}, \tag{8}$$

where $\underline{\mathbf{f}}$ and $\overline{\mathbf{f}}$ represent the minimum and maximum forces that the system can generate at pose $\mathbf{q}$, respectively.

At each pose, the available wrench set $\mathscr{W}_{avail}(\mathbf{q})$ in (8) can be generated through the convex hull method presented in [7]. In this method, the zonotope wrench set $\mathscr{W}_{avail}(q)$ is constructed using the convex hull of a set defined as

$$H = \left\{\mathbf{w} \in \mathbb{R}^n | \mathbf{w} = \sum_{i=1}^{m} f_i \mathbf{l}_i,\ f_i \in \left\{\underline{f_i}, \overline{f_i}\right\}\right\}, \tag{9}$$

where n is the number of degrees-of-freedom and m is the number of cables. The maximum and minimum mass for cable i is represented by $\underline{f_i}$ and $\overline{f_i}$, respectively.

The vector $\mathbf{l}_i$ corresponds to the ith column of the matrix $-L^T(\mathbf{q})$. The convex hull is constructed from the set of 2^m vertices representing the combinations of the minimum and maximum cable forces of the system. For highly redundant systems, a large number of these vertices may be located inside the generated convex hull. Each vertex represents a combination of a set of minimum and maximum cable forces (9).

If a pose is within the static workspace, then the gravity wrench $\mathbf{G}(\mathbf{q})$ is within the available wrench set $\mathscr{W}_{avail}(\mathbf{q})$. If the mass of the system is denoted as m_s, then the normalised gravity wrench can be defined as $\hat{\mathbf{G}}(\mathbf{q}) = \mathbf{G}(\mathbf{q})/m_s$. Allowing the mass of the system to vary, the vector $\Gamma(\mathbf{q}) = m\hat{\mathbf{G}}(\mathbf{q})$ is geometrically equivalent to a ray in $\mathbb{R}^n$. For each pose that is within the static workspace, there exists a section of the ray that lies within the available wrench set $\Gamma(\mathbf{q}) \subset \mathscr{W}_{avail}(\mathbf{q})$ defined by

$$\Gamma(\mathbf{q}) = \{m\hat{\mathbf{G}}(\mathbf{q}),\ m \in [m_{min}, m_{max}]\}. \tag{10}$$

The minimum mass $m_{min} \geq 0$ and maximum mass $m_{max} > 0$ can be determined by solving for the intersections between the ray $\Gamma(\mathbf{q})$ and the surfaces of $\mathscr{W}_{avail}(\mathbf{q})$.

The surfaces of $\mathscr{W}_{avail}(\mathbf{q})$ represent the minimum and maximum wrenches that could be produced by the vertices from (9). Each of the surfaces V_k can be defined by the parametric relationship

$$V_k = \mathbf{v_0} + s(\mathbf{v_1} - \mathbf{v_0}) + t(\mathbf{v_2} - \mathbf{v_0}), \tag{11}$$

where $\mathbf{v_0}$, $\mathbf{v_1}$ and $\mathbf{v_2}$ represent the vertices within $\mathscr{W}_{avail}(\mathbf{q})$ that define the surface V_k. Each of the vertices $\mathbf{v_0}$, $\mathbf{v_1}$ and $\mathbf{v_2}$ from (11) corresponds to a set of active muscles at maximum activation to form the vertex and can be defined as $\mathcal{M}_{v_0}(\mathbf{q})$, $\mathcal{M}_{v_1}(\mathbf{q})$ and $\mathcal{M}_{v_2}(\mathbf{q})$, respectively.

An intersection between the ray (10) and surface V_k from (11) exists if a solution to $\Gamma(\mathbf{q}) = V_k$ can be found under the constraints

$$m \geq 0,\ s \geq 0,\ t \geq 0,\ s + t \leq 1. \tag{12}$$

As a result, the set of muscles $\mathcal{M}(\mathbf{q})$ contribute to producing the maximum gravity wrench can be determined by

$$\mathcal{M}(\mathbf{q}) = \mathcal{M}_{v_0}(\mathbf{q}) \cup \mathcal{M}_{v_1}(\mathbf{q}) \cup \mathcal{M}_{v_2}(\mathbf{q}). \tag{13}$$

To illustrate the wrench set and the determination of the set of contributing muscles (*muscle function*), consider the example of the human shoulder system presented in Sect. 2.2 at the pose $\mathbf{q}_{abduction} = [\alpha\ \beta\ \gamma]^T = [0°\ 60°\ 0°]^T$. Figure 4(a) shows the forces that each muscle is capable of producing at pose $\mathbf{q}_{abduction}$.

Figure 4b shows the resulting wrench set $\mathscr{W}_{avail}(\mathbf{q}_{abduction})$ for the shoulder system at pose $\mathbf{q}_{abduction}$. The intersection between the ray $\Gamma(\mathbf{q}_{abduction})$ and the wrench set is also shown in the figure. In this example, the ray and wrench set intersected at $m = m_{max} = 39.11$ kg and the set of muscles that contribute to the intersecting surface is $\mathcal{M}(\mathbf{q}_{abduction}) = \{3, 4, 5, 6, 7, 12, 13, 14\}$. Furthermore,

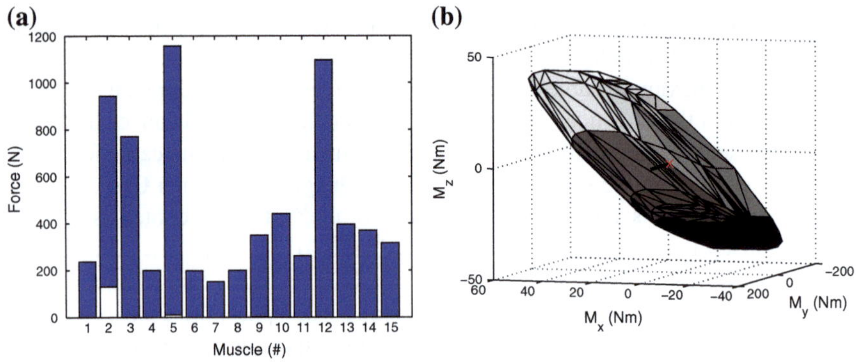

Fig. 4 The muscle wrench set $\mathscr{W}_{avail}(\mathbf{q}_{abduction})$ for the human shoulder at pose $\mathbf{q}_{abduction} = [0°60°0°]^T$ showing the solution to the muscle function for the gravity wrench. The solid bars in (a) represent the range of force that can be produced by the muscles in pose $\mathscr{W}_{avail}(\mathbf{q}_{abduction})$. **a** Muscle force ranges. **b** Resultant wrench set

from Fig. 4a it can be observed that within $\mathcal{M}(\mathbf{q}_{abduction})$ muscles number 3, 5 and 12 have the largest capability in producing the gravity wrench. In addition to producing the maximum gravity wrench, the muscles were also required to counteract the passive muscle force of muscle number 2. From the example, by observing set of muscles required in generating the maximum force in a particular direction of force (the gravity wrench $\hat{\mathbf{G}}$ in this study), the function of muscles can be determined.

4 Muscle Function over the Workspace

In Sect. 3, the method to determine the set of muscles required to sustain the gravity wrench was shown. By analysing the contribution of muscles for every pose, the muscle function for the static workspace can be studied. In this section, the function for different muscles on the workspace is analysed and compared to that from biomechanics studies. Figure 5 shows the poses that each muscle contributes to the maximum gravity wrench. From the muscle function cross-sections, it can be observed that different muscles contribute to gravity wrench in different poses within the static workspace.

It can be observed that the set of deltoid muscles (Fig. 5a, b, c) is a contributor to shoulder flexion (positive α direction and zero β), extension (negative α direction and zero β) and abduction (positive β direction and zero α) [14–16]. Furthermore, this is consistent with the literature that the deltoid anterior, deltoid middle and deltoid posterior muscle sections are primarily responsible for shoulder flexion, abduction and extension, respectively.

Furthermore, it can be observed that the set of pectorialis muscles (Fig. 5d, e, f) is predominantly responsible for flexion motion [17]. Additionally, it can be observed that the three subregions of the pectorialis muscle perform a very similar muscle

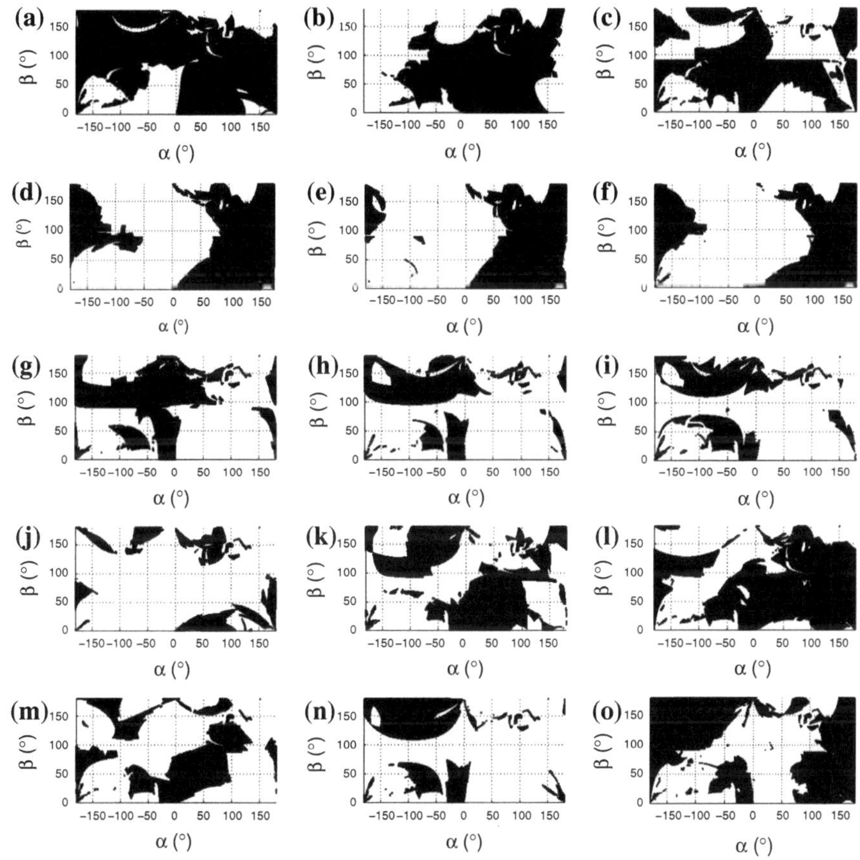

Fig. 5 Muscle function for the set of muscles in the human shoulder on the α-β cross section for zero rotation ($\gamma = 0°$). The shaded region corresponds to the poses in which the particular muscle is required to sustain the maximum gravity force. **a** Deltoid (anterior). **b** Deltoid (middle). **c** Deltoid (posterior). **d** Pectorialis (sternal). **e** Pectorialis (clavicular). **f** Pectorialis (Ribs). **g** Latissimi dorsi (thoracic). **h** Latissimi dorsi (lumbar). **i** Latissimi dorsi (iliac). **j** Coracobrachialis. **k** Infraspinatus. **l** Subscapularis. **m** Supraspinatus. **n** Teres major. **o** Teres minor

function. This is consistent with the fact that the three muscle subregions are arranged at similar locations on the shoulder [22].

Similarly, Fig. 5g, h and i show that the three subregions of the latissimi dorsi have a similar muscle function, and is known to be largely responsible for the extension motion of the shoulder [18]. Additionally, it can be observed that the latissimi dorsi muscles only contribute to small degrees of abduction. This result is also consistent with the literature where it has been observed that the strength of the muscles diminishes as the amount of abduction is increased. This indicates that the muscle force range and corresponding affect on the wrench polygon also diminishes with increasing abduction. The functions of muscles in Fig. 5j–o can be interpreted in the same manner.

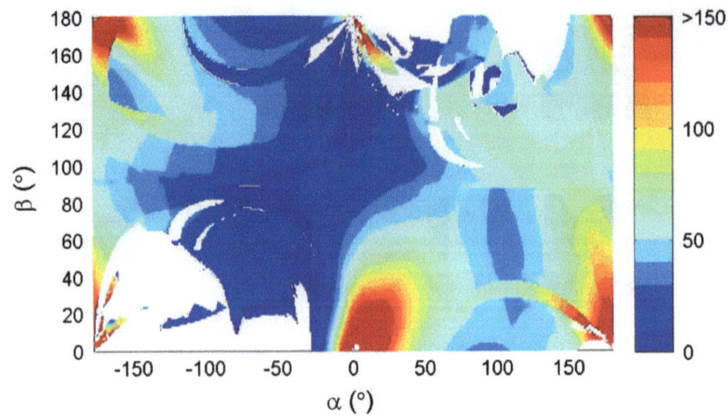

Fig. 6 The maximum mass that the shoulder muscles can sustain on the α-β cross section for zero rotation ($\gamma = 0°$)

Compared with previous biomechanics studies, the proposed approach provides a more complete description of muscle function over the entire workspace, not requiring any trajectory or inverse dynamics objective function to be specified. The muscle functions obtained represent the muscles that are responsible for producing the largest force in a particular direction, for example, the gravity wrench. In addition the function of each muscle, Fig. 6 shows the maximum mass that can be sustained for the gravity force over the α-β workspace cross-section for zero rotation $\gamma = 0°$.

It can be observed from Fig. 6 that the shoulder is capable of sustaining gravity forces from masses of few kilograms to above 150 kg. The capability of the shoulder to sustain mass is dependent on the shoulder pose as the moment arms vary for both the gravity force and the set of muscles. This approach in studying muscle function allows the capabilities of the shoulder to be studied for varying properties and attachment locations of the muscles.

In addition to the distribution of mass that can be obtained from the approach, the number of muscles required in producing the maximum mass is also beneficial in the study of a CDPM. Figure 7 shows the dimension of $\mathcal{M}(\mathbf{q})$, or the number of muscles required to generate the maximum mass, for the different poses within the workspace. It can be observed that for a majority of the α-β workspace cross-section, an average of 6–9 cables were required to produce the maximum gravity wrench. This describes the level of redundancy in generating the gravity wrench ray $\Gamma(\mathbf{q})$. Hence, for masses $m < m_{max}$, a subset of the muscles could be selected to counteract the gravity force.

It should be noted that in the proposed approach, the resulting muscle function shown is limited to one particular direction of force, the gravity force in this study. However, the selection of the gravity wrench is a natural choice in studying the function of the human shoulder. For example, many common actions performed by the upper arm is in the direction opposite to gravity, such as the swinging of the arm, to lift heavy objects or in exercises for the upper limb.

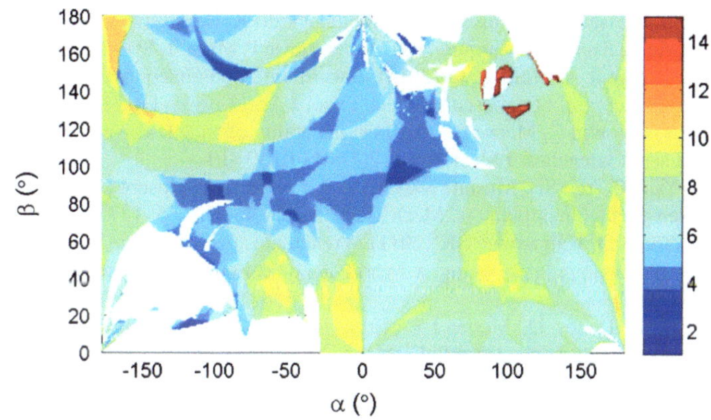

Fig. 7 The number of muscles required in sustaining the maximum mass on the α-β cross section for zero rotation ($\gamma = 0°$)

5 Conclusion

The analysis of cable function on the musculoskeletal static workspace of the human shoulder was performed. It was shown that the cable function for state dependent force generators could be determined by computing the set of cables required in balancing the maximum gravity force. The function of muscles for the human shoulder was performed to demonstrate the proposed method. The results were compared to the muscle function obtained from motion analysis performed in biomechanics studies, showing that the obtained results matched that from human studies. The proposed method provides a more complete description of muscle function than that of trajectory based approaches. Future work will focus on studying the muscle function for changes in actuator properties and different pathological conditions.

References

1. Albus JS, Bostelman RV, Dagalakis N (1993) The NIST robocrane. J Robot Syst 10(5):709–724
2. Williams RL II, Albus JS, Bostelman RV (2004) 3D cable-based cartesian metrology system. J Robot Syst 21(5):237–257
3. Lau D, Bhalerao K, Oetomo D, Halgamuge SK (2012) On the task specific evaluation and optimisation of cable-driven manipulators. In: Dai JS, Zoppi M, Kong X (eds) Advances in reconfigurable mechanisms and robots, vol I, chapter 63. Springer, London, pp 707–716
4. Lau D, Oetomo D, Halgamuge SK (2013) Generalized modeling of multilink cable-driven manipulators with arbitrary routing using the cable-routing matrix. IEEE Trans Robot 29(5):1102–1113
5. Kozuki T, Mizoguchi H, Asano Y, Osada M, Shirai T, Urata J, Nakanishi Y, Okada K, Inaba M (2012) Design methodology for thorax and shoulder of human mimetic musculoskeletal humanoid kenshiro: a thorax with rib like surface. In: Proceedings of IEEE/RSJ international conference on intelligent robots and systems, pp 3687–3692

6. Wittmeier S, Alessandro C, Bascarevic N, Dalamagkidis K, Devereux D, Diamond A, Jäntsch M, Jovanovic K, Knight R, Marques HG, Milosavljevic P, Mitra B, Svetozarevic B, Potkonjak V, Pfeifer R, Knoll A, Holland O (2013) Toward anthropomimetic robotics: development, simulation, and control of a musculoskeletal torso. J Artif Life 19(1):171–193
7. Bouchard S, Gosselin C, Moore B (2010) On the ability of a cable-driven robot to generate a prescribed set of wrenches. J Mech Robot 2(1):011 010/1–10
8. Lim WB, Yang G, Yeo SH, Mustafa SK (2011) A generic force-closure analysis algorithm for cable-driven parallel manipulators. Mech Mach Theory 46(9):1265–1275
9. Lau D, Oetomo D, Halgamuge SK (2011) Wrench-closure workspace generation for cable driven parallel manipulators using a hybrid analytical-numerical approach. J Mech Des 133(7):071 004/1–7
10. Gouttefarde M, Daney D, Merlet J-P (2011) Interval-analysis-based determination of the wrench-feasible workspace of parallel cable-driven robots. IEEE Trans Robot 27(1):1–13
11. Lau D, Eden J, Oetomo D, Halgamuge SK (2014) Musculoskeletal static workspace of the human shoulder as a cable-driven robot. IEEE/ASME Trans Mechatron 99:1–6
12. Tang X, Tang L, Wang J, Sun D (2013) Workspace quality analysis and application for a completely restrained 3-DOF planar cable-driven parallel manipulator. J Mech Sci Technol 27(8):2391–2399
13. Sasaki K, Neptune RR (2006) Differences in muscle function during walking and running at the same speed. J Biomech 39(11):2005–2013
14. Ringelberg JA (1985) EMG and force production of some human shoulder muscles during isometric abduction. J Biomech 18(12):939–947
15. Kronberg M, Brostrom L-A (1995) Electromyographic recordings in shoulder muscles during eccentric movements. Clin Orthop Relat R 314:143–151
16. Escamilla RF, Yamashiro K, Paulos L, Andrews JR (2009) Shoulder muscle activity and function in common shoulder rehabilitation exercises. Sports Med 39(8):663–685
17. Palastanga N, Field D, Soames RW (2006) Anatomy and human movement: structure and function, 5th edn. Butterworth-Heinemann Elsevier, Edinburgh
18. Gerling ME, Brown SH (2013) Architectural analysis and predicted functional capability of the human latissimus dorsi muscle. J Anat 223(2):112–122
19. Zajac FE (1989) Muscle and tendon: properties, models, scaling, and application to biomechanics and motor control. Crit Rev Biomed Eng 17(4):359–411
20. Martin CF, Schovanec L (1999) The control and mechanics of human movement systems. Prog Syst Control Theory 25:173–202
21. Vilimek M (2007) Musculotendon forces derived by different muscle models. Acta Bioeng Biomech 9(2):41–47
22. Holzbaur KRS, Murray WM, Delp SL (2005) A model of the upper extremity for simulating musculoskeletal surgery and analyzing neuromuscular control. Ann Biomed Eng 33(6):829–840
23. Delp SL, Anderson FC, Arnold AS, Loan P, Habib A, John CT, Guendelman E, Thelen DG (2007) OpenSim: open-source software to create and analyze dynamic simulations of movement. IEEE Trans Biomed Eng 54(11):1940–1950
24. Yu J, Ackland DC, Pandy MG (2011) Shoulder muscle function depends on elbow joint position: an illustration of dynamic coupling in the upper limb. J Biomech 44(10):1859–1868

A Reconfigurable Cable-Driven Parallel Robot for Sandblasting and Painting of Large Structures

Lorenzo Gagliardini, Stéphane Caro, Marc Gouttefarde, Philippe Wenger and Alexis Girin

Abstract The research work presented in this paper introduces a Reconfigurable Cable Driven Parallel Robot (RCDPR) to be employed in industrial operations on large structures. Compared to classic Cable-Driven Parallel Robots (CDPR), which have a fixed architecture, RCDPR can modify their geometric parameters to adapt their own characteristics. In this paper, a RCDPR is intended to paint and sandblast a large tubular structure. To reconfigure the CDPR from one side of the structure to another one, one or several cables are disconnected from their current anchor points and moved to new ones. This procedure is repeated until all the sides of the structure are sandblasted and painted. The analysed design procedure aims at defining the positions of the minimum number of anchor points required to complete the task at hand. The robot size is minimized as well.

1 Introduction

Over the last decades, several companies faced the necessity to manufacture novel large industrial structures. Surface finishing, e.g. painting and sandblasting, can be part of the manufacturing process of those structures. According to the different

L. Gagliardini · A. Girin
IRT Jules Verne, Chemin du Chaffault, 44340 Bouguenais, France
e-mail: lorenzo.gagliardini@irt-jules-verne.fr

A. Girin
e-mail: alexis.girin@irt-jules-verne.fr

S. Caro (✉) · P. Wenger
CNRS–IRCCyN, 1, rue de la Noë, 44321 Nantes Cedex 03, France
e-mail: Stephane.Caro@irccyn.ec-nantes.fr

P. Wenger
e-mail: Philippe.Wenger@irccyn.ec-nantes.fr

M. Gouttefarde
Laboratoire d'Informatique, de Robotique et de Micro-électronique de Montpellier (LIRMM–CNRS–UM2), 161, rue Ada, 34392 Montpellier Cedex 05, France
e-mail: marc.gouttefarde@lirmm.fr

A. Pott and T. Bruckmann (eds.), *Cable-Driven Parallel Robots*,
Mechanisms and Machine Science 32, DOI 10.1007/978-3-319-09489-2_20

structures, painting and sandblasting are usually performed by human operators, with the support of non-automatic machines. Alternative methods can improve the efficiency of these operations and release human operators from their unpleasant tasks. *Cable Driven Parallel Robots* (CDPR) are one of the possible solutions. By definition, CDPRs are parallel robots composed of a platform connected to a fixed base by cables (in this paper, the connection points between the cables and the base will be referred as anchor points). Sandblasting and painting tools can be embarked on the CDPR platform, which will follow the profile of the structure to be painted according to an off-line planned strategy.

Advantages of CDPRs are their wide workspace, the possibility to carry heavy loads and the simplicity of their mechanical components [1]. However, a limitation of CDPRs is the possible interferences between cables as well as between cables and the surrounding environment. Furthermore, the non-rigid nature of CDPR links requires a rigorous study of the force transmission characteristics.

The potentialities of CDPRs have already been proved in different industrial contexts [2, 3]. Other research studies are being performed in the framework of the European project *CableBot* [4]. Most of the previous works have been dedicated to CDPRs with a fixed architecture and a fixed geometry (cable layout). This type of robots cannot always guarantee good performances when installed in cluttered environments. In this context, *Reconfigurable Cable-Driven Parallel Robots* (RCDPR) should represent a better solution. Indeed, they can modify their geometric parameters in order to adapt their characteristics or to avoid cable collisions.

One of the first works related to CDPR reconfigurability was part of the *NIST RoboCrane* project [5]. Further studies on reconfigurability have been performed by Zhuo et al. [6], as well as by Izard et al. [7] and Rosati et al. [8]. Rosati suggested to add additional DoF to a classical CDPR (e.g., moving the cable anchor points on a rail) and optimize analytically the robot properties, such as the payload capability. This method has been proved to be efficient for planar robots. However, it cannot be easily applied to three-dimensional case studies, where the analytic solution of the problem is very difficult to define.

The research work presented in this paper focuses on the design of a RCDPR for sandblasting and painting of a three-dimensional tubular structure represented in Fig. 1. These operations are performed by appropriate tools embarked on the robot platform. The robot platform approaches each external side of the structure and the tools perform their work. Due to the structure complexity, reconfigurability is required in order to avoid cable collisions. Each external side of the structure is sandblasted and painted through a different configuration of the cable anchor points. To reconfigure the CDPR from one side of the structure to another one, one or several cables are disconnected from their current anchor points and moved to new ones. This procedure is repeated until all the sides of the structure are sandblasted and painted. The variables of the corresponding design problem are thus the Cartesian coordinates of the anchor points of the three required configurations associated to the paths $\mathscr{P}_1$, $\mathscr{P}_2$ and $\mathscr{P}_3$ illustrated in Fig. 2.

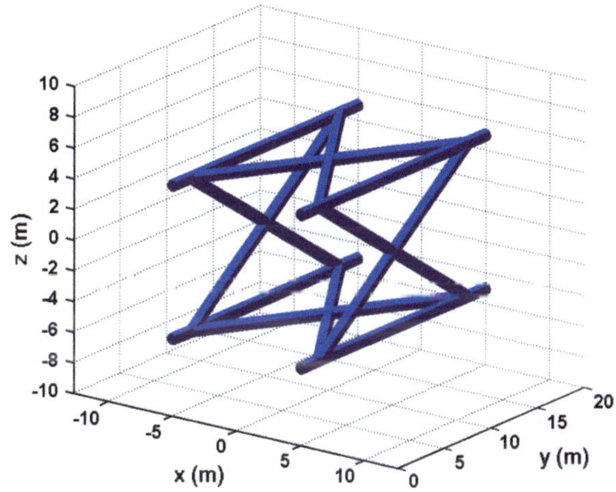

Fig. 1 Case study model. The structure is 20 m long, with a cross section of 10 m × 10 m

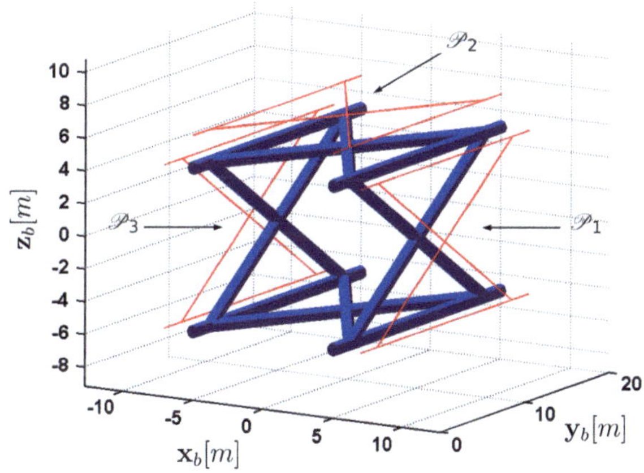

Fig. 2 Definition of the desired paths, $\mathscr{P}_1$, $\mathscr{P}_2$ and $\mathscr{P}_3$ of the platform CoM

In the present work, we aim at minimizing the total number of anchor points on the base, selecting the anchor point locations that can be shared between two or more configurations. Thereby, during a configuration change, not all the cable anchor points need to be modified. Furthermore, we also aim at minimizing the robot overall size. The feasibility of each configuration has to be guaranteed: cable interferences as well as cable collisions with the structure are not permitted. Moreover, a minimum platform pose precision is required.

This paper is organized as follows. Section 2 briefly introduces the industrial context and the problem at hand. Section 3 presents the RCDPR geometric, static and elastostatic models used in this paper. Section 4 provides a description of the selected design strategy. Section 5 presents the achieved results. Section 6 concludes this article.

2 Context and Problem Description

The structure selected for the given case study is 20 m long, with a cross section of 10 m × 10 m. The number of tubes to be painted is equal to twenty. Their diameter, ϕ_s, is equal to 0.8 m. The sandblasting and painting operations are realised indoor. The structure lies horizontally in order to reduce the dimensions of the painting workshop. The whole system can be described with respect to a fixed reference frame, $\mathscr{F}_b$, of origin O_b and axes $\mathbf{x}_b$, $\mathbf{y}_b$, $\mathbf{z}_b$, as illustrated in Fig. 2.

Sandblasting and painting tools are embarked on the RCDPR mobile platform. The CoM of the platform follows the profile of the structure tubes and the tools perform the required operations. The paths to be followed, $\mathscr{P}_1$, $\mathscr{P}_2$ and $\mathscr{P}_3$, are represented in Fig. 2. They are located at a distance of 2 m from the structure tubes. No path has been assigned to the lower external side of the structure, since it is sandblasted and painted from the ground.

In order to avoid collisions between the cables and the structure, the reconfigurability of the robot anchor point positions is necessary. Each external side of the structure should be painted by one and only one robot configuration. Three configurations are necessary to work at the exterior of the structure: configuration $\mathscr{C}_i$ being associated to path $\mathscr{P}_i$, $i = 1, 2$ and 3, in order not to interrupt the painting and sandblasting operations during their execution. Passing from a configuration to another, one or more cables are disconnected from their anchor points and connected to other anchor points located elsewhere. For each configuration, the locations of the cable exit points are defined as variables of the design problem. In the present work, the dimensions of the platform as well as the position of the cable connection points on the platform are fixed. They will be both detailed in Sect. 4.

A suspended and a fully constrained 8-cable CDPR architectures are considered. The suspended architecture is inspired by the CoGiRo CDPR prototype [9]. For the fully constrained configuration, note that 8 cables is the smallest possible even number of cables that can be used for the platform to be fully constrained by the cables. In the suspended architecture, the static equilibrium of the mobile platform is obtained thanks to the gravity force that plays the role of an additional cable pulling the mobile platform downward.

The RCDPR should be as cheap and simple as possible. For this reason, the minimization of the total number of cable anchor points is required. Consequently, the number of anchor point locations, shared by two or more configurations, should be maximized. The size of the robot is minimized as well, in order to reduce the dimensions of the sandblasting and painting workshop.

Since the sandblasting and painting operations are performed at low speed, the motion of the CDPR platform can be considered to be quasi-static. Hence, only the static equilibrium of the robot mobile platform will be considered. Collisions between the cables as well as collisions between the cables and the structure tubes should be avoided. Besides, the platform positioning precision is constrained as detailed in Sect. 4. Here, the cable mass is not considered.

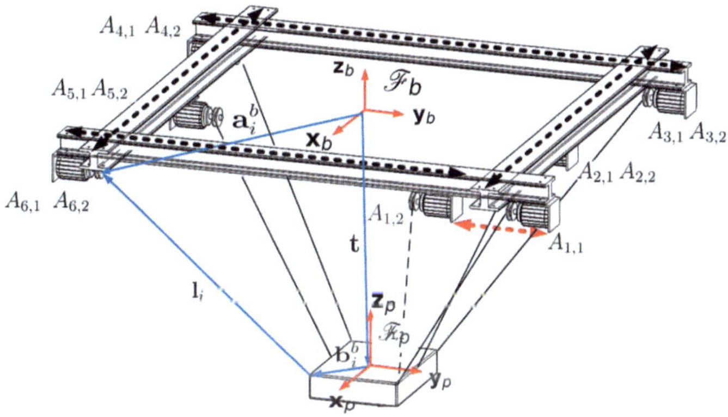

Fig. 3 Geometric model of a generic RCDPR. In this example, the anchor points are connected to a four-beam structure. Two configurations are illustrated. The anchor point $A_{1,1}$, belonging to the first configuration, is exchanged for the position $A_{1,2}$, belonging to the second configuration

3 RCDPR Kinetostatic Modeling

A RCDPR is mainly composed of a mobile platform connected to the base through a set of cables, as illustrated in Fig. 3. The connection points of the i-th cable on the platform are denoted as $B_{i,c}$, where c represents the configuration number. The position of each point $B_{i,c}$ is expressed by the vector $\mathbf{b}^p_{i,c}$ with respect to a local reference frame $\mathscr{F}_p$, attached to the platform and of origin O_p and axes $\mathbf{x}_p$, $\mathbf{y}_p$ and $\mathbf{z}_p$. O_p is the platform Center of Mass (CoM). For the cth configuration, the anchor point of the ith cable is denoted by $A_{i,c}$, $i = 1, \ldots, 8$. The Cartesian coordinates of each point $A_{i,c}$, with respect to $\mathscr{F}_b$, are given by the vector $\mathbf{a}^b_{i,c}$.

The pose of the platform, with respect to $\mathscr{F}_b$, is defined by the vector $\mathbf{p} = [\mathbf{t}, \boldsymbol{\Phi}]^{\mathrm{T}}$. The vector $\mathbf{t}$ represents the Cartesian coordinates of the platform CoM. The platform orientation is defined by the vector $\boldsymbol{\Phi}$, through the Euler angles ϕ, θ and ψ corresponding to rotations around $\mathbf{z}_b$, $\mathbf{x}_b$ and $\mathbf{y}_b$, respectively.

For a given configuration c, the vector directed along the cable from $B_{i,c}$ to $A_{i,c}$, expressed in $\mathscr{F}_b$, is defined as follows:

$$\mathbf{l}^b_{i,c} = \mathbf{a}^b_{i,c} - \mathbf{t} - \mathbf{R}\mathbf{b}^p_{i,c} \qquad i = 1, \ldots, 8 \tag{1}$$

where $\mathbf{R}$ is the rotation matrix defining the platform orientation:

$$\mathbf{R} = \mathbf{R}_z(\phi)\mathbf{R}_x(\theta)\mathbf{R}_y(\psi) = \begin{bmatrix} c\phi c\psi - s\phi s\theta s\psi & -s\phi c\theta & c\phi s\psi + s\phi s\theta c\psi \\ s\phi c\psi + c\phi s\theta s\psi & c\phi c\theta & s\phi s\psi - c\phi s\theta c\psi \\ -c\theta s\psi & s\theta & c\theta c\psi \end{bmatrix} \tag{2}$$

The unit vector $\mathbf{d}_{i,c}$ associated to each vector $\mathbf{l}_{i,c}$ is given by:

$$\mathbf{d}_{i,c} = \frac{\mathbf{l}_{i,c}}{\|\mathbf{l}_{i,c^b}\|_2}, \quad i = 1, \ldots, 8 \tag{3}$$

The ith cable exerts on the platform a wrench $\mathbf{w}_i$. This wrench is produced by a positive cable tension τ_i. All the cable tensions are collected in the vector $\boldsymbol{\tau} = [\tau_1, \ldots, \tau_8]^{\mathrm{T}}$. The static equilibrium of the platform is described by the following equation:

$$\mathbf{W}\boldsymbol{\tau} + \mathbf{w}_e = 0 \tag{4}$$

where $\mathbf{W}$ is the wrench matrix, defined as follows:

$$\mathbf{W} = \begin{bmatrix} \mathbf{d}^b_{1,c} & \mathbf{d}^b_{2,c} & \ldots & \mathbf{d}^b_{8,c} \\ \mathbf{R}\mathbf{b}^p_{1,c} \times \mathbf{d}^b_{1,c} & \mathbf{R}\mathbf{b}^p_{2,c} \times \mathbf{d}^b_{2,c} & \ldots & \mathbf{R}\mathbf{b}^p_{m,c} \times \mathbf{d}^b_{8,c} \end{bmatrix} \tag{5}$$

The vector $\mathbf{w}_e$ is the external wrench. It describes the wrench transmitted by the sandblasting or painting tools to the RCDPR platform and the weight of the platform and of the embarked tools. The weight of the platform and the embarked tools is considered to be equal to 60 Kg.

$$\mathbf{w}_e = [\mathbf{f}, \mathbf{m}]^{\mathrm{T}} = [f_x, f_y, f_z, m_x, m_y, m_z]^{\mathrm{T}} \tag{6}$$

The forces transmitted along the axes $\mathbf{x}_b$, $\mathbf{y}_b$ and $\mathbf{z}_b$ are represented by the components f_x, f_y and f_z, respectively. The maximum intensity of these components is considered to be equal to 50 N. The maximum values of the transmitted torques m_x, m_y and m_z, are equal to 7.5 Nm.

$$-50\,\mathrm{N} = f_{min} \le f_x, f_y, f_z \le f_{max} = 50\,\mathrm{N} \tag{7}$$

$$-7.5\,\mathrm{Nm} = m_{min} \le m_x, m_y, m_z \le m_{max} = 7.5\,\mathrm{Nm} \tag{8}$$

Cable tensions can be computed from Eq. (4):

$$\boldsymbol{\tau} = \boldsymbol{\tau}_n + \boldsymbol{\tau}_0 = \mathbf{W}^{\dagger}\mathbf{w}_e + \boldsymbol{\lambda}\mathbf{N}, \ \boldsymbol{\tau} \ge \mathbf{0} \tag{9}$$

$\mathbf{W}^{\dagger}$ is the Moore-Penrose generalized inverse of $\mathbf{W}$, $\boldsymbol{\lambda} \in \mathbb{R}^2$ and $\mathbf{N}$ is the null space of $\mathbf{W}$ [10].

The elastostatic model defines the relationship between the infinitesimal change $\delta\mathbf{p}$ in the platform pose and the corresponding infinitesimal change of the external wrench $\delta\mathbf{w}_e$:

$$\delta\mathbf{w}_e = \mathbf{K}\delta\mathbf{p} \tag{10}$$

The expression of the stiffness matrix $\mathbf{K}$ is [11]

$$\mathbf{K} = \sum_{i=1}^{m} k_i \begin{bmatrix} \mathbf{d}_{i,c}\mathbf{d}_{i,c}^{\mathrm{T}} & \mathbf{d}_{i,c}\mathbf{d}_{i,c}^{\mathrm{T}}\,\hat{\mathbf{b}}_{i,c}^{b\ \mathrm{T}} \\ \hat{\mathbf{b}}_{i,c}^{b}\,\mathbf{d}_{i,c}\mathbf{d}_{i,c}^{\mathrm{T}} & \hat{\mathbf{b}}_{i,c}^{b}\,\mathbf{d}_{i,c}\mathbf{d}_{i,c}^{\mathrm{T}}\,\hat{\mathbf{b}}_{i,c}^{b\ \mathrm{T}} \end{bmatrix} + \\ - \sum_{i=1}^{m} \frac{\tau_i}{\|\mathbf{l_i}\|} \begin{bmatrix} -\mathbf{I}_{3,3} + \mathbf{d}_{i,c}\mathbf{d}_{i,c}^{\mathrm{T}} & \left(-\mathbf{I}_{3,3} + \mathbf{d}_{i,c}\mathbf{d}_{i,c}^{\mathrm{T}}\right)\hat{\mathbf{b}}_{i,c}^{b\ \mathrm{T}} \\ \hat{\mathbf{b}}_{i,c}^{b}\left(-\mathbf{I}_{3,3} + \mathbf{d}_{i,c}\mathbf{d}_{i,c}^{\mathrm{T}}\right) & \left(\hat{\mathbf{b}}_{i,c}^{b}\left(-\mathbf{I}_{3,3} + \mathbf{d}_{i,c}\mathbf{d}_{i}^{\mathrm{T}}\right) + \hat{\mathbf{b}}_{i,c}^{b\ \mathrm{T}}\right)\hat{\mathbf{b}}_{i,c}^{b} \end{bmatrix} \tag{11}$$

where k_i denotes the ith cable longitudinal stiffness, $\mathbf{I}_{3\times 3}$ denotes the 3×3 identity matrix, $\hat{\mathbf{b}}_{i,c}^{b}$ and $\hat{\mathbf{d}}_{i,c}$ represent the cross product matrices of vectors $\mathbf{b}_{i,c}^{b}$ and $\mathbf{d}_{i,c}$, respectively. For a generalized 3-dimensional vector $\mathbf{r} = \left[r_x, r_y, r_z\right]^{\mathrm{T}}$, the cross product matrix is defined as follows:

$$\hat{\mathbf{r}} = \begin{bmatrix} 0 & -r_z & r_y \\ r_z & 0 & -r_x \\ -r_y & r_x & 0 \end{bmatrix} \tag{12}$$

4 Design Procedure

The design problem aims at identifying the locations of points $A_{i,c}$ for the configurations $\mathscr{C}_1$, $\mathscr{C}_2$ and $\mathscr{C}_3$. The number of cables, $m = 8$, the cable properties and the dimensions of the platform are given in Sect. 4.1. Those parameters are the same for the three robot configurations.

At first, in order to identify the set of feasible locations for the anchor points $A_{i,c}$, the three robot configurations are parameterized and analysed separately in Sects. 4.2–4.4. A set of anchor points will be considered feasible if the design constraints are satisfied along the whole path to be followed by the CoM of the platform. The design constraints are listed in Sect. 4.5. Note that each path $\mathscr{P}_i, i = 1, \ldots, 3$ is discretized into 38 points $\mathrm{P}_{j,i}, j = 1, \ldots, 38\ i = 1, \ldots, 3$

Once the set of feasible solutions has been obtained for each configuration, i.e., for each path, a list of robots with a minimum number of anchor points, n_c, is extracted from the list of feasible robots. Finally, the most compact robots from the list of robots with a minimum number of anchor points are optimal solutions.

4.1 Design Parameters

The RCDPR developed in this paper is composed of steel cables. The Young Modulus, E, is equal to 100 GPa. Their diameter, ϕ_c is equal to 4 mm, the stiffness coefficient, k_i, is equal to 252 KN/m. The maximum allowed tension in the cables, τ_{max}, is equal to 34950 N:

$$0 < \tau_i \leq \tau_{max}, \quad \forall i = 1, \ldots, 8 \tag{13}$$

l_p, w_p and h_p denote the length, width and height of the platform, respectively: $l_p = 30\,\text{cm}$, $w_p = 30\,\text{cm}$ and $h_p = 60\,\text{cm}$. The design (constant) parameter vector $\mathbf{q}$ is expressed as:

$$\mathbf{q} = \left[m, \phi_c, k_i, \tau_{max}, l_p, w_p, h_p\right]^{\mathrm{T}} \tag{14}$$

4.2 Configuration $\mathscr{C}_1$

A fully constrained architecture has been assigned to the configuration $\mathscr{C}_1$. As shown in [12], this type of robot architecture can assure the robot static equilibrium while minimizing its size. The anchor points $A_{i,1}$ have been arranged in a parallelepiped layout. The edges of the parallelepiped are aligned with the axes of frame $\mathscr{F}_b$. This layout can be fully described by means of five variables: u_1, u_2 and u_3 define the Cartesian coordinates of the parallelepiped centre; u_4 and u_5 represent the half-lengths of the parallelepiped along the axes $\mathbf{x}_b$ and $\mathbf{y}_b$, respectively. Therefore, the Cartesian coordinates of the anchor points $A_{i,1}$ are expressed as follows:

$$\mathbf{a}^b_{1,1} = [u_1 + u_4,\ u_2 + u_5,\ -u_3]^{\mathrm{T}} \quad \mathbf{a}^b_{2,1} = [u_1 + u_4,\ u_2 + u_5,\ u_3]^{\mathrm{T}} \tag{15}$$

$$\mathbf{a}^b_{3,1} = [u_1 - u_4,\ u_2 + u_5,\ -u_3]^{\mathrm{T}} \quad \mathbf{a}^b_{4,1} = [u_1 - u_4,\ u_2 + u_5,\ u_3]^{\mathrm{T}} \tag{16}$$

$$\mathbf{a}^b_{5,1} = [u_1 - u_4,\ u_2 - u_5,\ -u_3]^{\mathrm{T}} \quad \mathbf{a}^b_{6,1} = [u_1 - u_4,\ u_2 - u_5,\ u_3]^{\mathrm{T}} \tag{17}$$

$$\mathbf{a}^b_{7,1} = [u_1 + u_4,\ u_2 - u_5,\ -u_3]^{\mathrm{T}} \quad \mathbf{a}^b_{8,1} = [u_1 + u_4,\ u_2 - u_5,\ u_3]^{\mathrm{T}} \tag{18}$$

The layout of the first robot configuration is described in Fig. 4. The design variables of the design problem at hand are collected into the vector $\mathbf{x}_1$:

$$\mathbf{x}_1 = [u_1, u_2, u_3, u_4, u_5]^{\mathrm{T}} \tag{19}$$

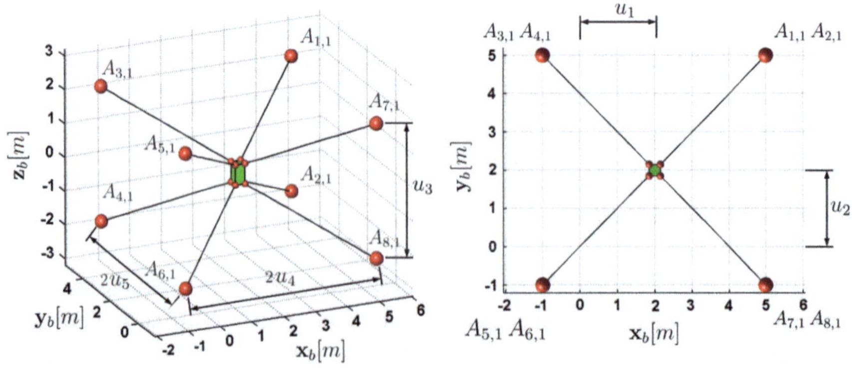

Fig. 4 Design variables parametrizing the configuration $\mathscr{C}_1$

Table 1 Design variables associated with configurations $\mathscr{C}_1$, $\mathscr{C}_2$ and $\mathscr{C}_3$

	Variables	Lower bounds	Upper bounds	Number of values
$\mathscr{C}_1$	u_1	5.5	7.5	9
	u_2	8.0	12.0	9
	u_3	6	10	5
	u_4	0.5	2.5	9
	u_5	10	14	5
$\mathscr{C}_2$	v_1	−1	1	9
	v_2	8.0	12.0	5
	v_3	7	11	9
	v_4	5	7.5	11
	v_5	10	14	5
$\mathscr{C}_3$	u_1	−7.5	−5.5	9
	u_2	8.0	12.0	9
	u_3	6	10	5
	u_4	0.5	2.5	9
	u_5	10	14	5

The Cartesian coordinates of the connection points of the cables to the platform, $B_{i,1}$, are expressed as:

$$\mathbf{b}^b_{1,1} = \frac{1}{2}\left[l_p, w_p, h_p\right]^{\mathrm{T}}, \qquad \mathbf{b}^b_{2,1} = \frac{1}{2}\left[l_p, w_p, -h_p\right]^{\mathrm{T}} \tag{20}$$

$$\mathbf{b}^b_{3,1} = \frac{1}{2}\left[-l_p, w_p, h_p\right]^{\mathrm{T}}, \qquad \mathbf{b}^b_{4,1} = \frac{1}{2}\left[-l_p, w_p, -h_p\right]^{\mathrm{T}} \tag{21}$$

$$\mathbf{b}^b_{5,1} = \frac{1}{2}\left[-l_p, -w_p, h_p\right]^{\mathrm{T}}, \qquad \mathbf{b}^b_{6,1} = \frac{1}{2}\left[-l_p, -w_p, -h_p\right]^{\mathrm{T}} \tag{22}$$

$$\mathbf{b}^b_{7,1} = \frac{1}{2}\left[l_p, -w_p, h_p\right]^{\mathrm{T}}, \qquad \mathbf{b}^b_{8,1} = \frac{1}{2}\left[l_p, -w_p, -h_p\right]^{\mathrm{T}} \tag{23}$$

A discretized set of design variables has been considered. The lower and upper bounds as well as the number of values for each variable are given in Table 1. 18,225 robot configurations have been generated with those values. It turns out that 4,576 configurations satisfy the design constraints expressed in Sect. 4.5 along the 38 discretized points of path $\mathscr{P}_1$.

4.3 *Configuration* $\mathscr{C}_2$

A suspended architecture has been attributed to the configuration $\mathscr{C}_2$ in order to avoid any possible collision with the tubular structure. The selected architecture is based on CoGiRo, a suspended CDPR designed and built in the framework of the ANR

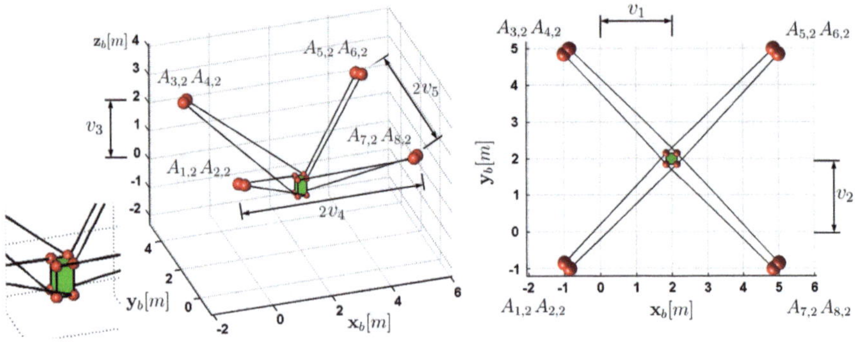

Fig. 5 Design variables parametrizing the configuration $\mathscr{C}_2$

CoGiRo project [9, 13]. An advantage of this architecture is the possibility to balance efficiently the external wrench throughout a very large part of the robot footprint. The anchor points $A_{i,2}$ have been arranged in a parallelepiped layout. The Cartesian coordinates $\mathbf{a}_{i,c}$ are defined as follows:

$$\mathbf{a}^b_{1,2} = \mathbf{a}^b_{2,2} = [v_1 - v_4,\ v_2 - v_5,\ v_3]^T \tag{24}$$

$$\mathbf{a}^b_{3,2} = \mathbf{a}^b_{4,2} = [v_1 - v_4,\ v_2 + v_5,\ v_3]^T \tag{25}$$

$$\mathbf{a}^b_{5,2} = \mathbf{a}^b_{6,2} = [v_1 + v_4,\ v_2 + v_5,\ v_3]^T \tag{26}$$

$$\mathbf{a}^b_{7,2} = \mathbf{a}^b_{7,2} = [v_1 + v_4,\ v_2 - v_5,\ v_3]^T \tag{27}$$

Variables $v_1 = 1, \ldots, 5$ assume the same geometric roles assigned to variables $u_1 = 1, \ldots, 5$. The layout of this configuration is illustrated in Fig. 5. The design variables of configuration $\mathscr{C}_2$ are collected into the vector $\mathbf{x}_2$:

$$\mathbf{x}_2 = [v_1, v_2, v_3, v_4, v_5]^T \tag{28}$$

Note that this architecture is composed of couples of anchor points theoretically connected to the same locations: $\{A_{1,2}, A_{2,2}\}$, $\{A_{3,2}, A_{4,2}\}$, $\{A_{5,2}, A_{6,2}\}$ and $\{A_{7,2}, A_{8,2}\}$. From a technical point of view, in order to avoid any cable interference, the coupled anchor points should be separated by a distance at least greater than the cable diameter. For the design problem at hand, this distance has been fixed to 5 mm.

The Cartesian coordinates of points $B_{i,2}$ are defined as:

$$\mathbf{b}^b_{1,2} = \frac{1}{2}\left[l_p, -w_p, h_p\right]^T, \quad \mathbf{b}^b_{2,2} = \frac{1}{2}\left[-l_p, w_p, -h_p\right]^T \tag{29}$$

$$\mathbf{b}^b_{3,2} = \frac{1}{2}\left[-l_p, -w_p, h_p\right]^{\mathrm{T}}, \quad \mathbf{b}^b_{4,2} = \frac{1}{2}\left[l_p, w_p, -h_p\right]^{\mathrm{T}} \tag{30}$$

$$\mathbf{b}^b_{5,2} = \frac{1}{2}\left[-l_p, w_p, h_p\right]^{\mathrm{T}}, \quad \mathbf{b}^b_{6,2} = \frac{1}{2}\left[l_p, -w_p, -h_p\right]^{\mathrm{T}} \tag{31}$$

$$\mathbf{b}^b_{7,2} = \frac{1}{2}\left[l_p, w_p, h_p\right]^{\mathrm{T}}, \quad \mathbf{b}^b_{8,2} = \frac{1}{2}\left[-l_p, -w_p, -h_p\right]^{\mathrm{T}} \tag{32}$$

Table 1 describes the lower and upper bounds as well as the number of values considered for the configuration $\mathscr{C}_2$. Combining these values, 22,275 configurations have been generated. Amongst these configurations, only 5,579 configurations are valid.

4.4 Configuration $\mathscr{C}_3$

The configuration $\mathscr{C}_3$ follows the path $\mathscr{P}_3$. This path is symmetric to the path $\mathscr{P}_1$ with respect to the plane $\mathbf{y}_b O \mathbf{z}_b$, O, $\mathbf{y}_b$ and $\mathbf{z}_b$ being the origin, the $\mathbf{y}$-axis and the $\mathbf{z}$-axis of the base frame. Considering the symmetry of the tubular structure, the robot architecture used for the configuration $\mathscr{C}_1$ has been assigned to the configuration $\mathscr{C}_3$. The discretized set of design variables chosen for the configuration $\mathscr{C}_3$ is described in Table 1. The design variables for the configuration $\mathscr{C}_3$ are collected into the vector $\mathbf{x}_3$:

$$\mathbf{x}_3 = [w_1, w_2, w_3, w_4, w_5]^{\mathrm{T}} \tag{33}$$

where the variables w_i amount to variables u_i described in Sect. 4.2. Therefore, the Cartesian coordinates of the anchor points $A_{i,3}$ are expressed as follows:

$$\mathbf{a}^b_{1,3} = [w_1 + w_4,\ w_2 + w_5,\ -w_3]^{\mathrm{T}} \quad \mathbf{a}^b_{2,3} = [w_1 + w_4,\ w_2 + w_5,\ w_3]^{\mathrm{T}} \tag{34}$$

$$\mathbf{a}^b_{3,3} = [w_1 - w_4,\ w_2 + w_5,\ -w_3]^{\mathrm{T}} \quad \mathbf{a}^b_{4,3} = [w_1 - w_4,\ w_2 + w_5,\ w_3]^{\mathrm{T}} \tag{35}$$

$$\mathbf{a}^b_{5,3} = [w_1 - w_4,\ w_2 - w_5,\ -w_3]^{\mathrm{T}} \quad \mathbf{a}^b_{6,3} = [w_1 - w_4,\ w_2 - w_5,\ w_3]^{\mathrm{T}} \tag{36}$$

$$\mathbf{a}^b_{7,3} = [w_1 + w_4,\ w_2 - w_5,\ -w_3]^{\mathrm{T}} \quad \mathbf{a}^b_{8,3} = [w_1 + w_4,\ w_2 - w_5,\ w_3]^{\mathrm{T}} \tag{37}$$

4.5 Constraints

The static equilibrium of the robot should be assured during the entire task execution, balancing any possible external wrench $\mathbf{w}_e$, according to the limits specified in Eqs. (7) and (8). This condition is verified for all the points $\mathscr{P}_i$, $i = 1, \ldots, 3$. The set of required external wrenches consists of a hyperrectangle, defined as $[\mathbf{w}]_r$. In order the robot to be in a static equilibrium in a given posture, the hyperrectangle of the required external wrench should be included inside the zonotope of the admissible wrench, $[\mathbf{w}]_a$, as follows:

$$\forall \mathbf{w}_e \in [\mathbf{w}]_r,\ \exists \boldsymbol{\tau} \in [\boldsymbol{\tau}] \text{ such that } \begin{cases} [\mathbf{w}]_r \subset [\mathbf{w}]_a \\ \mathbf{W}\boldsymbol{\tau} + \mathbf{w}_e = 0 \end{cases} \tag{38}$$

$[\mathbf{w}]_a$ represents the possible external wrenches that the platform can balance. $[\mathbf{w}]_a$ depends on the robot geometry and the cable tension limits, defined by Eq. (13). Any mobile platform pose respecting the previous condition is said to be *wrench feasible*. The set of wrench feasible poses represents the *Wrench Feasible Workspace* (WFW) of the given robot.

In [14], Gouttefarde et al. improved the wrench feasibility verification procedure defined by Bouchard et al. in [15]. Both the procedures lead to a set of inequalities:

$$\mathbf{C}\mathbf{w} \leq \mathbf{d}, \quad \forall \mathbf{w} \in [\mathbf{w}]_r \tag{39}$$

Cable interferences have to be avoided. The interference between the ith cable and the jth cable is verified analysing the distance $d^{cc}_{i,j}$ between them. The cables have been modelled as linear segments, neglecting the mass and elastic effects on the sagging phenomenon. A fast computation of $d^{cc}_{i,j}$ is realised through Lumelsy's approach [16]. For the constraint to be satisfied, distance $d^{cc}_{i,j}$ should be greater than the diameter of the cables, ϕ_c, i.e.,

$$d^{cc}_{i,j} \geq \phi_c \quad \forall i, j = 1, \ldots, m, \quad i \neq j \tag{40}$$

The same approach is used to detect collisions between the cables and the tubular structure. The collision between the ith cable and the kth structure tube does not occur when their distance $d^{cc}_{i,k}$ is greater than the sum of the cable and tube radii:

$$d^{cs}_{i,k} \geq \frac{(\phi_c + \phi_s)}{2} \quad \forall i = 1, \ldots, m, \quad \forall k = 1, \ldots, 12 \tag{41}$$

The RCDPR should respect a prescribed pose accuracy along all the paths. Inside the limits defined by the required wrench set, the maximum mobile platform linear displacements δt_x, δt_y and δt_z along $\mathbf{x}_b$, $\mathbf{y}_b$ and $\mathbf{z}_b$ axes should be lower than 5cm:

$$-5\,\text{cm} \leq \delta t_x, \delta t_y, \delta t_z \leq 5\,\text{cm} \tag{42}$$

The rotation displacements of the platform δr_x, δr_y and δr_z about axes $\mathbf{x}_b$, $\mathbf{y}_b$ and $\mathbf{z}_b$ should not be higher than 0.1rad:

$$-0.1\,\text{rad} \leq \delta r_x, \delta r_y, \delta r_z \leq 0.1\,\text{rad} \tag{43}$$

The linear and rotational displacements of the mobile platform are computed from Eq. (10).

4.6 Objective Functions

In Sects. 4.2–4.4, the feasible robot configurations associated to paths $\mathscr{P}_1$, $\mathscr{P}_2$ and $\mathscr{P}_3$ have been identified. For each path, a configuration is selected, aiming at minimizing the total number of anchor points required by the RCDPR to complete the task. These optimal solutions have been computed in two phases. At first, the 4,576 feasible robot configurations for path $\mathscr{P}_1$ are compared with the 5,579 feasible robot configurations for path $\mathscr{P}_2$. The sets of robot configurations that minimize the number, n_c, of required anchor points along the three paths are selected.

Then, the problem aims at minimizing the size of the robot, defined as the convex hull of the robot anchor points. The Cartesian coordinates of anchor point $A_{i,c}$ are defined as $\mathbf{a}_{i,c} = [a^x_{i,c}, a^y_{i,c}, a^z_{i,c}]^T$. The variables $\underline{s}_x$, $\underline{s}_y$ and $\underline{s}_z$ denote the smallest Cartesian coordinates of the robot anchor points along the axes $\mathbf{x}_b$, $\mathbf{y}_b$ and $\mathbf{z}_b$, respectively:

$$\underline{s}_x = \min a^x_{i,c}, \ \forall i = 1, \ldots, 8, \ c = 1, \ldots, 3 \tag{44}$$

$$\underline{s}_y = \min a^y_{i,c}, \ \forall i = 1, \ldots, 8, \ c = 1, \ldots, 3 \tag{45}$$

$$\underline{s}_z = \min a^z_{i,c}, \ \forall i = 1, \ldots, 8, \ c = 1, \ldots, 3 \tag{46}$$

The upper bounds on the Cartesian coordinates of the RCDPR anchor points, along the axes $\mathbf{x}_b$, $\mathbf{y}_b$, $\mathbf{z}_b$, are denoted by $\bar{s}_x$, $\bar{s}_y$ and $\bar{s}_z$, respectively.

$$\bar{s}_x = \max a^x_{i,c}, \ \forall i = 1, \ldots, 8, \ c = 1, \ldots, 3 \tag{47}$$

$$\bar{s}_y = \max a^y_{i,c}, \ \forall i = 1, \ldots, 8, \ c = 1, \ldots, 3 \tag{48}$$

$$\bar{s}_z = \max a^z_{i,c}, \ \forall i = 1, \ldots, 8, \ c = 1, \ldots, 3 \tag{49}$$

Hence, the objective function related to the size of the robot is expressed as follows:

$$\mathscr{V}_2 = (\bar{s}_x - \underline{s}_x)(\bar{s}_y - \underline{s}_y)(\bar{s}_z - \underline{s}_z) \tag{50}$$

4.7 Design Problem Formulation

The design problem of the CDPR is formulated as follows:

$$
\begin{aligned}
&\text{minimize} && \begin{cases} \mathscr{V}_1 = n_c \\ \mathscr{V}_2 = (\bar{s}_x - \underline{s}_x)(\bar{s}_y - \underline{s}_y)(\bar{s}_z - \underline{s}_z) \end{cases} \\
&\text{over} && \mathbf{x}_1, \mathbf{x}_2, \mathbf{x}_3 \\
&\text{subject to:} && \\
&\forall \mathrm{P}_{m,n},\ m = 1, \ldots, 38 \quad n = 1, \ldots, 3 && \begin{cases} \mathbf{Cw} \leq \mathbf{d}, \quad \forall \mathbf{w} \in [\mathbf{w}]_r \\ d_{i,j}^{cc} \geq \phi_c \quad \forall i, j = 1, \ldots, 8, \quad i \neq j \\ d_{i,k}^{cs} \geq \dfrac{(\phi_c + \phi_s)}{2} \quad \forall i = 1, \ldots, 8, \quad \forall k = 1, \ldots, 12 \\ -5\,\mathrm{cm} \leq \delta t_x, \delta t_y, \delta t_z \leq 5\,\mathrm{cm} \\ -0.1\,\mathrm{rad} \leq \delta r_x, \delta r_y, \delta r_z \leq 0.1\,\mathrm{rad} \end{cases}
\end{aligned}
\tag{51}
$$

5 Results

Following the procedure described in Sect. 4, a set of feasible configurations has been identified for $\mathscr{C}_l, l = 1, \ldots, 3$. 16,516 triplets of configurations minimize the total number of anchor points.

A generic CDPR composed of 8 cables requires 8 anchor points $A_i = 1, \ldots, 8$ on the base. It is the case for the fully-constrained configuration $\mathscr{C}_1$ described in Sect. 4.2. The suspended CDPR proposed in Sect. 4.3 presents 4 coincident couples of anchor points. Hence, the maximum total number of anchor points of the RCDPR is equal to 20. The best results provide a reduction of 4 points. Regarding the configurations $\mathscr{C}_1$ and $\mathscr{C}_2$, points $A_{5,2}$ and $A_{7,2}$ can be coincident with points $A_{3,1}$ and $A_{5,1}$, respectively. Alternatively, points $A_{5,2}$ and $A_{7,2}$ can be coincident with points $A_{1,1}$ and $A_{7,1}$. As far as configurations $\mathscr{C}_2$ and $\mathscr{C}_3$ are concerned, points $A_{1,2}$ and $A_{3,2}$ can be coincident with points $A_{8,3}$ and $A_{2,3}$, respectively. Likewise, points $A_{1,2}$ and $A_{3,2}$ can be coincident with points $A_{4,3}$ and $A_{6,3}$, respectively.

The total volume of the robot has been computed for the 16,516 configurations minimizing the total number of anchor points. 96 RCDPRs amongst the 16,516 robot configurations have the smallest size, this minimum size being equal to 5,104 m^3. An optimal solution is illustrated in Fig. 6. The corresponding optimal design parameters are given in Table 2.

Figure 7 illustrates the *minimum degree of constraint satisfaction* introduced in [17] and defined thereafter along the paths $\mathscr{P}_1$, $\mathscr{P}_2$ and $\mathscr{P}_3$, which is discretized into 388 points.

$$
s = \min_{i=1,\ldots,n} \left(\min_{j=1,\ldots,p} s_{j,i} \right) \tag{52}
$$

It turns out that mobile platform is in a static equilibrium along all the paths because the minimum degree of constraint satisfaction remains negative. The degree of constraint satisfaction $s_{j,i}$ is the signed distance from the jth vertex $\mathbf{w}_{e,j}$ of $[\mathbf{w}]_r$ to the

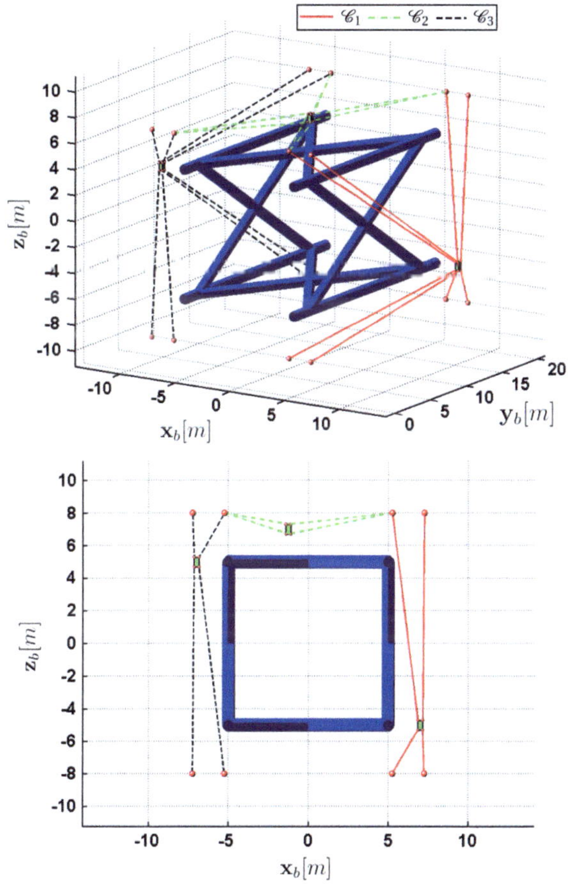

Fig. 6 Optimal reconfigurable cable-driven parallel robot

Table 2 Design parameters of the selected optimum RCDPR

Conf.	var.1	var.2	var.3	var.4	var.5
$\mathbf{x}_1$	6.25	10.0	8.0	1.0	11.0
$\mathbf{x}_3$	0	10.0	8.0	5.25	11.0
$\mathbf{x}_3$	-6.25	10.0	8.0	1.0	11.0

ith face of $[\mathbf{w}]_a$. $s_{j,i}$ is negative when a pose is wrench feasible. Configurations $\mathscr{C}_1$ and $\mathscr{C}_3$ maintain their degree of satisfaction lower than the 400 N. On the contrary, configuration $\mathscr{C}_2$ is often close to 0. The poses where s vanishes require that two cables of the suspended CDPR are slack.

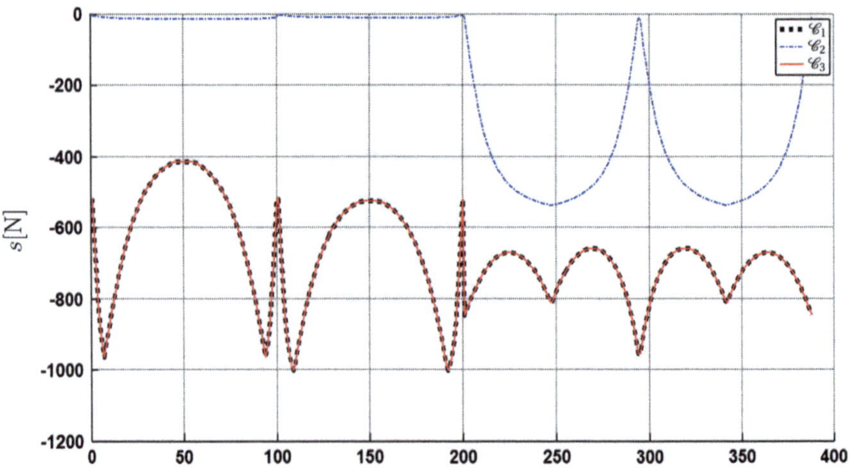

Fig. 7 Minimum degree of constraint satisfaction. The analysis has been performed by discretizing the paths $\mathscr{P}_1$, $\mathscr{P}_2$ and $\mathscr{P}_3$ into 388 points

6 Conclusions

This paper dealt with the design of a Reconfigurable Cable Driven Parallel Robot (RCDPR) for sandblasting and painting of large structures. A tubular structure of 20 m long and 10 m wide was considered. The design problem aimed at determining the optimal locations of the cable anchor points such that the number of anchor points is minimized and the RCDPR is as small as possible. The problem solving led to 96 optimal solutions.

A RCDPR was required in order to complete a given task in a complex environment. Unfortunately, the design strategy introduced in this paper is very time consuming. The whole procedure, performed on a Intel® Core™ i7-3630QM 2.40 GHz, required 19 h of computation, on Matlab® 2013a. Therefore, the development of more efficient strategies for the design of RCDPRs will be part of our future work. Moreover, the mass of the cables will be taken into account.

Acknowledgments This paper is part of the CAROCA project managed by IRT Jules Verne (French Institute in Research and Technology in Advanced Manufacturing Technologies for Composite, Metallic and Hybrid Structures). The authors wish to associate the industrial and academic partners of this project, namely, STX Europe, DCNS, AIRBUS, IRCCyN and LIRMM.

References

1. Gouttefarde M, Daney D, Merlet J-P (2011) Interval-analysis-based determination of the wrench-feasible workspace of parallel cable-driven robots. IEEE Trans Robotics 27(1):1–13

2. Bostelman R, Albus J, Dagalakis N, Jacoff A (1994) Applications of the NIST RoboCrane. In: Proceedings of the 5th international symposium on robotics and manufacturing, Maui, Aug 1994
3. Holland CS, Cannon DJ (2002) Cable array robot for material handling, March 2002
4. CableBOT. Available http://www.cablebot.eu/en/
5. Bostelman R, Jacoff A, Proctor F, Kramer T, Wavering A (2000) Cable-based reconfigurable machines for large scale manufacturing. In: Proceedings of the (2000) Japan-USA symposium on flexible automation. MI, July, Ann Arbor 2000
6. Zhou A, Tang CP, Krovi V (2012) Analysis framework for cooperating mobile cable robots. In: IEEE international conference on Robotics and Automation, pp 3128–3133, RiverCentre, Saint Paul, MN, May 2012
7. Izard J-B, Gouttefarde M, Tempier O, Baradat C (2013) A reconfigurable robot for cable-driven parallel robotic research and industrial scenario proofing. Cable-Driven Parallel Robots 12:135–148
8. Rosati G, Zanotto D, Agrawal SK (2011) On the design of adaptive cable-driven systems. J Mech Robotics 3(2):021004
9. Lamaury J, Gouttefarde M, Chemori A, Herve P-E (2013) Dual-space adaptive control of redundantly actuated cable-driven parallel robots. In: IEEE/RSJ international conference on intelligent robots and systems (IROS), pp 4879–4886, Tokyo, Japan, Nov 2013
10. Roberts RG, Graham T, Lippitt T (1998) On the inverse kinematics, statics, and fault tolerance of cable-suspended robots. J Robotic Syst 15(10):581–597
11. Behzadipour S, Khajepour A (2006) Stiffness of cable-based parallel manipulators with application to stability analysis. J Mech Des 128:303–310
12. Gagliardini L, Caro S, Gouttefarde M, Wenger P, Girin A (2014) Optimal design of cable-driven parallel robots for large industrial structures. In: IEEE international conference on robotics and automation (ICRA 2014), Hong Kong, China, May 2014
13. CoGiRo. Available http://www.lirmm.fr/cogiro/
14. Gouttefarde M, Krut S (2010) Characterization of parallel manipulator available wrench set facets. In: Advances in robot kinematics: motion in man and machine, pp 475–484. Piran, 2010
15. Bouchard S, Gosselin CM, Moore B (2008) On the ability of a cable-driven robot to generate a prescribed set of wrenches. In: Proceedings of the ASME 2008 international design engineering technical conferences & computers and information in engineering conferences, pp 1–12, Brooklyn, New York, USA, Aug 2008
16. Lumelsky VJ (1985) On fast computation of distance between line segments. Inf Process Lett 21:55–61
17. Guay F, Cardou P, Cruz AL, Caro S (2013) Measuring how well a structure supports varying external wrenches. In: The second conference on mechanisms, transmissions and applications, Bilbao, Spain, Oct 2013

ARACHNIS: Analysis of Robots Actuated by Cables with Handy and Neat Interface Software

Ana Lucia Cruz Ruiz, Stéphane Caro, Philippe Cardou and François Guay

Abstract This paper presents ARACHNIS, a graphical user interface for the analysis and parametric design of Cable Driven Parallel Robots (CDPRs). ARACHNIS takes as inputs the design parameters of the robot, the task specifications, and returns a visualisation of the CDPR Wrench Feasible Workspace (WFW) and Interference-Free Constant Orientation Workspace (IFCOW). The WFW is traced from the capacity margin, a measure of the robustness of the equilibrium of the robot. Interferences between the moving parts of a CDPR are also determined by an existing technique for tracing the interference-free workspace of such robots. Finally, the WFW and the IFCOW of a planar cable-driven parallel robot and of a spatial cable-driven parallel robot are plotted in order to demonstrate the potential of ARACHNIS.

Keywords Workspace · Wrench-feasible workspace · Interference · Graphical user interface · Capacity margin · Static equilibrium · Cable-driven robot · Wire-driven robot

A.L.C. Ruiz
IRCCyN, École Centrale de Nantes, 1 rue de la Noë, 44321 Nantes, France
e-mail: analu.610@gmail.com

S. Caro
CNRS–IRCCyN, 1 rue de la Noë, 44321 Nantes, France
e-mail: stephane.caro@irccyn.ec-nantes.fr

P. Cardou (✉) · F. Guay
Laboratoire de robotique, Département de génie mécanique, Université Laval,
Quebec City, QC, Canada
e-mail: pcardou@gmc.ulaval.ca

F. Guay
e-mail: francois.guay.2@ulaval.ca

A. Pott and T. Bruckmann (eds.), *Cable-Driven Parallel Robots*,
Mechanisms and Machine Science 32, DOI 10.1007/978-3-319-09489-2_21

1 Introduction

Cable-Driven Parallel Robots (CDPRs) may be seen as Gough-Stewart platforms in which the prismatic actuators are replaced with cables. Hence, the cables are connected between the moving platform and fixed eyelets. The platform moves around by increasing or decreasing the lengths of the different cables in concert while preventing any cable from becoming slack [1]. This architecture allows for marked advantages over conventional robots, such as the possibility of a larger workspace and higher accelerations, thanks to lighter mechanical components. CPDRs also suffer from important drawbacks, however. First, there is the possibility of interferences among the cables or between a cable and the moving platform. Furthermore, the non-rigid nature of the CPDR links further restrict its workspace, and requires a rigorous study of its force transmission characteristics.

Nevertheless, CDPRs have been used, often with success, in many areas such as robotic cranes, cable-driven cameras, medicine, automatic painting and rescue operations, to name but a few. The reader is referred to a recent literature survey on the topic [2] for a more thorough account of these applications. These CDPRs and others were developed independently, in most cases, meaning that their designers usually started the analysis and design process development "from scratch". As CDPRs become more popular, there is an obvious need to automate their design and analysis, to accelerate their development and to provide a common working ground to the engineers involved.

To partly answer this need, we propose the graphical user interface ARACHNIS, or Analysis of Robots Actuated by Cables with Handy and Neat Interface Software. In short, ARACHNIS is dedicated to the parameter-based analysis and design of CDPRs. It allows the designer to enter the design parameters of the robot, to specify the task that the robot should perform in terms of force and moment ranges, and to assess the performance of the robot through the visualization of its Wrench-Feasible Workspace (WFW) [3, 4] and its Interference-Free Constant Orientation Workspace (IFCOW) [5]. ARACHNIS seems to be similar in its functions to a previously developed interface named WireCenter [6]. From the little information we could gather on WireCenter, it seems that the two interfaces differ mainly in that WireCenter allows to trace the wrench-closure workspace (WCW), while ARACHNIS allows to trace the WFW.

This workspace was chosen to analyze the static-equilibrium of the robot because it applies to a wider range of tasks than the WCW [7] or the Static Workspace (SW) [8]. In fact, the WCW and the SW can be shown to be special cases of the WFW. To the knowledge of the author, two methods have been reported to trace the WFW of a spatial parallel robot driven by m cables. Both of these methods are purely numerical, since, as was pointed out in [9], computing symbolic expressions of the WFW-boundaries is too complex to be practical in the general case. This spurred other researchers to simply use a brute-force method [10] to trace the WFW as a cloud of points. Others resorted to interval arithmetics to trace the WFW as a set of boxes that are either completely inside the workspace or not completely outside of it [4]. Both of these methods are relatively slow and generally yield poor renderings of the workspace boundary. In ARACHNIS, we resort to an alternative method that was recently proposed by the authors [11], and which, if far from perfect, appears be somewhat faster and to yield smoother estimates of the WFW than the two existing ones. The IFCOW, on the other hand, is determined according to the method proposed in [12].

The paper is organized as follows: Sect. 2 describes the kinetostatic model of a cable-driven parallel mechanism. Section 3 introduces the capacity margin as a measure of the robustness of CDPR equilibrium. Section 4 explains how the capacity margin is used to trace the WFW. Section 5 presents the polygons of the IFCOW where interferences occur between the cables and the moving platform. Section 6 describes the ARACHNIS interface. A planar cable-driven parallel mechanism and a spatial cable-driven parallel mechanism are studied in Sects. 7 and 8, respectively. Finally, we present our conclusions and future work in Sect. 9.

2 Kinetostatic Model of a Cable-Driven Parallel Robot

Formally, a cable-driven parallel robot consists of a moving platform connected to a fixed base by a set of m cables acting in parallel. The winding and unwinding of these cables on fixed spools controls the platform motion in space.

To mathematically model this mechanism, let us attach frame $\mathscr{P}$ to the moving platform, and frame $\mathscr{F}$ to the fixed frame, as shown in Fig. 1. The origins of these frames are P and O, respectively. The platform pose may thus be represented by the vector $\mathbf{p}$ pointing from O to P, and by the rotation matrix $\mathbf{Q}$ rotating $\mathscr{F}$ onto $\mathscr{P}$.

The ith cable is modeled as a straight-line segment from the attachment point onto the moving platform, R_i, to the attachment point onto the fixed frame, A_i. The

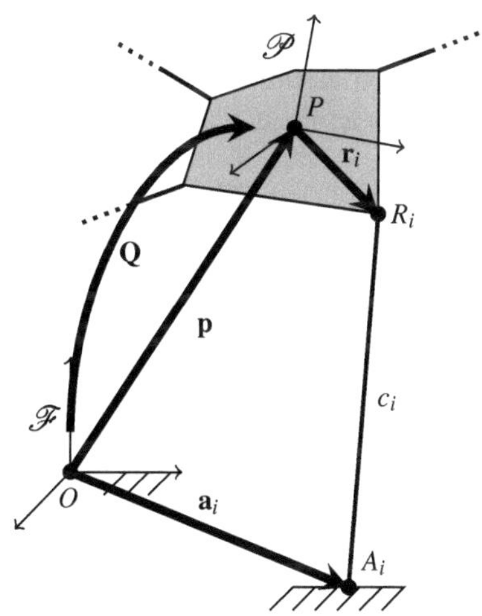

Fig. 1 The kinetostatic model of a cable-driven parallel mechanism

positions of these attachment points are respectively given by $\mathbf{r}_i$, expressed in frame $\mathscr{P}$, and $\mathbf{a}_i$, expressed in frame $\mathscr{F}$. The ith cable may thus be represented by the vector

$$\mathbf{c}_i \equiv \mathbf{a}_i - \mathbf{p} - \mathbf{Q}\mathbf{r}_i, \tag{1}$$

and the corresponding cable length is

$$c_i \equiv \|\mathbf{c}_i\|_2, \; i = 1, \ldots, m. \tag{2}$$

Let us now assume that external forces are applied on the moving platform. These forces are equivalent to the resultant force-moment system $\mathbf{w}_e \equiv [\mathbf{f}_e^T \; (1/r)\mathbf{n}_e^T]^T$, the force $\mathbf{f}_e \in \mathbb{R}^3$ being applied at P, the moment being $\mathbf{n}_e \in \mathbb{R}^3$, and r being a positive length used to render $\mathbf{w}_e$ dimensionally homogeneous. Here, we define this length quite arbitrarily as $r^2 = (1/m)\sum_{i=1}^{m} \|\mathbf{r}_i\|_2^2$. For the moving platform to remain in equilibrium, there must be a combination of cable tensions $\mathbf{t} \equiv [t_1 \; \cdots \; t_m]^T$ that balances the system of external forces. By the application of the Newton-Euler equations to the moving platform, we obtain

$$\mathbf{W}\mathbf{t} + \mathbf{w}_e = \mathbf{0}_6, \tag{3}$$

where $\mathbf{0}_6$ is the six-dimensional null vector and $\mathbf{W}$ is the Jacobian matrix of the mechanism at this particular pose. This matrix may be computed as

$$\mathbf{W} = \begin{bmatrix} \mathbf{c}_1/c_1 & \cdots & \mathbf{c}_m/c_m \\ (\mathbf{Q}\mathbf{r}_1) \times \mathbf{c}_1/(r\,c_1) & \cdots & (\mathbf{Q}\mathbf{r}_m) \times \mathbf{c}_m/(r\,c_m) \end{bmatrix}. \tag{4}$$

To this standard kinetostatic model of a cable-driven parallel robot, let us add limits to the involved forces and moments. Indeed, for a given application, we may safely assume that the designer of a cable-driven parallel mechanism can define a set of external force-moment systems $\mathbf{w}_e$ that can be applied on the mobile platform. We further assume that this set is a box $\mathscr{W}_e$ in $\mathbb{R}^6$. For reasons that will later become apparent, let us give the vertex representation of this box, namely,

$$\mathscr{W}_e = \{\mathbf{w}_e \in \mathbb{R}^6 : \; \mathbf{w}_e = \sum_{j=1}^{n} \alpha_j \mathbf{w}_{e,j}, \; \sum_{j=1}^{n} \alpha_j = 1, \; \alpha_j \geq 0, \; j = 1, \ldots, n\}, \tag{5}$$

where $\mathbf{w}_{e,j}$, $j = 1, \ldots, 64$ are the vertices of $\mathscr{W}_e$.

The cable tensions t_i, $i = 1, \ldots, m$, are also limited by the load capacities of the motors and by the strengths of the cables and spool transmission elements. From the components he/she selected, the designer should thus be able to assess the lower and upper tension bounds, respectively $\underline{\mathbf{t}}$ and $\overline{\mathbf{t}}$. Formally, these define the m-dimensional box of feasible tensions

$$\mathscr{T} = \{\mathbf{t} \in \mathbb{R}^m : \; \underline{\mathbf{t}} \leq \mathbf{t} \leq \overline{\mathbf{t}}\}. \tag{6}$$

3 The Capacity Margin as a Measure of the Robustness of Equilibrium

According to the model of a cable-driven parallel robot presented above, the moving platform is in equilibrium for every possible external force-moment system $\mathbf{w}_e$ if and only if, in every case, the Newton-Euler equilibrium, Eq. (3), can be satisfied by a set of feasible cable tensions $\mathbf{t}$. Such a pose of "complete" equilibrium of the moving platform is often called a "feasible pose" in the scientific literature. The set of all feasible poses is called the "wrench-feasible workspace" (WFW), and is formally expressed in Definition 1.

Definition 1 The wrench-feasible wokspace $\mathscr{F}$ is defined as

$$\mathscr{F} = \{(\mathbf{p}, \mathbf{Q}) \in \mathbb{R}^3 \times SO(3) : \forall \mathbf{w}_e \in \mathscr{W}_e, \exists \mathbf{t} \in \mathscr{T}, \ \mathbf{W}\mathbf{t} + \mathbf{w}_e = \mathbf{0}_6\}, \qquad (7)$$

where $SO(3)$ is the group of proper rotation matrices, $\mathscr{W}_e$ is defined in Eq. (5), $\mathscr{T}$ is defined in Eq. (6), and $\mathbf{W}$ is defined in Eq. (4).

The capacity margin was recently introduced by the same authors in [11] under the name *minimum degree of constraint satisfaction*. We believe that the name *capacity margin* describes better this index, if only for the sake of brevity, but for other reasons that will be made apparent from its definition.

This definition stems from the geometric interpretation of the wrench-feasible workspace, which is represented in Fig. 2. The two sets involved in the definition of $\mathscr{F}$, $\mathscr{W}_e$ and $\mathscr{T}$, live in different spaces and have different numbers of dimensions. In this paper, we assume the number of dimensions of $\mathscr{W}_e$ to be six, although the proposed method applies equally well to lower dimensional cases. The number of cables m corresponds to the number of dimensions of $\mathscr{T}$. For the sake of illustration, the dimensions of $\mathscr{W}_e$ and $\mathscr{T}$ in Fig. 2 were chosen to be two and three, respectively. The two sets are connected by Eq. (3), by which the tensions, $\mathbf{t}$, are linearly mapped onto the wrench space via the transformation $-\mathbf{W}$. Hence, the box $\mathscr{T}$ of feasible tensions becomes the zonotope $\mathscr{W}_t$ in the wrench space.

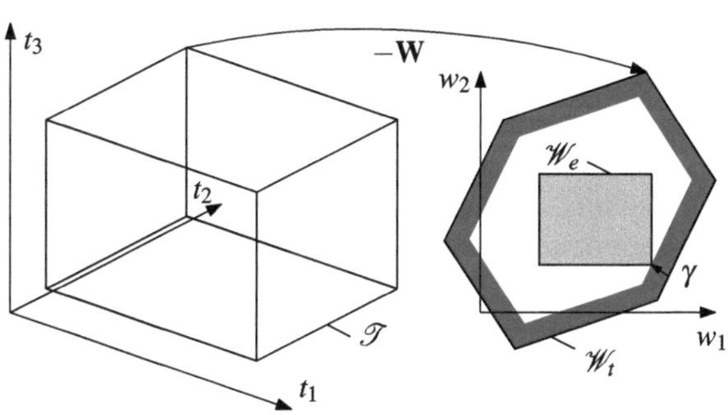

Fig. 2 An analog geometric representation of the capacity margin γ

From the definition of the wrench-feasible workspace, this zonotope should include the set $\mathscr{W}_e$ of external wrenches. Indeed, for an external wrench $\mathbf{w}_e$ to be sustained by the moving platform, there must exist a set of tensions in $\mathscr{T}$ such that $\mathbf{w}_e = -\mathbf{W}\mathbf{t}$. Geometrically, this means that the point $\mathbf{w}_e$ should be contained in $\mathscr{W}_t$. Therefore, for any external wrench to be sustained, we should have

$$\mathscr{W}_e \subseteq \mathscr{W}_t, \tag{8}$$

which gives us a necessary and sufficient condition for a pose $(\mathbf{p}, \mathbf{Q})$ to be wrench-feasible.

This binary condition does not tell us how far is a feasible pose from being unfeasible. Such a question is important to the designer, as an answer would indicate the robustness of the equilibrium. This is precisely what the capacity margin represents.

In Fig. 2, the capacity margin is geometrically represented by γ, the signed width of the margin between the boundary of $\mathscr{W}_t$ and $\mathscr{W}_e$. This width is taken positive when the margin is inside $\mathscr{W}_t$, and negative when it is outside. It may thus be seen as the degree of inclusion of $\mathscr{W}_e$ within $\mathscr{W}_t$.

Mathematically, we compute γ as follows:

1. Compute the facet representation of $\mathscr{W}_t$ using the hyperplane shifting method [3]. We thus obtain $\mathbf{a}_l$ and b_l, $l = 1, \ldots, p$ in the equation $\mathscr{W}_t = \{\mathbf{w} \in \mathbb{R}^6 : \mathbf{a}_l^T \mathbf{w} \leq b_l, l = 1, \ldots, p\}$.
2. Compute the coordinate $\gamma_{j,l}$ of each vertex $\mathbf{w}_{e,j}$ of $\mathscr{W}_e$ with each facet $(\mathbf{a}_l, b_l)$ of $\mathscr{W}_t$. This coordinate is taken from the facet in its normal direction $\mathbf{a}_l$, which gives $\gamma_{i,j} = (b_l - \mathbf{w}_{e,j}^T \mathbf{a}_l)/\|\mathbf{a}_l\|_2$, $j = 1, \ldots, n$, $l = 1, \ldots, p$.
3. Keep the minimum of all coordinates as a worst case scenario: $\gamma = \min_{j=1,\ldots,n} \min_{l=1,\ldots,p} \gamma_{j,l}$.

4 Tracing the Wrench-Feasible Workspace Using the Capacity Margin

The capacity margin γ indicates the *degree of inclusion* of the set $\mathscr{W}_e$ of external wrenches into the set $\mathscr{W}_t$ of feasible wrenches. γ is a greater *positive* scalar as $\mathscr{W}_e$ is further inside $\mathscr{W}_t$ opposed to being a greater *negative* scalar as at least a part of $\mathscr{W}_e$ is further outside $\mathscr{W}_t$. Consequently, γ is zero when $\mathscr{W}_e \subseteq \mathscr{W}_t$ and a point of $\mathscr{W}_e$ is on the boundary of $\mathscr{W}_t$. From Definition 1, this latter case corresponds to a pose that is on the limit of feasibility. Hence, the boundary of the wrench-feasible workspace may simply be expressed as

$$\gamma(\mathbf{p}, \mathbf{Q}) = 0. \tag{9}$$

In this paper, we use this fact to trace the boundary of the wrench-feasible workspace as the zero *isosurface* or *isocontour* of the capacity margin. To compute the zero isocontour of a function, one first needs to evaluate it at a grid of points covering its domain. The next step consists in interpolating the between adjacent points of the grid in search of roots of the function. This process is readily implemented in virtually all scientific computation packages. In `Matlab`, for instance, the functions `isosurface` and `contour` can be used to trace the isosurface and isocontour of a function, respectively. In `Maple`, the analogous functions are `implicitplot3d` and `implicitplot`.

5 Tracing the Interference Polygons

The wrench-feasible workspace alone is not sufficient to design a cable-driven parallel robot. Interferences between the moving parts must also be considered, either by shear intuition or by systematic analysis. A technique for such analysis of the interference-free workspace was proposed in [12], and was implemented in ARACHNIS. Let us note, however, that the technique only applies to a constant orientation of the moving platform. Moreover, it is developed under the assumptions that the cables are line segments and that the moving platform is a convex polyhedron represented by its edges.

In [12], it was shown that the region of the Cartesian space where a pair of cables collide is composed of two disjoint polygons in space. Mathematically, the two polygons corresponding to an interference between the ith and jth cables may be expressed as

$$\mathscr{C}_{i,j} = \{\mathbf{p} \in \mathbb{R}^3 : \mathbf{p} = \mathbf{a}_j - \mathbf{Q}\mathbf{r}_i + \alpha(\mathbf{a}_j - \mathbf{a}_i) + \beta\mathbf{Q}(\mathbf{r}_j - \mathbf{r}_i),\ \alpha, \beta \geq 0\}, \quad (10a)$$

$$\mathscr{C}_{j,i} = \{\mathbf{p} \in \mathbb{R}^3 : \mathbf{p} = \mathbf{a}_i - \mathbf{Q}\mathbf{r}_j + \alpha(\mathbf{a}_i - \mathbf{a}_j) + \beta\mathbf{Q}(\mathbf{r}_i - \mathbf{r}_j),\ \alpha, \beta \geq 0\}. \quad (10b)$$

Interferences can also occur between a cable and the moving platform. The associated interference regions are computed in [12]. This is done by tracing the interference region between the ith cable and the edge connecting the jth and kth vertices of the polyhedron-shaped moving-platform. Let the positions of these vertices be represented by $\mathbf{v}_j$ and $\mathbf{v}_k$, respectively, in the mobile frame $\mathscr{P}$. In this case, the interference region $\mathscr{E}_{i,j,k}$ takes the form of a polygon in space, that is,

$$\begin{aligned}\mathscr{E}_{i,j,k} = \{&\mathbf{p} \in \mathbb{R}^3 : \mathbf{p} = \mathbf{a}_i - \mathbf{Q}\mathbf{r}_i + \nu_j\mathbf{Q}(\mathbf{v}_j - \mathbf{b}_i) + \nu_k\mathbf{Q}(\mathbf{v}_k - \mathbf{b}_i),\\ &\nu_j, \nu_k \leq 0,\ \nu_j + \nu_k \leq -1\}.\end{aligned} \quad (11)$$

Notice that the regions $\mathscr{C}_{i,j}$ and $\mathscr{E}_{i,j,k}$ are unbounded in some directions. In ARACHNIS, we assume that the portion of interest of the Cartesian workspace is a box containing all the fixed points A_i, $i = 1, \ldots, m$. In the proposed interface, we thus trace the intersections of this box with the interference regions $\mathscr{C}_{i,j}$, $i, j = 1, \ldots, m$, and $\mathscr{E}_{i,j,k}$, $i = 1, \ldots, m$, $j, k = 1, \ldots, q$, where q is the number of vertices of the moving platform.

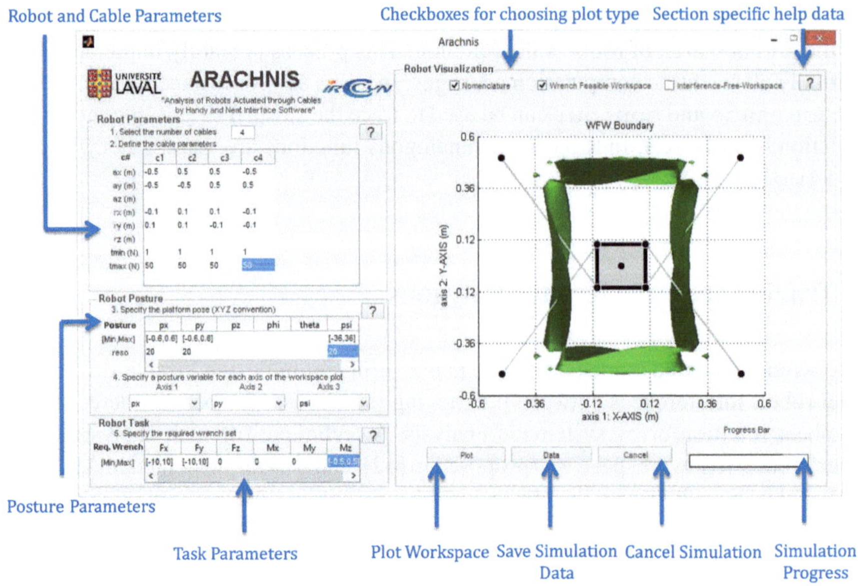

Fig. 3 ARACHNIS and its main features

6 The ARACHNIS Interface

ARACHNIS, or "Analysis of Robots Actuated by Cables with Handy and Neat Interface Software", is a graphical user interface developed in MATLAB to automate the design and the analysis of CDPRs. At the conceptual design stage, this tool aids the designer in choosing the design that best meets two requirements: equilibrium throughout the workspace and interference avoidance.

The interface provides the degree to which these requirements are satisfied for a particular design through the generation of the WFW and the IFCOW. As shown in Fig. 3, it is divided into four main areas: Robot Parameters, Robot Posture, Robot Task and Robot Visualization.

The section *Robot Parameters* is dedicated to the specification of the geometry of the platform and base, cable tension limits and cable arrangement. The geometry of the platform is described through the Cartesian coordinates of the cable anchor points on the platform (rx, ry and rz) expressed in the platform frame and the geometry of the base is described through the Cartesian coordinates of the cable anchor points on the base (ax, ay and az) expressed in the base frame. The tension limits are specified via the minimum and maximum tensions allowed in each cable.

In the section *Robot Posture*, the poses required from the platform during a specific task are specified through positions and orientations with respect to the X, Y and Z axes (*px*, *py*, *pz*, *phi*, *theta*, *psi*). The angles *phi*, *theta* and *psi* are the Euler angles specifying the attitude of the platform frame with respect to the fixed frame according to the *XYZ* convention. Fixed positions and orientations are specified by single numbers, while parameters that vary are specified by intervals and resolutions. Each varying parameter is then assigned to a workspace axis through the pop-up menus.

By allowing any combination and assignation of parameters to the workspace axes, ARACHNIS is able to generate any WFW from two to six dimensions. The WFW above three dimensions cannot be traced in three-dimensional space, of course. In such cases, three of the axes are selected to trace the workspace. The poses contained in the WFW are then the poses for which the full range of poses in the remaining axes are wrench-feasible poses.

The section *Robot Task* consists in describing the prescribed robot task by specifying intervals of forces and moments along each dimension (*Fx*, *Fy*, *Fz*, *Mx*, *My*, *Mz*). The forces are assumed to be applied at the origin of the moving-platform frame, but the components of the forces and moments are parallel to the axes of the fixed frame.

The section *Robot Visualization* consists of a graphical area to visualize the CDPR geometry, its WFW and its IFCOW. This information can be hidden or shown by first clicking the corresponding checkboxes and then clicking the *plot* button. This section also contains a *data* button, which generates an Excel spreadsheet containing the information on the interface and the simulation data, a *cancel* button to stop the workspace computation, and a progress bar to indicate the simulation progress. ARACHNIS also contains *Help* buttons and error messages customized for each section to guide the user as he or she inputs the parameters.

7 Case Study 1: A Planar Cable-Driven Parallel Mechanism

This case study demonstrates the capacity of ARACHNIS to generate the wrench feasible workspace (WFW) of planar cable-driven mechanisms. The mechanism under study is the eight-cable robot shown in Fig. 4, which has previously been analyzed in [13]. As in [13], the tension limits were set to 10 and 1,000 N and no external wrench was applied. The *total orientation workspace* in Fig. 4 was computed by displacing the robot in a 1.2 m × 1.2 m × 120° grid using a resolution of 80 points per axis.

The green iso-surface corresponding to $\gamma = 0$ represents the frontier of the WFW. Inside this boundary the robot can safely perform the task. We can acquire a more quantitative view of the stability of the robot by slicing the total orientation workspace and generating the *constant orientation workspace* shown in Fig. 4. Each contour in the figure represents a value of γ. Again, the contour where $\gamma = 0$ defines the stability frontier. The higher this value, the greater the robustness of the moving-platform equilibrium. Notice that these results match those obtained in [13]. Other examples of the workspaces generated by ARACHNIS were done by the same authors in [11], which match the workspaces in Ref. [4]. The computation time reported in [4] was 51 s, while that obtained with ARACHNIS was 6 s. This

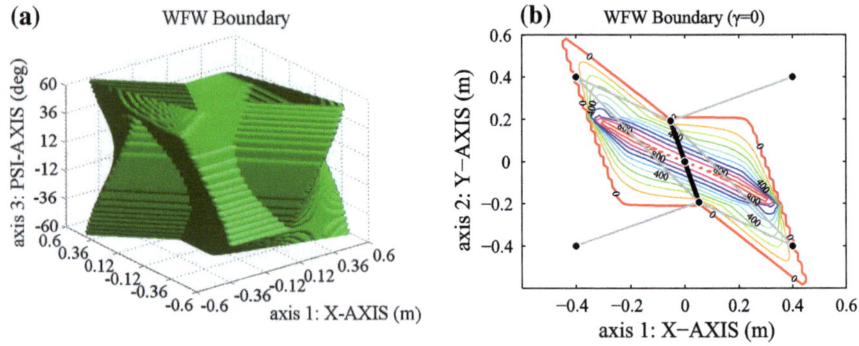

Fig. 4 Wrench feasible workspace (WFW) of a planar robot. **a** Total orientation WFW. **b** Constant orientation WFW: 60°

time difference is significant, and cannot be attributed only to the different personal computers that were used in the two studies. Notice that all the simulations were run on a Dell Inspiron 15 r laptop (Intel Core i7-3537U 2.50 GHz).

8 Case Study 2: A Spatial Cable-Driven Parallel Mechanism

This case study demonstrates the capacity of ARACHNIS to generate workspaces of spatial cable-driven mechanisms. The mechanism under study is the spatial robot with eight cables displayed in Fig. 5, which has also been previously analyzed in [4]. The tension limits for all cables were set to 1 and 540 N. The externally applied forces along each dimension varied from -10 to 10 N, while the moments varied from -0.5 to 0.5 N m.

The 5D WFW workspace of this robot was generated by translating it along a 1 m × 1 m × 1 m grid. At each point on this grid, the robot was rotated around the X and Y axes through orientations ranging from -15 to 15°, while retaining a fixed orientation with respect to the Z-axis. The capacity margin was then evaluated at each of these poses. A set of values γ was thus obtained for every position on a 3D grid. The minimum value from each set was selected and assigned to its corresponding position on the grid. Finally, the values were interpolated, creating the boundary shown Fig. 5. As in the previous section, we computed the 2D slices of this higher dimensional workspace to assess quantitatively the performance of the robot. A slice along the XZ plane is also shown in Fig. 5.

These results also match those obtained in [4] by means of interval arithmetics. An argument often made in favor of interval arithmetics is the guarantee of exactness that comes with its results. In the example shown in [4], however, the volume occupied by the uncertainty boxes—boxes that have not been determined to be fully inside or outside the WFW—amounts to 60 % of the total workspace volume. In these

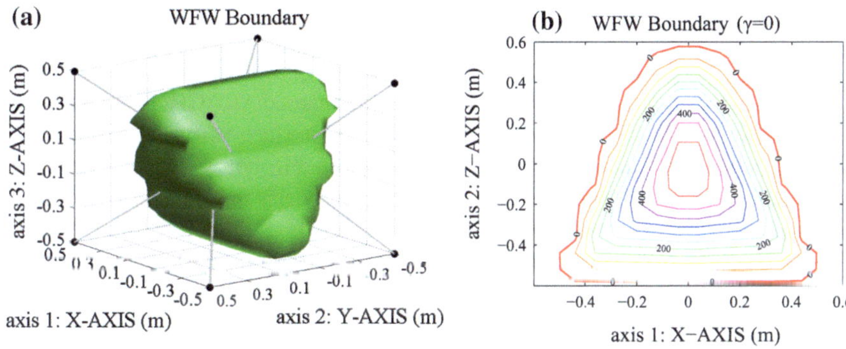

Fig. 5 Wrench feasible workspace (WFW) of a spatial robot. **a** 5D WFW. **b** Slice: Plane XZ

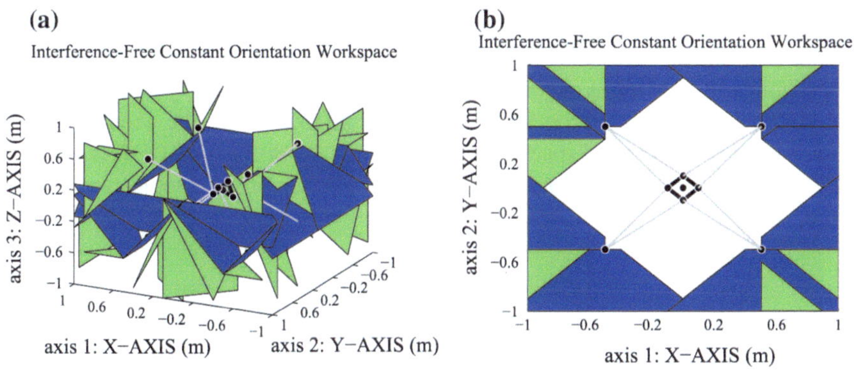

Fig. 6 Interference-free constant orientation workspace (IFCOW) of a spatial robot. **a** Isometric view. **b** *Top* view

conditions, there is no guarantee on the boxes of interest, i.e., those that are close to the boundary of the WFW. Hence, for this example, interval arithmetics offers no more guarantee than any other method. This case study also highlights the computational efficiency of ARACHNIS. The workspace was computed in 225 s, outperforming the algorithm presented in [4] that provided a computation time equal to 1,567 s. The interference-free constant orientation workspace was also computed through the method described in Sect. 5 and is displayed in Fig. 6. The green polygons symbolize the interference regions between cables, while the blue ones represent the interference regions between a cable and a moving-platform edge. The results show that this robot has an interference free workspace within a box of 1 m × 1 m × 1 m.

9 Conclusions

In this paper, we presented a graphical user interface named ARACHNIS, or Analysis of Robots Actuated by Cables with Handy and Neat Interface Software. ARACHNIS allows the design and analysis of Cable Driven Parallel Robots (CDPRs). It requires the designer to enter the parameters of the robot, to specify the loads involved in the robot task, and to assess the performance of the robot through the visualization of its Wrench Feasible Workspace (WFW) and its Interference-Free Constant Orientation Workspace (IFCOW). The capacity margin was first defined as a measure of the robustness of the moving-platform equilibrium and was used to trace the wrench-feasible workspace of the CDPRs. Interferences between the moving parts have also been considered. An existing technique for tracing the interference-free workspace of CDPRs was summarized and implemented in ARACHNIS. The WFW and the IFCOW of a planar cable-driven parallel mechanism and of a spatial cable-driven parallel mechanism have been plotted with ARACHNIS. It turns out that ARACHNIS is very competitive in terms of computation time for tracing the WFW and the IFCOW of CDPRs. The use of the proposed capacity margin to compute the wrench-closure workspace, the static workspace and the dynamic workspace of CDPRs and the conversion of ARACHNIS into a stand-alone interface are the subjects of future work.

References

1. Roberts RG, Graham T, Lippitt T (1998) On the inverse kinematics, statics, and fault tolerance of cable-suspended robots. J Robot Syst 15(10):581–597
2. Gosselin C (2014) Cable-driven parallel mechanisms: state of the art and perspectives. J Jpn Soc Mech Eng 1(1):1–17
3. Bouchard S, Gosselin C, Moore B (2010) On the ability of a cable-driven robot to generate a prescribed set of wrenches. ASME J Mech Robot 2(1):011010
4. Gouttefarde M, Daney D, Merlet JP (2011) Interval-analysis-based determination of the wrench-feasible workspace of parallel cable-driven robots. IEEE Trans Robot 27(1):1–13
5. Merlet JP (2004) Analysis of the influence of wires interference on the workspace of wire robots. In: Advances in robot kinematics. Sestri Levante, Italy, pp 211–218
6. Pott A, Meyer C, Verl A (2010) Large-scale assembly of solar power plants with parallel cable robots. In: 41st international symposium on robotics and 6th German conference on robotics. Munich, Germany, pp 999–1004
7. Gouttefarde M, Gosselin C (2006) Analysis of the wrench-closure workspace of planar parallel cable-driven mechanisms. IEEE Trans Robot 22(3):434–445. doi:10.1109/TRO.2006.870638
8. Stump E, Kumar V (2006) Workspaces of cable-actuated parallel manipulators. J Mech Des 128. doi:10.1115/1.2121741
9. Bosscher P, Riechel AT, Ebert-Uphoff I (2006) Wrench-feasible workspace generation for cable-driven robots. IEEE Trans Robot 22(5):890–902
10. Azizian K, Cardou P (2013) The dimensional synthesis of spatial cable-driven parallel mechanisms. ASME J Mech Robot 5(4):044502
11. Guay F, Cardou P, Ruiz ALC, Caro S (2013) Measuring how well a structure supports varying external wrenches. In: 2nd conference on mechanisms, transmissions and applications. Bilbao, Spain

12. Perreault S, Cardou P, Gosselin C, Otis M (2010) Geometric determination of the interference-free constant-orientation workspace of parallel cable-driven mechanisms. ASME J Mech Robot 2(3):031016
13. Verhoeven R (2004) Analysis of the workspace of tendon-based stewart platforms. Ph.D. thesis, University of Duisburg-Essen

Upper Limb Rehabilitation Using a Planar Cable-Driven Parallel Robot with Various Rehabilitation Strategies

XueJun Jin, Dae Ik Jun, Xuemei Jin, Jeongan Seon, Andreas Pott, Sukho Park, Jong-Oh Park and Seong Young Ko

Abstract Robotic technology became an important tool for rehabilitation especially for stroke patients. This paper presents development of three degrees-of-freedom cable-driven parallel robot (CDPR) for upper limb rehabilitation. Main features of the proposed rehabilitation robot are to provide relatively large workspace and to be less dangerous especially in the situation of robot's malfunction owing to its reduced inertia of a moving part. In addition, the cable-driven rehabilitation robot has many advantages such as transportability, low cost, low actuation power, safeness, large workspace and so on. In this paper, we analyzed the patient's joint movement during the passive rehabilitation using the developed CDPR. In addition, the paper presents the several types of rehabilitation therapy strategies and their implementation using the proposed CDPR system.

Keywords Planar cable-driven parallel robot · Rehabilitation · Upper limb

---

X. Jin · D.I. Jun · X. Jin
Department of Mechanical Engineering, Graduate School of Chonnam National University, Gwangju, South Korea

J. Seon
Robot Research Initiative, Chonnam National University, Gwangju, South Korea

S. Park · J.-O. Park (✉) · S.Y. Ko (✉)
School of Mechanical Engineering, Chonnam National University, Gwangju, South Korea
e-mail: jop@jnu.ac.kr

S. Park
e-mail: spark@jnu.ac.kr

S.Y. Ko
e-mail: sko@jnu.ac.kr

A. Pott
Fraunhofer Institute for Manufacturing Engineering and Automation IPA, Stuttgart, Germany
e-mail: Andreas.Pott@ipa.fraunhofer.de

A. Pott and T. Bruckmann (eds.), *Cable-Driven Parallel Robots*,
Mechanisms and Machine Science 32, DOI 10.1007/978-3-319-09489-2_22

1 Introduction

Stroke is one of the most frequent causes of death worldwide and it is ranked top two in Korea. In addition, many stroke survivors are permanently disabled. According to the national health insurance claim data from 2000 to 2010, the number of stroke hospitalizations grew by a compound annual growth at a rate of 6.4 %, and health care expenditures showed continuing growth at a rapid rate of 13.2 % [1].

Rehabilitation of post-stroke hemiplegic subjects aims to reduce motor impairment, to improve functional use of their limbs, and thus to help them early reintegration into social and domestic life. The rehabilitation is performed by moving their upper and/or lower limbs during some amount of time. The rehabilitation using robots can be more efficient than that helped by therapists since the robots can spend more time in the therapy than the human therapists. There has been a substantial evolution in robotic technology for the upper limb rehabilitation over the last several decades, which has created a variety of opportunities and challenges.

Robot-aided rehabilitation therapy has two main advantages comparing to a traditional one. First, the therapy is carried on autonomously by the robot although it is programmed by a therapist. Thus, the therapy can last longer than a traditional therapy done by the human therapist. Secondly, the patient's strength exerting on the end-effector of the robot and its end-effector's motion can be measured. Thus, performance of the therapy, i.e. patients' motion improvement, can be recorded and qualified.

According to a common classification suggested by [2], there are two kinds of upper limb rehabilitation robotic devices: the exoskeleton type devices and the end-effector type ones as shown in Fig. 1.

The most successful example of an end-effector type is the MIT-Manus [3], developed by Massachusetts Institute of Technology (MIT), USA. The MIT-Manus robot consists of two degrees-of-freedom (DOFs) serial robot for elbow and forearm

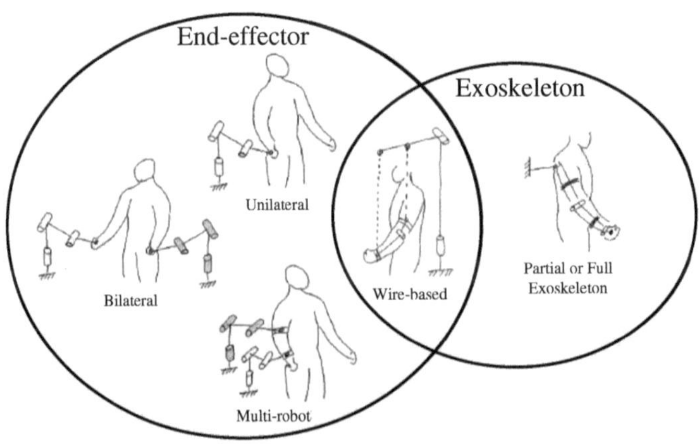

Fig. 1 Main configuration categories of upper limb rehabilitation robot [2]

motion. A 6-DOF force sensor is attached to the robot to control the motion by a personal computer. The robot spans the workspace of 381 mm × 457.2 mm, and produces the impedance up to 4.2 N/mm, which can displaya virtual wall effectively. Some disadvantages of this robot are (a) that this robot's workspace cannot be changed and (b) the robot's speed needs to be restricted low due to possibility of the robot's hitting the patient.

Many research groups have developed exoskeleton-type robots, such as ARM in III [4] which is one of the most advanced robotic arms available today. Its active 6-DOF motion allows the practice of a wide range of realistic arm movements in even severely impaired subjects. The ARMin has undergone pilot clinical testing in brain injured subjects, and generates several modes of passive and active-assisted movement. It also provides a virtual environment to motivate the exercises in the form of simple games or activities and daily living (ADL) tasks. The structure of exoskeleton robots resembles the human arm anatomy, where the mechanical rotation axes correspond with the human rotation axes. Consequently, the patient's arm is fixed to the exoskeleton at several points. This type of robotic devices can provide the rehabilitation of individual joints. However, its number of DOFs is in general higher than the end-effector type systems and each link of the robot system should be adjusted to the corresponding patient arm segment. At last, in the case of the robot's malfunction, it may be severely harmful to the patient's limbs.

Compared with other rehabilitation robots, a cable-driven parallel robot (CDPR) offers the advantages of a low-cost mechanical structure, also low actuation power, intrinsically safe treatment due to the low inertia, high acceptability by the patients, who are not comfortable to an "industrial-like" robot. We are currently developing a planar CDPR for the upper limb rehabilitation. The CDPRs are a type of parallel manipulators wherein its end-effector is supported in parallel by multiple cables that are controlled by multiple tensioning actuators, so-called winches. Several examples of the CDPRs can be found in the rehabilitation field, e.g. MACARM [5], NeReBot [6] and STRING-MAN [7]. The cable rehabilitation robot in this paper is a planar and fully constrained cable robot.

In addition, the developed rehabilitation robot is integrated with virtual reality (VR) to provide the virtual games that are designed to motivate the patients and to engage them in a period of intensive training. VR-based rehabilitation can also improve the efficiency of a physical therapist by providing tools for better assessment and remote monitoring of multiple patients [8].

The paper is organized as following. At first, we briefly introduce the rehabilitation robots and advantages of the cable-driven parallel robot in this chapter. Chapter 2 provides the overview of a prototype of our robotic system. In Chap. 3, the workspace analysis using an optical tracking system and analysis of patient's joint movement are given. In Chap. 4, four different rehabilitation modes are defined and explained briefly. Finally, the conclusion and future work are described in Chap. 5.

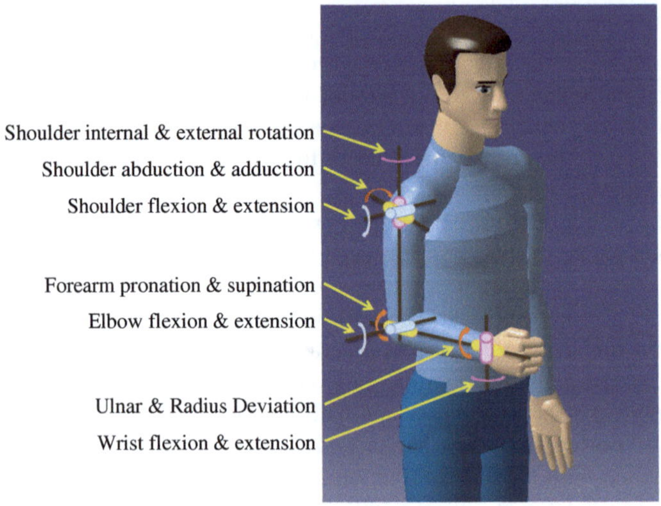

Fig. 2 Upper limb motion

2 Mechanical Design

2.1 Design Scheme

The robot design affects the efficiency of the rehabilitation. According to the clinical trials data, the planar robot inspired by 2D gravity-compensated therapy is more effective than 3-D spatial therapy [9]. As a first prototype, we developed a desk-shaped planar CDPR. That means that the robot is integrated to the desk where patients sit closely in front of it. Patients are supposed to leans forward to the desk during the therapy.

2.2 Upper Limb Motion

As shown in Fig. 2, a human upper limb consists of seven DOFs; 3 DOFs in the shoulder joint, 2 DOFs in the elbow joint and 2 DOFs in the wrist joint. The basic motions of the upper limb can be categorized into eight individual motions [10]; shoulder vertical flexion/extension, shoulder horizontal flexion/extension, shoulder adduction/abduction, shoulder internal/external rotation, elbow flexion/extension, forearm supination/ pronation, wrist flexion/extension, and wrist ulnar/ radial deviation. The activities of daily living (ADL) for the upper limb are the combination of these basic motions. Although fingers also have lots of DOFs, their motions are not considered in this study.

Table 1 Dimensions of rehabilitation system

	Dimensions (mm)
Width	1,200
Length	600
Height	700

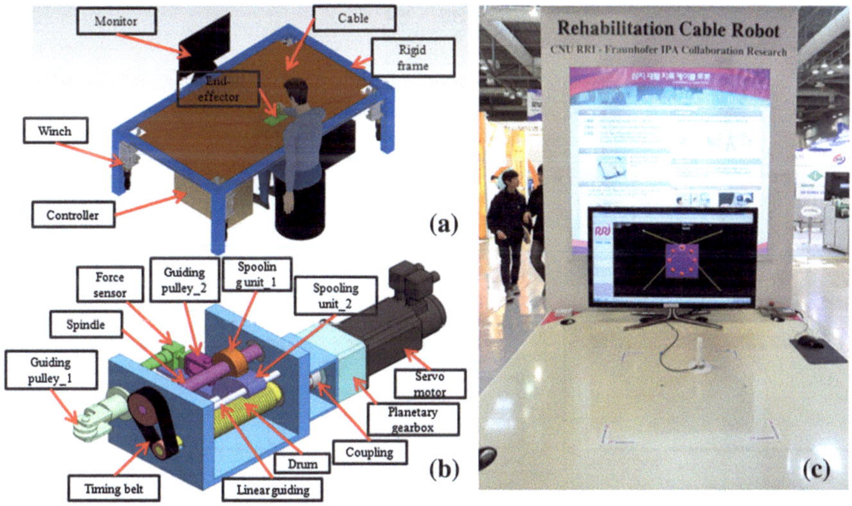

Fig. 3 A planar cable-driven parallel robot. **a** Concept of planar cable-driven parallel robot. **b** Winch used in this system. **c** Prototype of planar cable-driven parallel robot

2.3 Prototype and Its Size

In order to provide sufficient workspace for Korean patients, the dimensions of our planar CDPR are selected as in Table 1 based on database of the average size of Koreans [11]. The width of the frame was chosen to be 2 or 3 times larger than the average shoulder width. The length was chosen to be a bit longer than the length of the upper limb. The height was chosen similarly to normal desk.

Figure 3a, c show a conceptual design and a prototype of the developed rehabilitation robot. Similarly to a normal planar CDPR, the developed system is composed of a rigid frame, four winches, an end-effector (handle), cables, a PC as a controller, and the monitor. The winches for controlling of cable lengths are mounted beneath the desk for the better look. The patient sits near the desk and grabs the end-effector with his/her either hand. Figure 3b shows assembly of the winch, whose basic concept is same as the winch explained in [12].

2.4 System Communication Structure and Winches

The planar CDPR is composed of a rigid frame, four winches that control the cable lengths, a low-level position controller, a PC for a high-level control, an end-effector,

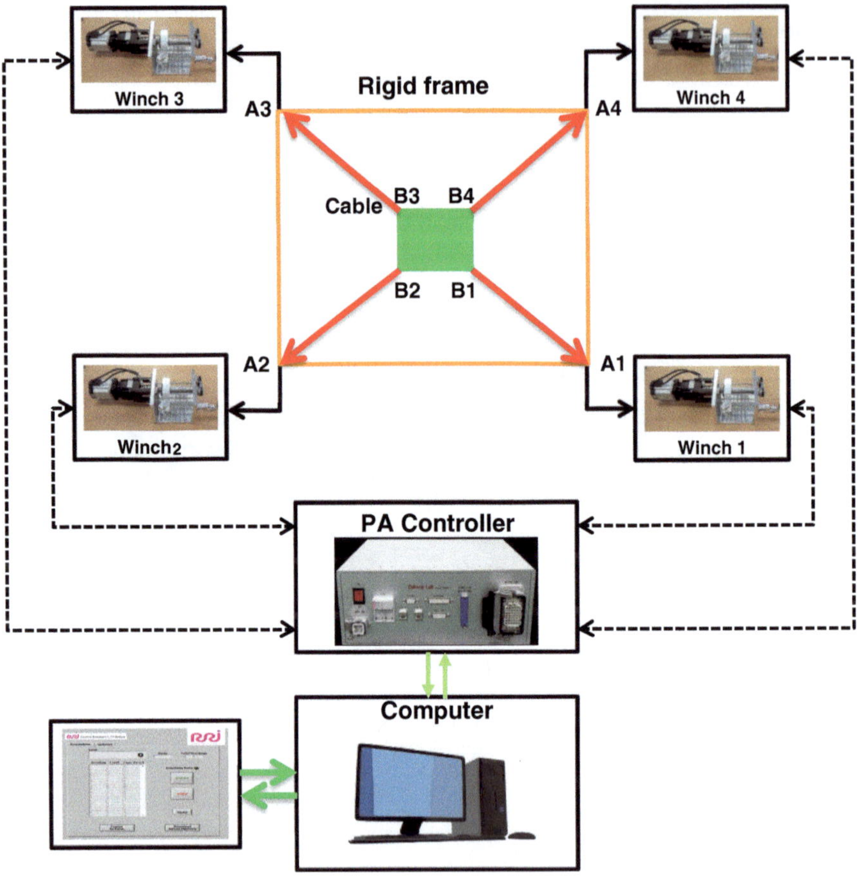

Fig. 4 System communication structure of the planar cable-driven robot

as shown in Fig. 4. For the low level control, a PA controller from Precise Automation Ltd., USA is used to control the motors and set-point values of motors are transferred to PC with a sampling time of 5 ms for monitoring the status of the controller.

3 Motion Analysis

3.1 Experimental Setup

In this section, we discuss the upper limb motion during the rehabilitation therapy using the developed robotic system. For this purpose, we attached ball markers at the four different positions on the upper limbs to measure three-dimensional (3-D) positions of shoulder/elbow/wrist joints as shown in Figs. 5 and 6.

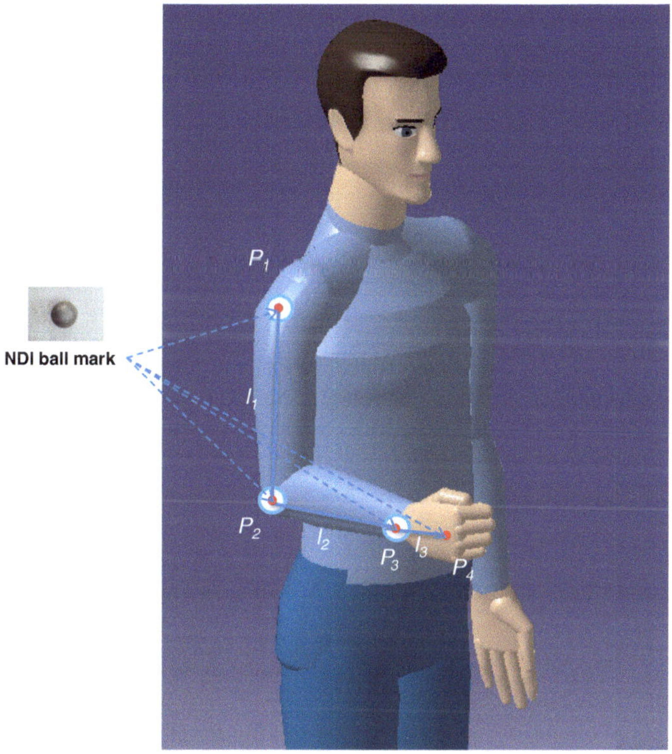

Fig. 5 Markers' positions for motion analysis of upper limb

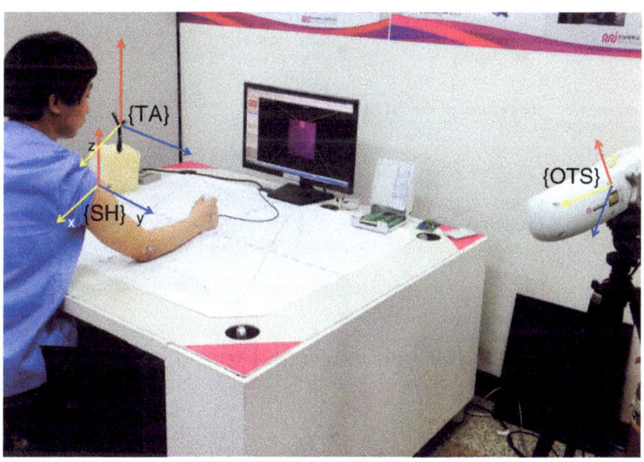

Fig. 6 Experimental setup and the coordinate systems

As depicted in Fig. 6, the patient sits down in front of the system and grabs the end-effector gently. The end-effector of robot moves to follow a circle at the constant speed. At the upper limb, four infra-red (IR) reflective ball markers of optical tracking system (OTS) from NDI Ltd. are attached at the joints i.e. shoulder (P_1), elbow (P_2), wrist (P_3) joints and at the hand (P_4) to measure their 3D position according to time.

3.2 Experiment Result and Discussion

After measuring the 3D positions, first we transform the positions relative to the optical tracking system coordinate frame {OTS} into the positions relative to the shoulder coordinate frame {SH} as in (1). The rotational matrix of the shoulder coordinate frame {SH} is calibrated before the measuring process, and the origin position of {SH} is set as the 3D position of shoulder joints. The vector of each link of the upper limbs has been defined in (2).

$$^{SH}P_i = (^{OTS}T_{SH})^{-1}\,{}^{OTS}P_i \tag{1}$$

$$^{SH}V_i = {}^{SH}P_{i+1} - {}^{SH}P_i \tag{2}$$

where $^{SH}V_1$, $^{SH}V_2$ and $^{SH}V_3$ indicate the positions of the upper arm, forearm, and hand distally, respectively.

The angle of elbow, the shoulder flexion and the shoulder adduction are defined and calculated as following:

$$\theta_{elbow} = \cos^{-1}\left(\frac{^{SH}V_1 \cdot {}^{SH}V_2}{|^{SH}V_1|\,|^{SH}V_2|}\right) \tag{3}$$

$$\theta_{flexion} = \text{atan2}\left(-{}^{SH}V_{1y}, {}^{SH}V_{1x}\right) \tag{4}$$

$$\theta_{adduction} = \cos^{-1}\left(\frac{^{SH}V_1 \cdot -\hat{z}}{|^{SH}V_1|\,|-\hat{z}|}\right) \tag{5}$$

where $-\hat{z}$ indicates the negative z-axis.

Figure 7a shows traces of four ball markers. From this figure, we can see the data has been measured reasonably.

Figure 7b and Table 2 show calculated joint angle ranges of upper limb joints when the subjects hold end-effector that is following a circle trajectory (R = 150 mm). The experiment results show that using the developed system we can produces the elbow motion of 53.2 ± 5.7, the shoulder flexion of 91.6 ± 5.2, and the shoulder adduction of $32.8 \pm 5.5°$. Measurement and analysis of the joint motions are performed here to see the generated joint motion is acceptable for the upper limb rehabilitation. The

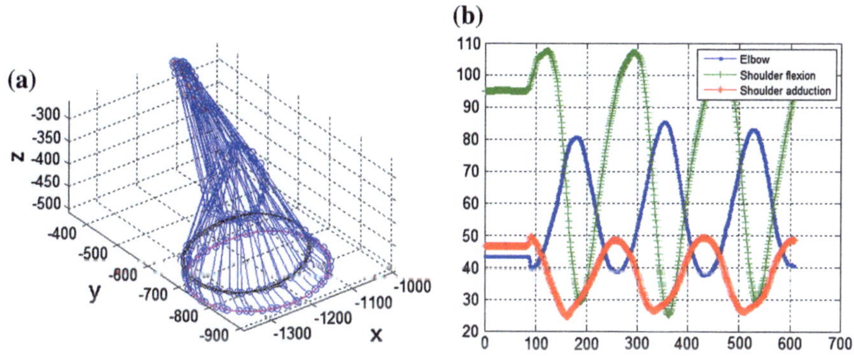

Fig. 7 **a** Plot of ball mark trajectory. **b** Plot of upper limb joint angle

Table 2 Measured range of joint motions

Subject	Links Lengths (mm)			Joint Angles (min/max/range) (degree)		
	Upper arm (L_1)	Forearm (L_2)	Hand (L_3)	Elbow angle	Shoulder flexion	Shoulder adduction
S1	231	244	82	38/85/47	20/108/88	25/50/25
S2	252	272	74	35/90/55	10/110/100	23/53/30
S3	266	241	83	33/95/62	15/105/90	28/58/30
S4	245	255	68	28/80/52	22/115/93	22/60/38
S5	277	246	82	39/89/50	24/111/87	16/57/41
Min	231	241	68	28	10	16
Max	277	272	83	95	115	60
Range	–	–	–	53.2 ± 5.7	91.6 ± 5.2	32.8 ± 5.5

motion range of the shoulder flexion is sufficient since it produces almost similar to the joint range; however the others are required to further improvement. The measurement and analysis capability may be helpful to the surgeon if it is integrated to the rehabilitation system to provide the surgeon with the patient joint motion in real time.

4 Rehabilitation Therapy Strategies Using the Planar CDPR

We consider four basic therapy strategies for the rehabilitation using the developed planar CDPR: preprogrammed mode which the robotic device moves the patient's arm, active assistant mode which the patient executes the end-effector moving along the guidance line and the robot just help the patient move along guidance line, active range of motion which the patient attempts to move and the robot provides no help but

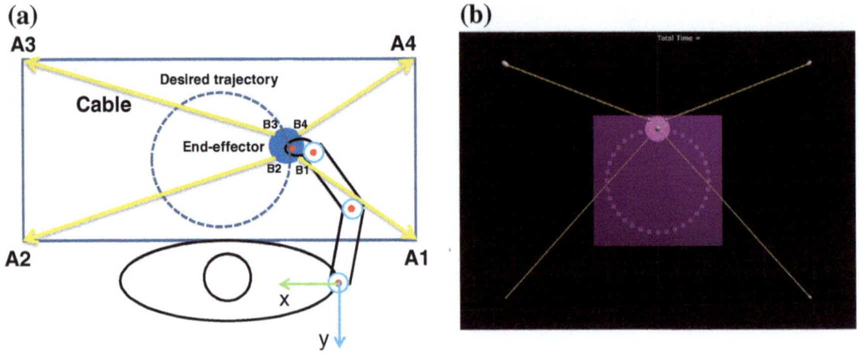

Fig. 8 Rehabilitation strategy in preprogrammed mode. **a** The system configuration. **b** The graphic user interface

there have virtual wall protect patient safety, and progressive resistance the patient is required to perform an exercise against an resistive force provided by the robot [13].

4.1 Preprogrammed Mode (Passive Assistant Mode)

As a first step of rehabilitation, our system provides the preprogrammed motion. Since at the beginning of the rehabilitation patients do not have enough muscle power to move even their own limbs and they have pains in their joints during motion, the robot needs to help their motion forcedly with moderate speed. In this way, they can regain functionalities of their joints in the upper limb.

We developed the trajectory of various shapes in planar motion. Figure 8a shows one example trajectory, the circle, and Fig. 8b shows the graphic user interface to guide the patient to follow the path.

4.2 Position-Based Admittance Control

To implement the passive rehabilitation, a 6-DOF force/torques sensor (Mini40 from ATI industrial automation co.) is attached to the end-effector. Admittance control modulates the position trajectory as a function of the force measured [14]. The admittance is in general defined as:

$$\frac{\dot{p}(s)}{f(s)} = \frac{1}{M_a s + c_a + \frac{k_a}{s}} \tag{6}$$

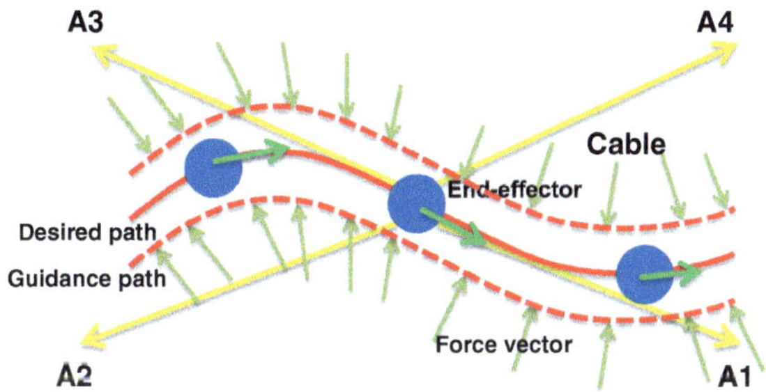

Fig. 9 Active assistant mode

where p is the Cartesian position of the end-effector and f represents force and torque of the force sensor, M_a, c_a and k_a are virtual inertia, virtual damping and virtual stiffness components of admittance control. The desired position that corresponds to the input force can be expressed as:

$$p_d(s) = \frac{f_d(s)}{M_a s^2 + c_a s + k_a} \tag{7}$$

4.3 Active Assistant Mode

In the active assistant mode, one desired path line and two guide-path lines that are offset from the original desired path are constructed. When the patient moves the end-effector along/near the desired path, he/she can move the robot freely in the admittance mode control, in which the force exerted by the patient is used to generate the robot's desired velocity. In this way, the patient moves his/her upper limb as a therapist designed. The conceptual diagram of the active assistant mode is shown in Fig. 9.

As depicted in Fig. 9, the active assistant mode can be considered as the combination of two virtual walls that are placed closely. To implement the admittance control and at the same time to provide the virtual wall perception, we need to modify the input command, i.e. f in (6). Consider the situation where the end-effect are located near the virtual wall as in Fig. 10, where the origin point is drawn in cyan dashed line and the virtual wall is drawn in red dotted line. Then, F_p is the force applied by the patient to move the end-effector, and F_{vw} is the reaction force from the virtual wall. The magnitude and the direction of the force is directly relate to the position and speed of the end-effector, and the spring constant (k) and damper-constant (c), as in (8).

$$F_p = m_a \ddot{p} + c_a \dot{p} + k_a p \tag{8}$$

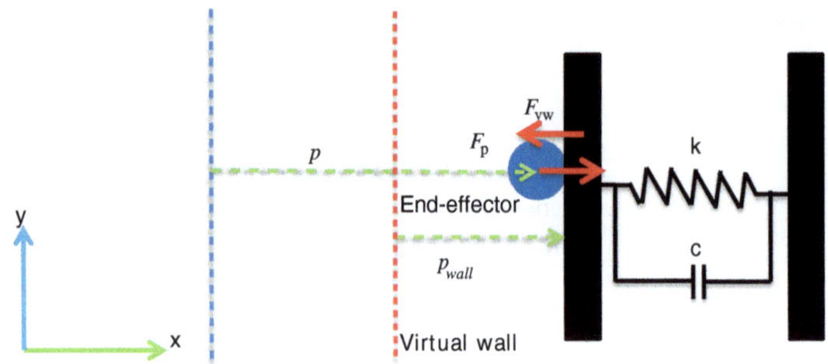

Fig. 10 Virtual wall spring-mass-damper model

where m_a, c_a and k_a are the virtual mass, the virtual damping and the virtual spring of the admittance control. The position p is the displacement from the closest points on the virtual wall to the position of the end-effector.

When the end-effector gets contacted to the virtual wall, the virtual contact force will be considered. If the virtual reactive force is defined as in (9), the total force is also calculated as in (10).

$$F_{vw} = c\ \dot{p}_{wall} + k\ p_{wall} \tag{9}$$

$$F_{total} = F_p + \lambda F_{vw} \tag{10}$$

where

$$\lambda = \begin{cases} 1; & p(x, y) \geq p_{wall}(x, y) \\ 0; & p(x, y) < p_{wall}(x, y) \end{cases}$$

4.4 Active Range of Motion

In this mode, we utilized only one virtual wall instead of two. The virtual wall can be constructed in any shapes. However, the basic geometry—e.g. line, rectangular, circle is considered because the function can be obtained straight forwardly. The algorithm related to this mode is mostly same as the control strategy described in Sects. 4.2 and 4.3.

These walls bound the desired working area and allow the end-effector works inside (Fig. 11). Moreover the virtual wall can protect patient from injury.

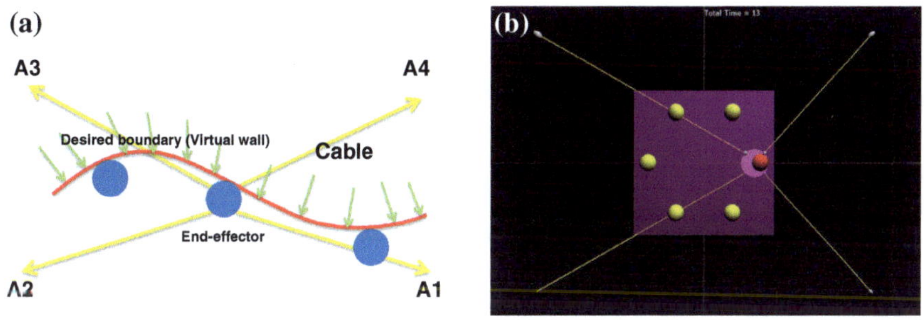

Fig. 11 Active range of motion. **a** The concept. **b** The graphic user interface

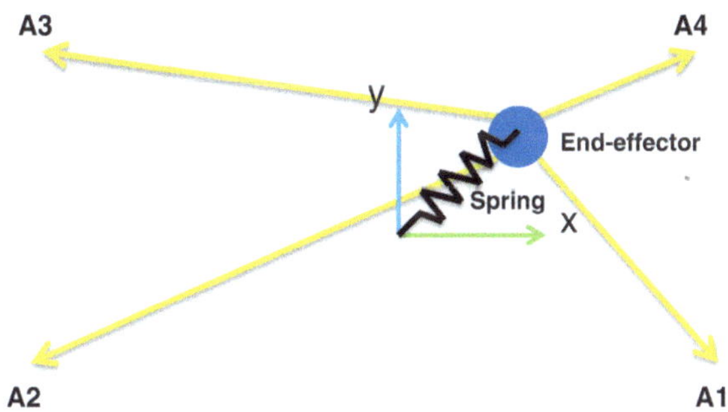

Fig. 12 Progressive resistance mode

4.5 Progressive Resistance Mode

As a last step, if patients get recovered well, they need to increase their muscle power. For this purpose, the progressive resistance mode is also provided, where the robotic system resists from moving thus the patient needs to execute larger force than the others. As shown in Fig. 12 when the patient appliesthe force to move the end-effector, the resistive reaction force is applied as in (11). This is considered as the situation where the virtual wall is as small as the point. The calculated resistive force can replace the virtual wall force F_{vw} in (10).

$$F_{re} = k(\Delta p) \tag{11}$$

4.6 The Virtual Rehabilitation Therapy Game

Three games were developed to improve the effectiveness of the developed rehabilitation robot [15]. The graphic user interface (GUI) is developed so as to be intuitive and easy to understand as shown in Fig. 13. The game is designed to provide the force feedback to the player as well.

Fig. 13 Virtual rehabilitation therapy games, **a** painting game, **b** pong game, **c** ball game

While the patient plays the game, the simulation transparently measures and stores several parameters in the database: recording time until completing the exercise, errors between the ideal path and the path taken by the patient. Through several modes of feedback, the patient's will be encouraged and put some effort to improve the result data through several modes of feedback. A congratulatory word is displayed when the patient completes an exercise.

5 Conclusion and Future Research

This paper presents the design of a planar cable-driven parallel robot for the upper limb rehabilitation. This robot's specific design is decided using the limb information from Korean size database. To analyze the real motion of joints, we measured the 3-dimensional positions of the joints using an optical tracking system and analyze their angle variations. Our robot is developed to deliver four different rehabilitation modes using a cable-driven parallel robot.

In the future, a larger number of experiments will be done to quantify the role of the upper limb joint angle more systematically. The clinical application is finally envisaged for an extensive use of the machine to neurological patients (such as chronic post-stroke, inpatients, and brain injured people).

Acknowledgments This research was supported by Leading Foreign Research Institute Recruitment Program through the National Research Foundation of Korea (NRF) funded by the Ministry of Education, Science and Technology (MEST) (2012K1A4A3026740).

References

1. Kwon YD, Chang H, Choi YJ, Yoon SS (2012) Nationwide trends in stroke hospitalization over the past decade. J Korean Med Assoc 55:1014–1025
2. Loureiro RC, Smith TA (2011) Design of the ROBIN system: whole-arm multi-model sensorimotor environment for the rehabilitation of brain injuries while sitting or standing. In: IEEE international conference on rehabilitation robotics, pp 1–6

3. Krebs HI, Ferraro M, Buerger SP, Newbery MJ, Makiyama A, Sandmann M, Lynch D, Volpe BT, Hogan N (2004) Rehabilitation robotics: pilot trial of a spatial extension for MIT-Manus. J NeuroEng Rehabil 1:5
4. Nef T, Mihelj M, Kiefer G, Perndl C, Muller R, Riener R (2007) ARMin-Exoskeleton for arm therapy in stroke patients. In: IEEE 10th international conference on rehabilitation robotics (ICORR), pp 68–74
5. Mayhew D, Bachrach B, Rymer WZ, Beer RF (2005) Development of the MACARM-a novel cable robot for upper limb neurorehabilitation. In: International conference on rehabilitation robotics, pp 299–302
6. Rosati G, Gallina P, Masiero S (2007) Design, implementation and clinical tests of a wire-based robot for neurorehabilitation. IEEE Trans Neural Syst Rehabil Eng 15:560–569
7. Surdilovic D, Bernhardt R (2004) STRING-MAN: a new wire robot for gait rehabilitation. In: IEEE international conference on robotics and automation, pp 2031–2036
8. Boian R, Sharma A, Han C, Merians A, Burdea G, Adamovich S, Recce M, Tremaine M, Poizner H (2002) Virtual reality-based post-stroke hand rehabilitation. Stud Health Technol Inf 85:64–70
9. Conroy SS et al (2011) Effect of gravity on robot-assisted motor training after chronic stroke: a randomizedTrial. Arch Phys Med Rehabil 92:1754–1761
10. Perry JC, Rosen J, Burns S (2007) Upper-limb powered exoskeleton design. IEEE/ASME Trans Mechatron 12:408–417
11. Size Korea. Available at http://sizekorea.kats.go.kr/
12. Jin X, Jun DI, Pott A, Park S, Park J-O, Ko SY (2013) Four-cable-driven parallel robot. In: International conference on control, automation and systems, pp 879–883
13. Prange GB, Jannink MJ, Groothuis-Oudshoorn CG, Hermens HJ, IJzerman MJ (2006) Systematic review of the effect of robot-aided therapy on recovery of the hemiparetic arm after stroke. J Rehabil Res 43:171–184
14. Zeng G, Hemami A (1997) An overview of robot force control. Robotica 15:473–482
15. Halton J (2008) Virtual rehabilitation with video games: a new frontier for occupational therapy. Occup Ther Now 9:12–14

Erratum to: Haptic Interaction with a Cable-Driven Parallel Robot Using Admittance Control

Werner Kraus, Alexander Mangold, Wei Yang Ho and Andreas Pott

Erratum to: Chapter 'Haptic Interaction with a Cable-Driven Parallel Robot Using Admittance Control' in: A. Pott and T. Bruckmann (eds.), *Cable-Driven Parallel Robots*, Mechanisms and Machine Science 32, DOI 10.1007/978-3-319-09489-2_14

In the published version of chapter 'Haptic Interaction with a Cable-Driven Parallel Robot Using Admittance Control', sequence of the authors is incorrect and the correct sequence is Werner Kraus, Alexander Mangold, Wei Yang Ho, Andreas Pott.

The online version of the original chapter can be found under
DOI 10.1007/978-3-319-09489-2_14

W. Kraus · A. Mangold · W. Yang Ho (✉) · A. Pott
Fraunhofer Institute for Manufacturing Engineering and Automation IPA, Stuttgart, Germany
e-mail: WeiYang.Ho@ipa.fraunhofer.de

W. Kraus
e-mail: wek@ipa.fhg.de

A. Mangold
e-mail: mangold.alexander@googlemail.com

A. Pott
e-mail: asp@ipa.fhg.de

A. Pott and T. Bruckmann (eds.), *Cable-Driven Parallel Robots*,
Mechanisms and Machine Science 32, DOI 10.1007/978-3-319-09489-2_23

Author Index

A. Pott and T. Bruckmann (eds.), *Cable-Driven Parallel Robots*,
Mechanisms and Machine Science 32, DOI 10.1007/978-3-319-09489-2

MIX
Papier aus verantwortungsvollen Quellen
Paper from responsible sources
FSC® C105338

If you have any concerns about our products,
you can contact us on
ProductSafety@springernature.com

In case Publisher is established outside the EU,
the EU authorized representative is:
Springer Nature Customer Service Center GmbH
Europaplatz 3, 69115 Heidelberg, Germany

Printed by Libri Plureos GmbH
in Hamburg, Germany